Butterflies of Cyprus

A Field Guide and Distribution Atlas

FSC
www.fsc.org
MIX
Paper from
responsible sources
FSC® C022174

2022

For our respective understanding and patient wives,
Jane and Roi

Butterflies of Cyprus

A Field Guide and Distribution Atlas

Eddie John and Christodoulos Makris

CABI

CABI is a trading name of CAB International

CABI
Nosworthy Way
Wallingford
Oxfordshire OX10 8DE
UK

Tel: +44 (0)1491 832111
E-mail: info@cabi.org
Website: www.cabi.org

CABI
WeWork
One Lincoln St
24th Floor
Boston, MA 02111
USA

Tel: +1 (617)682-9015
E-mail: cabi-nao@cabi.org

A catalogue record for this book is available from the British Library, London, UK.

ISBN-13: 9781800621251 (hardback)
9781800621268 (ePDF)
9781800621275 (ePub)

DOI: 10.1079/9781800621275.0000

Commissioning Editor: Ward Cooper
Editorial Assistant: Emma McCann
Production Editor: James Bishop

Typeset by SPi, Pondicherry, India
Printed and bound in the UK by Severn, Gloucester

Contents

Foreword

Islands are very special in many ways, and for threatened biodiversity in particular. Although islands contribute only 6.7% of the world's land surface area, they are home to roughly 20% of the Earth's biodiversity, 50% of the world's threatened species and 75% of the known extinctions. The Mediterranean Basin with its thousands of islands is a major hot-spot of biodiversity on Earth, with more than 25,000 species of flowering plants, and harbours 75% of the total European insect fauna. Cyprus is the third-largest island in the Mediterranean Sea and is at the crossroads of three continents, leading to a unique mixture of biodiversity. Unlike most other Mediterranean Islands, Cyprus has not been connected to any continent during the Pleistocene glacial cycles that led to sea-level drops of up to 130 metres compared with current levels. This explains the relatively low number of 56 butterfly species known from Cyprus, but also the existence of several endemic taxa, whether called species or subspecies.

The butterfly fauna of Cyprus is relatively well known, also due to the passionate research of two entomologists who joined forces to write this book: the Cypriot Christodoulos Makris, who published the first book on the butterflies of Cyprus almost 20 years ago, and the British lepidopterist Eddie John, who not only visited Cyprus many times during the past 20 years, but also founded the Cyprus Butterfly Study Group (CBSG) and relentlessly compiled records from other lepidopterists, both residents and visitors, that helped to increase our knowledge on the distribution, the biology and the ecology of butterflies in Cyprus.

This book is an amazing step forward which compiles the results of all those efforts into a truly comprehensive piece of work, with a wealth of information, which is of importance not only to students of the butterflies of Cyprus but also to those of the Western Palearctic Region in general. It contains 15 chapters, which include various general aspects such as biogeography, biology, habitats, behaviour, phenology and conservation.

Chapter 12 deals with each species in detail and contains very precise distribution maps based on the UTM coordinate system with an accuracy of 5 km^2. It also features stunning photos of all species (some including their early stages) and larval hostplants, which not only illustrate the diversity of butterflies, but also are a visual treat and whet the appetite to study them further and in even more detail.

The appetite for more has already increased in the author of this Foreword who has only been to Cyprus once, during a short stay in September 2013 by invitation of Elli Tzirkalli, on the occasion of her PhD defence. Elli has contributed a section on the new Cyprus Butterfly Monitoring Scheme, which she initiated as a means to understand better the dynamics of butterfly populations on the island. And this is of the utmost importance to safeguard its endemics and other native species that might harbour additional endemic genetic lineages. The recent arrival and possible establishment of the Lime Swallowtail *Papilio demoleus* in Cyprus illustrates that island faunas are subject to new arrivals, some of which can become invasive and pose a risk to native biodiversity.

Certainly, this book will stimulate further research in Cyprus and help to increase awareness about the conservation of its butterflies, which are an important ecological indicator group.

Dr Martin Wiemers
Head of the Ecology Section
Senckenberg German Entomological Institute (SDEI)
Müncheberg, December 2021

Preface

The concept for this book was initially discussed during a meeting between the authors at Pano Lefkara in 2013. The intervening period saw work on other projects, but provided further opportunities for data gathering before resumption of work on the final draft. Fortuitously, in terms of timing, 2021 saw the arrival and probable establishment in Cyprus of the island's first new butterfly species (*Papilio demoleus*) in recent history. Others will follow.

Our aim was to present a reference book and field guide that would enable easy recognition of the island's butterflies, while also presenting, for the first time, detailed distribution maps at 5 km^2 UTM square resolution, much of the data collated from both authors' fieldwork over decades. Of similar value, records from nearly 300 contributors, many of them members of the Cyprus Butterfly Study Group, further improved our knowledge of butterfly distribution within the island.

An important consideration was to present data thought to be of likely value to present and future researchers. With this in mind, the book details work by earlier lepidopterists and, where applicable, species are introduced with details of historical notes specific to Cyprus. Moreover, we wished to supplement this with a comprehensive bibliography relating to published works relevant to Cyprus and nearby Mediterranean countries, requiring the text to be heavily referenced.

The book summarizes decades of research into Cyprus's butterflies, complementing the seminal work of the second author's publication *Butterflies of Cyprus*, published in 2003.

Eddie John and Christodoulos Makris

A brief note about the authors

As discussed further in Chapter 2 (A Review of Butterfly Research in Cyprus), the authors have been engaged in the study of the butterfly fauna of Cyprus over a long period.

The second author, in particular, has spent a vast amount of time on fieldwork, spanning over 30 years. During this time, his interests have expanded to include the study of many insect Orders and he has published numerous papers, covering a variety of Orders, and is currently working on the Buprestidae (Coleoptera) and Chrysididae (Hymenoptera) of the island. He has collaborated with many international researchers and to date has discovered 17 species of insect new to science.

During a period of residency in Cyprus, the first author began recording the island's butterflies over 20 years ago, a practice continued throughout 15 later visits to Cyprus, with a heavy emphasis on the collation of butterfly distributional data from all parts of the island. This led to the authoring or co-authoring of over 60 papers and articles. In 2016 he co-edited (with David Sparrow) the 895-page *An Introduction to the Wildlife of Cyprus* and in 2020 co-authored and edited *Butterflies of the Levant – Vol. II* (208 pp.) with Dubi Benyamini. He is a Fellow of the Royal Entomological Society and a Fellow of the Linnean Society.

Acknowledgements

From the first author

Everyone benefits from having a mentor and the name of mine will certainly be familiar to many, especially those working in the field of academy. I refer to Professor Roger Dennis, whose unstinting encouragement helped steer me from an early age ('early', in terms of entomological writing), by first assisting with a paper written 20 years ago with the second author, on the elusive *Libythea celtis*. Unfailingly, since those naïve days, Roger has readily been available for advice, his communications invariably touched with humour, often at times of great personal tribulation for him. Alongside Roger, the late Torben Larsen was a further, constant source of inspiration. He remains well known as one of the principal authorities on butterflies in the region and his books and near-countless email messages have been referred to on many an occasion during the preparation of this book.

Another stalwart of the lepidopterological world, Rob Parker, whose name appears numerous times in this book, must also be acknowledged. At a time when very little authoritative, documented information was available on the island's butterflies, Rob brought order where there was chaos and in his 1983 paper constructed a species list that (new arrivals excepted) remains accurate today. The distribution maps appearing in this book follow a format not dissimilar to those devised by Rob in the days of true 'cut and paste'.

Reliable distribution maps lend assistance to those involved in conservation work, a discipline that is receiving increasing attention in Cyprus (Chapter 15). In addition to the extensive data garnered over decades by the authors, we are most grateful to the following for contributing records, either through direct reporting, by means of published material, or (very occasionally) in confirmed postings on websites:

Diederik van den Abbeele, Tim Adriaens, Jonathan Agius, Giannakis Antoniou, Marc Arbuckle, Guy Arien, Aristos Aristophanous, Marios Aristophanous, *Ray Atkinson*, Kevin Austin, Hasan Bağlar, Karl Bailey, Mike Bailey, Steve Baines, the late Colin Baker, Colin Baker (Port Talbot), Koos Ballintijn, Andrea Barden, Ian Barton, Onat Başbay, Ted Benton, Dubi Benyamini, Domenique Bernard, Jan Boeren, Greet Boomahuer, Luca Boscain, Marc Botham, Nick Bowles, R.F. Bretherton, Paul Brock, *Angela Brooks*, Nick Brown, Jennifer Brown, Trevor Bucknall, J.A. Bucknill, Clive Burrows, Steve Cale, Mark Calway, Jon Campbell, Sue Campbell, Aldo Catania, Geoff Cave, Jean-François Charmeux, Pambos Christodoulou, *Yiannis Christofides*, Andrew Cockcroft, Philip Collins, Simon Coombes, *Nigel Cottle*, Jurgen Couckuyt, Terry Coult, Mattijs Courbois, John Coutsis, Simon Cox, Alan Creaser, Linda Crossley, Andrew Cunningham, Simon Curson, Charles Cuthbert, Pawel Czudec, Mike Darlow, *Judy Dawes*, Mike Dawson, Steve Day, the late Philip Daykin, *Rosemarie Delaney*, R.C. Dening, Erik van Dijk, Arie van Dijk, Andreas Dimitropoulos, *Dorinda Drury*, Bob Duckhouse, *Dave Duke*, *Mark Easterbrook*, Steve Eckersley, Nico Elfferich, Brian Ellis, Olkan Erguler, Ian Everson, Paul Fisher, *Peter Flint*, Andy Foster, Francois Fournier, Bernard Fransen, Bob Frost, Daren Frost, Jürgen Fuchs, Brian Garner, Martin Gascoigne-Pees, Graham Gavaghan, Gabriel Georghiou, Graham Giles, Uffe Gjøl Sørensen, Bayram Göçmen, Valérie Goethals, James Gould, Keith Grant, Andy Grant, Zeki Gürsel, *Daniel Haines*, *Hilary Haines*, David Hall, Brian Hallam, Tamas Hapka, Andy Harding, *Eddie Harding*, Roger Harding, Mike Hardman, Miles Harkin, Daintre Hart, Les Hatton, Will Hawkes, *Ken Hearne*, Keith Heaven, John Hemmings, Bill Hipkiss, Vic Hitchings, Ralph Hobbs, Joanne Hodgkins, Ortwin Hoffmann, Peter Hofmann, Donna Hollinger, Allen Holmes, Chris Holt, *Gary Hulbert*, Hilary Humphrey, Mike Hunter, Lee Hurrell, Tony Hutchinson, Michel Huysmans, Barbara Jackson, Peter Jackson, L.A. Janzon, Klaas de Jong, Pat Jordan, Jacek Kazimierczak, Roger Kent, John Kinsey, Geert de Knijf, Ahmet Koçak, Peter Krijnen, Tom Nygaard Kristensen, Otakar Kudrna, Antoni Kwiczala, Bob Lambert, Antony Lamsdell, *Chris Lamsdell*, Jani Lassensen, *Ann Lawrence*, Judith Layzell, Roy Ledgerton, Ronny Leestmans, Stefan Lewandowski, Paul Lister, *Mike Lloyd*, Jim Lyes, Yiannakis Makriyianni, Luc Manil, *Shane Marshall*, Jean Massry, *Alison McArthur*, Derek McEwan, Karen McLeod, Jan Mees, Kris Mercer, Xavier Mérit, Benoît Méry, Paul Millard, Jon Millo, Tony Mills, Nick Morgan, Gerhard Müller, Ken Namba, Pavlos Neophytou, Ken Noble, Laurie Oakes, Özge Ogden-Fuller, Alain Olivier, *Alan Outen*, Uku Paal, *Athina Papatheodoulou*, *Rob Parker*, Kath Patrick, Nigel Peace, Nick Pegler, Phil Pemberton, Roger Perkins, Christoffe Perrett, Martha Petrou, John Phillips, Simon Plat, *Francis Plowman*, Dave Plowman, Michaelangelo Pontikis, Theodoulos Poullis, Karl Proctor, Chris Pye, Angela Pye, Patricia Radnor, Lois Rancilhac, Chris Raper, Julian Rawlins, Valéry Rebizant, Adrian Riley, Barry Robinson, Jean Robinson, Bourke Roosma, Matthew Rowlings, Peter Rowlings, Bert de Ruiter, Peter Russell, Paul Sammut, Lawrie Sampson, Matthias Sanetra, Phil Saunders, Michael

Schmitz, the late Dietmar Schuphan, Klaus Schurian, Tony Seguna, Klaus Siedle, Michael Seizmair, Alan Showler, Th. Siakides, Nejat Sik, Malcolm Simpson, David Simpson, Bjarne Skule, Ben Smit, Matt Smith, *David Sparrow, Rosalyn Sparrow*, Spyros Spyrou, Sarah Stille, Fitz Stöckl, Jos Stratford, Bob Swann, Tom Tams, *Sami Tamson*, David Taylor, Mark Telfer, John Tennent, Bethan Thomas, *Lyndon Thomas*, Paul Threapleton, David Tomlinson, Vic Tucker, Elli Tzirkalli, Philippe Van de Velde, William van der Velden, Lieuwe van Welie, Merijn van Leeuwen, Carlo van Seggelen, Bart Vanholder, Pepijn Venhuis, Rudi Verovnik, Esko Viitanen, Andrew Vincent, Dave Walker, Frank Walsh, Clive Walton, Jef Van de Water, Rhona Watson, David Webb, Brian Weeks, Jaap Westra, *David Whaley, Roger White*, Steven Whitnall, Wez Whittaker, Sven Wiessner, Phil Wilkins, Don Wilks, Jason Wilson, Josef Wimmer, Rupert Witherow, C.G.M. de Worms, Heather Young.

While the extent and depth of reporting has varied, many have contributed data both extensively and routinely to the first author over long periods; names of these recorders are shown in italics in the above list and we are especially grateful for their contributions.

Sadly, a number are no longer alive to note our appreciation of their assistance, not least the late Lyndon Thomas (cousin to the first author) and the highly respected, most helpful friend and expert lepidopterist, Torben Larsen, but it is fitting that their contributions are acknowledged.

The first author expresses his sincere appreciation to friends for their hospitality or with whom he has spent time in the field, including Angela Brooks, Yiannis Christofides, Nigel Cottle, Judy Dawes, Mark Easterbrook, Martin Gascoigne-Pees, David Hall, Ann Lawrence, Alison McArthur, Christodoulos and Roi Makris, Özge Özden-Fuller, Peter Russell, David and Rosalyn Sparrow, Lyndon and Pat Thomas, Elli Tzirkalli and David Whaley.

This section would be incomplete without a specific mention of the island's eminent entomologist and second author, Christodoulos Makris, whose help in the field and contribution to the book have been invaluable.

The first author is particularly grateful to David and Rosalyn Sparrow for their hospitality, generosity and entertaining company on numerous field explorations, especially for uncomplainingly tolerating the less hospitable areas of the Mesaoria Plain and the occasional ill-directed turn on unmarked mountain tracks, when the navigator guessed incorrectly!

Photographs

While the second author, Christodoulos Makris, has provided the majority, we are grateful to the following for generously allowing use of their photographs:

Marios Aristophanous, Ray Atkinson, Hasan Bağlar, the late Colin Baker, Onat Başbay, Ahmet Baytaş, Clive Burrows, Richard Somers Cox, Charles Cuthbert, Yiannis Christofides, Jeannette Dewaerheijt, Michael Field, Martin Gascoigne-Pees, Andy Grant, Georgios Hadjikyriakou, Daniel and Hilary Haines, Will Hawkes, Antonis Konis, Bob Lambert, Mike

Lloyd, Richard Lyszkowski, Alison McArthur, Yiannakis Makriyianni, Jean Massry, Nigel Peace, David Penney, Adrian Riley, Gerrit Ros, Mudar Salimeh, the late Dietmar Schuphan, Erdem Seven, Cristina Sevilleja, Anthony Simmons, Matt Smith, David and Rosalyn Sparrow, Sami Tamson, David Taylor, David Tomlinson and Elli Tzirkalli.

Ian Johnson (Pemberley Publishing) is thanked for granting permission to use various photographs previously used by the authors in papers published in *Entomologist's Gazette.*

Cover photograph credits:

Front cover:
Glaucopsyche paphos (Paphos Blue). *Photo*: Christodoulos Makris

Back cover (clockwise):
Allancastria cerisyi cypria (Eastern Festoon). *Photo*: Christodoulos Makris
Hipparchia cypriensis (Cyprus Grayling). *Photo*: Christodoulos Makris
Hipparchia syriaca cypriaca (Eastern Rock Grayling). *Photo*: Christodoulos Makris
Hyponephele lupina cypriaca (Oriental Meadow Brown). *Photo*: David & Rosalyn Sparrow
Chazara briseis larnacana (The Hermit). *Photo*: Christodoulos Makris
Maniola cypricola (Cyprus Meadow Brown). *Photo*: Christodoulos Makris

Wing diagrams

The authors are grateful to Bethan Wallace BA (Biological Sciences), University of Oxford, MPhil (Zoology), University of Cambridge, for providing the well-prepared wing diagrams.

Cyprus Map

Charalambos Alexandrou, Antonis Horattas and Takis Tsintides (Department of Forests, Nicosia) are thanked for permitting use and adaptation of the map originally prepared for Makris (2003).

Foreword

We express our thanks to Martin Wiemers, Head of Ecology, Senckenberg German Entomological Institute (SDEI), for kindly writing the Foreword to this book and for suggesting a project for future research in Cyprus (see Chapter 15).

Personal communications

The following people very kindly responded to enquiries relating to specific species and are acknowledged by means of a 'personal communication' in the text:

Aristos Aristophanous, Ray Atkinson, Karl Bailey, Hasan Bağlar, Onat Başbay, Dubi Benyamini, Ian Bloomer, Nick Bowles, Yiannis Christofides, Nigel Cottle, John Coutsis, Rienk de Jong, Vlad Dincă, Peter Flint, Andy Grant, Daniel and Hilary Haines, Mike Hardman, Jack Harrison, Ken Hearne, Allen Holmes, Rieks van Klinken, Otakar Kudrna (the late), Torben Larsen (the late), Mike Lloyd, Alison McArthur, Yiannakis Makriyianni, Alireza Naderi, Vazrick Nazari, Francis Plowman, Matt Rowlings, Peter Russell, Andrew Rumfitt, Mudar Salimeh, Klaus Schurian, Matt Smith, David and Rosalyn Sparrow, Chris van Swaay, Lyndon Thomas (the late), Elli Tzirkalli, Dick Vane-Wright, David Whaley, Roger White, Martin Wiemers and Mahran Zeity.

Papilio demoleus

We are also grateful to those who reported and assisted with investigations that helped track the initial dispersal of the recently arrived *Papilio demoleus*: Kostas Aristeidou, Feriha Bağlar, Hasan Bağlar, Irene Eliades, George Konstantinou, Giorgos Koronides, Hatice Mustafa, Filiz Oskay, Andrew Rumfitt, Roger White and Turan Huseyin Zeki.

Gegenes nostrodamus investigations

Many discussions have taken place in recent years on the vexed question of the likely presence of *G. nostrodamus* in Cyprus. While this will not be resolved to our full satisfaction until actual specimens are available for analysis, those principally involved in our investigations are sincerely thanked for their assistance. They include: John Coutsis (Greece), Richard Somers Cocks (UK), Charles Cuthbert (UK), Andy Foster (UK), James Hogan (OUNHM), Lee Hurrell (UK), Rienk de Jong (Naturalis Biodiversity Center, Leiden, The Netherlands), David Lees (BMNH), Alison McArthur (UK), Lazaros Pamperis (Greece), Rob Parker (UK) and Tim Termaat (The Netherlands).

At CABI, it has been a pleasure corresponding and working with Ward Cooper (Commissioning Editor), Emma McCann (Editorial Assistant), Alison Smith (Managing Editor), James Bishop (Production Editor), Valerie Porter (Copy Editor) and Megan Blackmore-Sly (Marketing Assistant).

We are most grateful to all for the assistance given.

Eddie John, Vale of Glamorgan, UK

1 Introduction

Origins

Butterflies (and moths) belong to the order Lepidoptera, a name derived from the Greek *lepis* (scale) and *ptera* (wings), i.e. insects having wings covered with scales. These near-microscopic scales, arranged in rows and present in many thousands on the wing membranes, create coloration and patterning ranging from the sombre brown of some species to the startlingly iridescent blue of others, enabling recognition not just by ourselves but, more importantly, by conspecific species.

Traditionally, the order Lepidoptera was broken down into the non-standard divisions Rhopalocera (butterflies) and Heterocera (moths), these being the only insects in the animal kingdom (Kingdom Animalia) to have scaled wings. Until very recently, members of the 'true' butterflies, i.e. those of the closely related family divisions Papilionidae, Pieridae, Nymphalidae, Lycaenidae and Riodinidae, were traditionally grouped together into the superfamily Papilionoidea. A second group, the more primitive moth-like skippers, formed a separate superfamily (Hesperioidea), a group that might reasonably be viewed as day-active moths, due to their manner of flight and posture when at rest. Another group, with a single genus *Macrosoma* comprising 36 nocturnal Neotropical 'butterfly-moths', formed a third superfamily, the Hedyloidea found in Central and Southern America. The controversial Hedylidae have been subjected to much molecular phylogenetic analysis, which has supported a sister relationship with Hesperiidae (Wahlberg *et al.*, 2005; McCullagh *et al.*, 2020; see also Kawahara and Breinholt, 2014). Thus, the three superfamilies, comprising *ca.* 19,000 species worldwide, are now regarded as belonging to a single superfamily, Papilionoidea (Nieukerken *et al.*, 2011; Simonsen *et al.*, 2012; Wiemers *et al.*, 2018).

DOI: 10.1079/9781800621275.0001

Figs 1.1 and 1.2. Detail of hindwing underside of *Papilio machaon* (Swallowtail) (left) and of the hindwing underside ocelli of *Vanessa cardui* (Painted Lady) (right). *Christodoulos Makris*

It is believed that butterflies originated from moths and co-evolved with plants, with the oldest butterfly fossil record (a hesperiid, *Protocoeliades kristenseni* de Jong, 2016) dating back about 55 MYA (de Jong, 2016, 2017) and fossils of flowering plants (angiosperms) dating to the early Cretaceous, about 140 MYA (Wing and Boucher, 1998). Contemporaneous evolution would therefore suggest that Lepidoptera have been in existence for well over 140 million years; indeed, the earliest known fossil believed to belong to the Lepidoptera (a primitive moth, *Archaeolepis mane* Whalley, 1986), dates to the Lower Jurassic, *ca.* 190 MYA (Grimaldi and Engel, 2005), although butterflies might have originated 70 MYA (Vane-Wright, 2004) or 100 MYA (Heikkilä *et al.*, 2012). For further information on the fossil record and butterfly-plant evolution, the reader is referred to Dennis (2020: 50).

Classification and Nomenclature

Butterflies have long fascinated collectors, with the oldest remaining specimens dating back over 300 years (Salmon, 2000). The earliest known *illustration* of a butterfly is that of a regular migrant to Cyprus, *Danaus chrysippus* (Linnaeus, 1758) (Plain Tiger or African Queen), several specimens of which were depicted on a painting in the Tomb of Nebamun in Luxor, Egypt and illustrated in Larsen (1984a). Sadly, although still conserved in the numerous natural history museums of the world, the legacy of many a collecting expedition may no longer be readily accessible to a casual visitor. However, secure behind closed doors remote from today's far less formal displays, the all-pervading aroma of paradichlorbenzene (formerly used to protect stored specimens against attack by species of, for example, *Anthrenus* beetle) may still signal the presence of mahogany cabinets filled with drawers of well-preserved butterfly specimens from earlier times. Nowadays, with digital imaging, butterfly photography has risen to the fore in a way unimagined by early collectors, but without their dedicated efforts, continued by today's researchers still working on actual specimens, the identification and describing of the majority of known butterflies (as well as other fauna and flora) would have been an impossible task.

The assembly and study, both of plants and of animals, led to John Ray, a British naturalist, establishing a classification system extended by Carl von Linné (better known as Linnaeus) who, in 1758, introduced the binomial (binominal) system for the naming of all known animal species. Linnaeus included all butterflies under a single genus name, '*Papilio*'; the concept of family divisions developed later, with *Papilio* being retained only for the Swallowtail family. The Red Admiral butterfly, so-named by James Petiver (1663–1718), was among a number of butterflies already known to 16th century lepidopterists by their vernacular names (only), and was given the binomial name of *Papilio atalanta* ('*Papilio*' relating to the genus, '*atalanta*' to the species) by Linnaeus along with the year (1758) in which the species was originally described. With the introduction of family divisions, *Papilio atalanta* Linnaeus, 1758 was listed within the Nymphalidae and became known as *Vanessa atalanta* (Linnaeus, 1758), the transfer of the butterfly into the genus *Vanessa* requiring the insertion of brackets to denote a change of generic name. It carries the same Latinized scientific name today.

Whereas vernacular names may differ from country to country, the scientific name is invariably understood, although such names are not necessarily unchanging or unchallenged. In the earlier example, *Vanessa atalanta* (nowadays regarded as a stable scientific name) became *Pyrameis* before reverting to *Vanessa*. Meticulous examination of specimens and the greater use of molecular techniques continue to reassign some species, as with certain Hesperiidae (Zhang *et al.*, 2020), but this is no guarantee of the universal acceptance of such changes, for example Coutsis and Benyamini (2021) in respect of Hesperiidae. In simple terms, the result is occasionally one of fluidity and lack of uniformity in the nomenclatural system. The species' entries that follow later refer to some of the earlier scientific names by which butterflies have been known, but these are by no means intended to be comprehensive.

European Butterfly Species: How Many Are There?

Dennis (1997) drew attention to the increasing number of butterfly species added to the European list, illustrated by the 374 species that appeared in Higgins and Riley (1983), 440 in Karsholt and Razowski (1996), rising to 491 in Kudrna (1996). Kudrna (2002: 308) listed 486, reducing to 441 in Kudrna *et al.* (2011: 484), increasing slightly to 445 species in Kudrna *et al.* (2015: 552) and to 446 in Kudrna (2019: 282). The number of butterfly taxa in Europe is currently given as 496 by Wiemers *et al.* (2018, 2020). However, the definition of 'Europe' may not correspond precisely among all authors; see Dennis (1997) and Wiemers *et al.* (2018) for further information.

Occasionally, species new to Europe are added, for example the accidentally introduced lycaenid *Cacyreus marshalli* Butler, 1898 (Geranium Bronze), but often the increases have resulted from elevation of subspecies to species rank, not least among species found in southern Europe, which were stated to show 'disproportionate increases in rare and endemic species' when compared with northern Europe (Dennis, 1997). Understandably, the same author called for an urgent revision of higher and lower European butterfly taxa, without which he believed the validity and objectivity of taxonomic practices could be called into question. Sbordoni *et al.* (2018: 1) also drew attention to overzealous taxonomic over-splitting (see also Kristensen *et al.*, 2007).

Furthermore, even where a species' status is not in question, this is no guarantee of universal agreement on the use of the *generic* name. Examples relevant to the Cyprus list may be found in recent publications; for example, Van Swaay *et al.* (2010a) and Verovnik and Wiemers (2010) referred to *Apharitis acamas* and *Pseudophilotes vicrama*, whereas Kudrna (2002: 108, 284; 2019: 105, 246) elected for *Cigaritis acamas* and *Scolitantides vicrama*, respectively. Wiemers *et al.* (2018) listed *Cigaritis acamas* and *Pseudophilotes vicrama*. Therefore, even within Europe, there is no standard comprehensive taxonomic list accepted by all researchers.

For more than two centuries the descriptions (and hence classification) of taxa were based almost exclusively on morphological characters, for example wing shape, markings and genitalia (Ackery *et al.*, 1999). However, Descimon and Mallet (2009: 219) reminded us that 'The tools used in making a decision on the rank of taxa ... include morphological, chromosomal (karyological), molecular, and ecological characters.' In respect of the last-mentioned, larval hostplant, habitat and phenology might be added (Dincă *et al.*, 2021). In stating that 'the whole is better than the sum of its parts', Wahlberg *et al.* (2005) expressed similar caution. Hołyński (2010) and Simonsen *et al.* (2012) concurred. Descimon and Mallet (2009: 222) added the caveat that of the 440 European butterfly species (referring to the list by Karsholt and Razowski, 1996) around 16% are known to hybridize with at least one other species in the wild, giving rise to fertile progeny in about half. Descimon and Mallet (2009: 222) further pointed out that in one group (European *Hipparchia*) only ten species were listed by Higgins and Riley (1970), whereas (largely resulting from elevation to species' rank) Tolman and Lewington (1997) listed 19, some of which have since been reduced to their subspecific status; so the picture is at best blurred. Mallet (2007) speculated that the elevation from subspecies to full species might even be driven by legislation that favours the conservation of taxa recognized as full species rather than subspecies, the latter being seen as possibly of less value. To quote from Kudrna *et al.* (2011: 25), 'Therefore a stable taxonomy of European butterflies is not in sight and a careful comparison of molecular and non-molecular data (i.e. integrative taxonomy) will remain indispensable for many years to come.' However, this is a highly complex issue beyond the scope of this book and certainly not a subject to be treated lightly.

2 A Review of Butterfly Research in Cyprus

Early Research

Rebel (1939) reported that Dr Theodor Kotschy, 'a botanist and scientific traveller', visited Cyprus in 1841 and while there also collected Lepidoptera, but no information was published on his findings. The first known concerted attempt at recording the island's butterflies was arranged by an Austrian lepidopterist, Julius Lederer, who sent a collector, Franz Zach, to Cyprus in 1853. After a four-week journey, Zach arrived at Larnaka on 7 May. Lederer (1855) noted (translated from the original German) that the area was 'bare, arid and the ground chalky, almost without vegetation; it is therefore not to be wondered at that ... outside some common butterflies and common ones throughout Europe [Zach] finds almost nothing. An excursion into the interior of the island also yielded an unsatisfactory result ... there was nothing to be found here either; ... the tour of the northern edge of the island had a pathetic result and so the enumeration of the butterflies collected on Cyprus has hardly any other interest than that for the geographical distribution of these animals.' The 38 species of butterfly (plus 'varieties') listed by Lederer (1855) are shown in Table 2.1. The list includes a number of species that have never been confirmed for the island, leading to a suspicion that mixing of specimens from Cyprus and Lebanon[1] appears probable. Wiltshire (1948) and Parker (1983) also expressed doubt about some of Zach's claims. Moreover, the exclusion from Table 2.1 of some very common Cyprus species, such as *Colias*

[1] On 22 July, Zach arrived in Beirut where he was somewhat more successful, collecting 50 butterfly species; a greater number might have been expected from a mainland country, but travel was far more difficult in 1853. Note that in the same account, Lederer (1855) also listed many more species collected in Turkey by Albert Kindermann from 1848 to 1850.

DOI: 10.1079/9781800621275.0002

Table 2.1. Species (including 'varieties') recorded in Cyprus by Franz Zach in 1853 and documented by Lederer (1855)

Family	Recorded as	Current name/species present
Papilionidae	*Papilio Machaon* L.	*Papilio machaon* Linnaeus, 1758
	Thais Cerysii God.	*Allancastria cerisyi cypria* Stichel, 1907
Pieridae	*Pieris Brassicae* L.	*Pieris brassicae* (Linnaeus, 1758)
	Pieris Rapae L.	*Pieris rapae* (Linnaeus, 1758)
	Pieris Daplidice L.	*Pontia daplidice* (Linnaeus, 1758)
	Anthocharis Ausonia Esp.	*Euchloe ausonia* (Hübner, [1804])
	Rhodocera Cleopatra L.	*Gonepteryx cleopatra* (Linnaeus, 1767)
Lycaenidae	*Polyommatus Phlaeas* L.	*Lycaena phlaeas* (Linnaeus, [1760])
	Lycaena Boetica L.	*Lampides boeticus* (Linnaeus, 1767)
	Lycaena Telicanus Herbst.	*Leptotes pirithous* (Linnaeus, 1767)
	Lycaena Balcanica Freyer	*Tarucus balkanicus* (Freyer, 1844)
	Lycaena Trochylus Freyer	*Freyeria trochylus* (Freyer, 1844)
	Lycaena Lysimon Hb.	*Zizeeria knysna* (Trimen, 1862)[1]
	Lycaena Cyllarus Fab.	*Glaucopsyche alexis* (Poda, 1761)[2]
	Lycaena Alexis Hufn. (*agestis* S. V.)	*Aricia agestis* ([Denis & Schiffermüller], 1775)
	Lycaena Icarus (*alexis* S. V.)	*Polyommatus icarus* (Rottemburg, 1775)
Nymphalidae	*Libythea Celtis* Fab.	*Libythea celtis* (Laicharting, 1782)
	Limenitis Camilla S. V.	*Limenitis reducta* Staudinger, 1901
	Melitaea Phoebe S. V.	*Melitaea phoebe* ([Denis & Schiffermüller], 1775)[3]
	Vanessa Cardui L.	*Vanessa cardui* (Linnaeus, 1758)
	Vanessa Atalanta L.	*Vanessa atalanta* (Linnaeus, 1758)
	Vanessa Polychloros L.	*Nymphalis polychloros* (Linnaeus, 1758)
	Vanessa Triangulum Fab.	*Polygonia egea* (Cramer, 1775)[4]
	Danais Chrysippus L.	*Danaus chrysippus* (Linnaeus, 1758)
	Satyrus Hermione L.	*Hipparchia syriaca cypriaca* (Staudinger, 1879)
	Satyrus Briseis L.	*Chazara briseis larnacana* (Oberthür, 1909)
	Satyrus Anthelea Hb.	*Pseudochazara anthelea*
	Pararga Roxellana Fab.	*Kirinia roxelana*
	Pararga Maera L. Var. *adrasta* Esp.	*Lasiommata maera maera* (Linnaeus, 1758)
	Pararga Megaera L.	*Lasiommata megera* (Linnaeus, 1767)
	Coenonympha Pamphilus L.	*Coenonympha pamphilus* (Hübner, 1824)[5]
	Coenonympha Pamphilus Var. *Cyllus* Esp.	*Coenonympha pamphilus*[6]
Hesperiidae	*Hesperia Malvarum* O.	*Carcharodus alceae* (Esper, 1780)
	Hesperia Marrubii Rb.	*Carcharodus baeticus* (Rambur, 1839)[7]
	Hesperia Alveus Hb.	*Pyrgus alveus* (Hübner, [1803])[8]
	Hesperia Alveolus Hb.	*Pyrgus malvae* (Linnaeus, 1758)[9]
	Hesperia Eucrate Esp.	*Spialia sertorius* (Hoffmansegg, 1804)[10]

Continued

Table 2.1. Continued.

Family	Recorded as	Current name/species present
	Hesperia Eucrate Var. *orbifer*	*Spialia orbifer* (Hübner, [1823])[11]
	Hesperia Actaeon Esp.	*Thymelicus acteon* (Rottemburg, 1775)
	Hesperia Nostradamus Fab.	*Gegenes nostrodamus* (Fabricius, 1793)[12]

[1]*Zizeeria karsandra* (Moore, 1865) in Cyprus (see species' entry);
[2]Misidentification for *Glaucopsyche paphos* Chapman, 1920 (see Rebel, 1939);
[3]Knapweed Fritillary, unconfirmed for Cyprus;
[4]Southern Comma, unconfirmed for Cyprus (see Chapter 13);
[5,6]Small Heath, since unconfirmed for Cyprus, other than for an anecdotal report by Wilson (in Turner, 1920) of one taken in Kyrenia in 1918 ([6] listed by Lederer as '*Cyllus*' [*sic*] = *Papilio lyllus* Esper, 1805);
[7]Southern Marbled Skipper, a species of the western Mediterranean;
[8]Large Grizzled Skipper, not known for the Levant;
[9]Grizzled Skipper, not known for the Levant;
[10]Red-underwing Skipper, a species of the western Mediterranean;
[11]Orbed Red-underwing Skipper (see Chapter 13);
[12]Questionable because the very similar *Gegenes pumilio* was not recorded, but see Chapter 13.

croceus (Geoffroy, 1785) (Clouded Yellow), *Hipparchia cypriensis* (Holik, 1949) (Cyprus Grayling) and especially *Maniola cypricola* (Graves, 1928) (Cyprus Meadow Brown), is puzzling. Although the last two mentioned species were not known at the time as Cyprus endemics, it is inexplicable that no species of *Maniola* was listed.

Staudinger (1878), in his extensive paper on the butterflies of Asia Minor, made frequent use of the words 'auch auf Cypern gefunden' (also found on Cyprus), but without providing any substantive data. He appears to have relied heavily, if not exclusively, on Lederer's (1855) publication; however, such was his contribution that several species or subspecies listed in this book bear his name as the describing authority.

Sir John Bucknill (Judge Advocate of Cyprus, who wrote mainly on matters ornithological), expanded on a list of known butterflies for Cyprus (Bucknill, 1911), previously documented in Hutchinson and Cobham (1909), which Bucknill updated and modified in later years in the same handbook series. As sources, Bucknill used his personal observations, along with those 'from the notes and collections of Major Bolton, Messrs. C. Noble, N. L. Petrides, J. W. Smallwood, the Rev. H. J. Smale and others'. 'Others' included Lederer (1855). Among some understandable misidentifications, *Euchloe ausonia* (Hübner, [1804]) was listed by Bucknill as *Euchloe belia*[2] (a synonym for *Euchloe crameri* Butler, 1869), *Maniola cypricola* listed as '*Epinephele jurtina L. form telmessia*', *Hyponephele lupina* (Costa, 1836) as *Epinephele lycaon* and *Apharitis acamas* (Klug, 1834) as *Cigaritis zohra* (Donzel, 1847).

[2] Specimens of '*Euchloe belia*' from Cyprus, sent by Otto Bang-Hass to Hans Rebel, could not be separated from *Euchloe ausonia* (Rebel, 1916), under which name Rebel included the species in his 1939 paper.

Of particular note is that Bucknill and the earlier observers referred to above were unable to confirm a number of earlier listings by Lederer (1855) and Staudinger (1881), including that of *Coenonympha pamphilus*, which is known from Lebanon (Larsen 1974: 139), increasing speculation that specimens collected by Zach from Cyprus and Beirut were mixed. Rebel (1939), however, commented (of *C. pamphilus*) that 'It seems likely that the species has lasted 80 years from the fauna [but] is gradually disappearing' (translated from the original German). Bucknill's (1911) compilation included *Gonepteryx rhamni* (Linnaeus, 1758), which was one of just three species of butterfly noted by an ornithological visitor (Guillemard, 1888); this was likely to have been *Gonepteryx cleopatra* (Linnaeus, 1767), although *rhamni* is present in nearby mainland countries. Bucknill invented many imaginative common names for the butterflies he encountered and it is a little sad that some, like 'All-the-gifts-Fritillary' (Cardinal, *Argynnis pandora* ([Denis & Schiffermüller], 1775), although cumbersome, failed to endure.

At around the end of the 19th century and the beginning of the 20th century, Henry J. Turner (1920) noted that Cyprus was visited by various collectors such as C.H. Glazner, D.M.A. Bate and T. Bainbrigge-Fletcher. Dorothea Bate, a palaeontologist at the Natural History Museum (BMNH) in London, visited a number of Mediterranean islands; her collected specimens from Cyprus are housed at the BMNH.

During this period, detailed studies were undertaken on individual species by, for example, Stichel (1907: *Zerynthia cerisyi cypria*), Oberthür (1909: *Satyrus briseis larnacana*), Bethune-Baker (1913: *Chilades galba*), Rebel (1916: *Satyrus anthelea acamanthis*), Chapman (1920: *Glaucopsyche paphos*), Riley (1921: *Epinephele lupinus cypriaca*), Riley (1925: *Apharitis acamas cypriaca*) and Graves (1928: *Epinephele cypricola*).

In Turner's (1920) highly detailed contribution, he drew heavily on Bucknill (1911), as well as on material collected on his (Turner's) behalf over a 3-year period by G.F. Wilson of the Chief Secretary's Office in Nicosia, and on notes made by A.E. Gibbs and T. Bainbrigge-Fletcher. Turner appears not to have visited Cyprus.

In June and July 1939, Hakan Lindberg (incidentally) collected Lepidoptera, although the emphasis seems to have been on acquiring representatives belonging to the orders Hemiptera and Coleoptera. Whilst numerous moths were recorded as new to Cyprus, his butterfly captures, listed in Hackman (1952), revealed nothing of significance.

A discussion on early contributions to our knowledge of the butterfly fauna of Cyprus would be incomplete without mention of two highly influential Cypriot entomologists, George Mavromoustakis (1898–1968) and Themistokles Shiakides (1904–1998). The former, although principally specializing in Hymenoptera, submitted large numbers of specimens of Lepidoptera to the BMNH and the Naturhistorische Museum in Vienna; the Agricultural University of Athens was the beneficiary of Shiakides' private collection.

After a three-season stay in Cyprus while serving in the RAF, dedicated research by Rob Parker (Fig. 2.1), both in Cyprus and later upon his return

Fig. 2.1. Rob Parker photographing *Pseudophilotes vicrama* (Eastern Baton Blue), Zakaki, 8 April 2006. *Christodoulos Makris*

to the UK, led to a substantive review of the island's butterflies (Parker, 1983). Excepting the later addition of new species to the Cyprus list, and other than the reassignment of certain species to differently named genera along with the elevation or sinking of some subspecies, Parker's list remains accurate 40 years later. Parker also provided the first 10 km^2 UTM square distribution maps for the island's butterflies, based upon his own observations and information gleaned from published documents. In the introduction to his paper, Parker (1983) stated, 'A distribution map is provided for each species and, although the records are sparse at present, these can form a basis for further research.' Indeed, they have.

Various other publications on Cyprus included those by Bretherton (1954, 1974), de Worms (1967), Tennent (1985), Dening (1985, 1987), Wimmer (1985), Hofmann and Rose (1987), Manil (1990) – a significant paper that also included 10 km^2 UTM square distribution maps – and Fuchs (1996).

For additional information on early collectors and researchers, the reader is referred to Rebel (1916, 1939), Parker (1983) and Makris (2003: 79). An updated, comprehensive bibliography relating to recent research on the butterflies of Cyprus is included in this book.

It is perhaps appropriate at this point to inform readers of the authors' experience in matters relating to the butterfly fauna of Cyprus.

Christodoulos Makris, the second author and a resident of Cyprus, has been studying the wildlife of the island since 1989. He published the island's first authoritative book on butterflies, *Butterflies of Cyprus* in 2003 (329 pp.), providing ecological information, photographs, distribution and phenological data for all the species of Cyprus. He was first to confirm the breeding in Cyprus of the species *Aporia crataegi* (Linnaeus, 1758) (Black-veined White), *Pontia chloridice* (Hübner, [1813]) (Small Bath White), *Libythea celtis* (Laicharting, [1782]) (Nettle-tree Butterfly) and *Danaus chrysippus* (Plain Tiger).

The first author, Eddie John, arrived in Cyprus in 1997 for a 14-month work-related period. During his stay, he began recording the island's butterflies and published the results in John (2000); this further updated the distribution maps and phenological data gathered by Parker (1983) and Manil (1990). He founded the Cyprus Butterfly Recording Scheme in 2000 and was initiator and co-founder of the Cyprus Butterfly Study Group in 2012. Further biographical details on both authors may be found in the Preface.

Changes to the Cyprus checklist

Earlier lists of species included a number of misidentified taxa that were corrected by Parker (1983), who contributed a species' list numbering 50. Changes to Parker's list have occurred since then and are listed chronologically:

- Manil (1990), in an update to Parker (1983), included *Colias erate* (Esper, [1805]) (Eastern Pale Clouded Yellow); this followed the publication of a record from Cyprus by Hofmann and Rose (1987).
- Makris (2003) included updated 10 km^2 UTM square distribution maps (provided by the first author) and, with the addition of *Catopsilia florella* (Fabricius, 1775) (African Migrant) and *Deudorix livia* (Klug, 1834) (Pomegranate Hairstreak), the latter first documented by Dennis *et al.* (2000), increased the species count to 53. (See species' entries for details.)
- John *et al.* (2006) and John (2007) reduced the species count to 52, with the removal of *C. erate* (see entry under *Colias croceus* for an explanation).
- John *et al.* (2011), with the addition of *Chazara persephone* (Hübner, [1805]) (Great Steppe Grayling), once again increased the list to 53 species, following a sighting of this species by Daniel and Hilary Haines in 2010.
- The previously unpublished observation of *Issoria lathonia* (Linnaeus, 1758) (Queen of Spain Fritillary) in May 2016 by Yiannakis Makriyianni further increased the species' list to 54.
- The anticipated arrival of *Papilio demoleus* (Linnaeus, 1758) (Lime Swallowtail) in August 2021 and first reported in John *et al.* (2021b), brought the total to 55 species.
- Also in 2021, the reported discovery of a museum specimen of *Hypolimnas misippus* (Linnaeus, 1764) (False Plain Tiger) captured in Cyprus by Prof. Bytinski-Salz in 1936, requires inclusion here (see species entry for details). The current checklist now stands at 56 species.

However, the species number of 56 includes one taxon thought to be extinct, and other very rare migrants or possible adventives, explaining why Van Swaay *et al.* (2010a), using data provided by the authors, took a more circumspect view in listing 48 species for Cyprus in their European Red List.

Brief mention should also be made of observations made over 73 collecting days by Ayberk and Akkuzu (2007) in northern Cyprus between 2000 and 2004. Apparent unfamiliarity with the species encountered meant that in total, 133 butterflies were taken for later identification, following which the authors claimed to have added no fewer than 19 species of butterfly to the Cyprus list. To achieve a 35% (at the time) increase in total species count from a small area of a relatively well-studied island seems quite remarkable, inevitably bringing into question the reliability of determinations. Additionally, the lack of observations/capture of several very common species, for example *Pieris rapae* (Linnaeus, 1758) and *Maniola cypricola* (but with declared captures of the very similar *Pieris napi* (Linnaeus, 1758) and *Maniola jurtina* (Linnaeus, 1758), respectively), would appear to suggest that the claim of 19 species new to Cyprus requires some very careful scrutiny. Along with colleagues, the authors of this book have spent much time in northern Cyprus in recent years without adding to the species list. Other than the historic inclusion of *H. misippus* and the recent observations of *C. persephone* in 2010, *I. lathonia* in 2016 and *P. demoleus* in 2021, no incontestable additions to the list have been reported to the Cyprus Butterfly Recording Scheme (CBRS) since its inception in 2000, although *Gegenes nostrodamus* (Fabricius, 1793) (Mediterranean Skipper) is another likely addition. See also Chapter 13.

3 Biogeography

Positioned in the far eastern end of the Mediterranean basin, Cyprus lies at the juncture of Africa, Asia and Europe, with Turkey 75 km to the north, Syria and Lebanon 105 km to the east and Egypt 380 km to the south (Sparrow and John, 2016: 3). The composition of the island's butterflies reflects elements of temperate parts of the Palaearctic and Asia and, to the south, the Afrotropical Region.

The Tropical Butterflies of Cyprus

Table 3.1 is based on Larsen (2015) and shows the tropical element in Cyprus to comprise 14 species, i.e. 25% of the total. Larsen (2015) estimated there to be *ca*. 210 butterfly species in his definition of the Levant[3], of which 27 (13%) were described as tropical (see also 'Biogeography' in John *et al*., 2010). In percentage terms, therefore, Cyprus potentially has experienced almost twice the number of tropical butterflies compared with the Levant as a whole. In common with the remainder of the Levant, the majority of the tropical species have African affinities. Larsen (1986) had included *Libythea celtis* in an earlier review of tropical species of the Mediterranean, as 'it does seem to be a Palaearctic representative of an essentially tropical genus'. This has been omitted from Table 3.1.

[3] Turkey (south of the Taurus Mountains, i.e. mainly Antalya and Hatay Provinces), Syria, Lebanon, Cyprus, Israel, Jordan, Sinai and the northernmost parts of Upper Egypt between the Suez Canal and the Nile Delta, and coastal areas west to Mersa Matruh.

DOI: 10.1079/9781800621275.0003

Table 3.1. The tropical butterflies of Cyprus (based on Larsen, 2015).

Family	Species	Biogeographical category	Migratory capability
Papilionidae	*Papilio demoleus*	Oriental	**
Pieridae	*Catopsilia florella*	Afrotropical	***
Lycaenidae	*Deudorix livia*	Afrotropical	**
	Lampides boeticus	Palaeotropical	***
	Leptotes pirithous	Afrotropical	***
	Azanus jesous	African/Indian	**
	Freyeria trochylus	African/Indian	–
	Zizeeria karsandra	Oriental	–
Nymphalidae	*Danaus chrysippus*	Palaeotropical	***
	Charaxes jasius	Afrotropical	–
	Ypthima asterope	African/Indian	–
Hesperiidae	*Gegenes pumilio*	African/Indian	–
	Pelopidas thrax	African/Indian	**

***Regular long-distance migrant, often in masses.
**Regular local and occasionally long-distance migrant.
–Completely non-migratory, sedentary species.

The Eremic Butterflies of Cyprus

Larsen (2015) considered certain other species to be of eremic origin and 'not readily assigned to any of the major biogeographical regions. Some are derived from tropical elements, others from Palaearctic elements.' Table 3.2 includes one species (*Gegenes nostrodamus*) strongly believed to be present on the island, but yet to be confirmed (see Chapter 13).

Table 3.2. The eremic butterflies of Cyprus (based on Larsen, 2015).

Family	Species	Distribution
Lycaenidae	*Apharitis acamas*	All regions of the Levant
	Tarucus balkanicus	Throughout the Levant, except Egypt (other than Sinai)
	Luthrodes galba	Northern Levant (absent from Egypt and Sinai)
Hesperiidae	(*Gegenes nostrodamus*)	Circum-Mediterranean, but unconfirmed (as yet) from Cyprus

4 The Life Cycle of the Butterfly

Throughout the various growth stages, all Pterygota (winged insects) undergo marked changes in form or structure, known as the process of metamorphosis. Butterflies are holometabolous insects, i.e. they undergo complete metamorphosis in the development from egg to adult stage, passing through intermediate larval and pupal stages. All butterflies, regardless of size or family, develop into adults via the egg (ovum), caterpillar (larva) and chrysalis (pupa), each of these stages being dramatically different from that of the preceding, with growth occurring only at the larval stage. This contrasts with certain other invertebrate orders such as Odonata (dragonflies), which have an incomplete (hemimetabolous) metamorphosis in which there is no intermediate pupal stage. The time spent in each stage may vary from species to species of butterfly and is determined by hostplant availability, overwintering behaviour and voltinism, and may be influenced by the effects of weather, especially drought. In some multiple-brooded species such as *Papilio machaon* Linnaeus, 1758 (Swallowtail), the life cycle may be completed in about a month in summer (second author, personal observation).

Mating

Mating usually occurs shortly after the butterfly emerges from the pupa, although where overwintering is in the adult stage, mating may be delayed until the following spring. Males often emerge earlier than females (protandry) and, according to species, adopt different ways for locating potential mates (see Chapter 6). Competition for acceptance by a female may be intense, as demonstrated in Figs 4.1 and 4.2.

DOI: 10.1079/9781800621275.0004

Fig. 4.1. Three ♂ *Papilio machaon* (Swallowtail) in competition to mate with a single ♀, 17 April 2018. Courtesy of *Matt Smith*

Fig. 4.2. Eleven ♂ *Zizeeria karsandra* (Dark Grass Blue) in competition to mate with two ♀♀, 10 September 2016. *Christodoulos Makris*

Fig. 4.3. Rejection of an approaching *Pieris brassicae* (Large White) ♂ (L) by a previously mated ♀ (R). Courtesy of *David and Rosalyn Sparrow*

Once mated, females of some species display rejection behaviour by opening wings and elevating the tip of the abdomen in order to deter further attempts at mating (Fig. 4.3) or by fluttering of wings (see Wickman, 2009).

After mating, a period of days is usually required to allow for egg maturation, whereupon females usually set off in search of egg-laying sites without further delay. However, there are exceptions. In Cyprus, for example, species of *Hipparchia* ('graylings') having just one brood per year, usually mate after emergence in spring and aestivate throughout the hottest of the summer months before any egg-laying takes place (see Chapter 6 and species entries).

Ovum (Egg)

Whereas some butterflies are fastidious about their selection of hostplants, taking much time in selecting the correct plant in the right condition and growing in a suitable position, other species select twigs or even rocks near the hostplant on which to oviposit. Eggs may be placed singly or in clusters or batches, depending on the species. Yet others scatter eggs randomly while flying over the hostplant, although this is not known for any Cyprus species. Egg size, shape, colour and texture vary from species to species; some are smooth and globular (Papilionidae, Figs 4.4 and 4.5; see also Chapter 12, Fig. 12.18), some spindle-shaped and ribbed (Pieridae, Figs 4.6 and 4.7), others domed and ridged (Satyrinae, Fig. 4.8) or doughnut shaped and reticulated (Lycaenidae, Fig. 4.9). Hatching time varies from about 1 to 2 weeks, or may be many months where overwintering is in the egg stage. For those species that hibernate in this way, there is yet more variation, with some species' eggs showing little or no

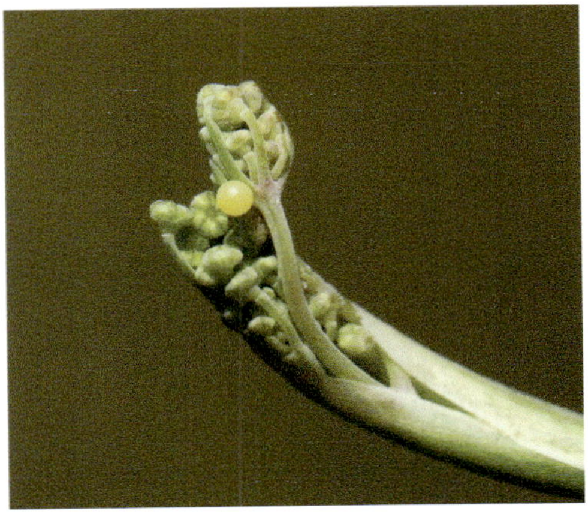

Fig. 4.4. *Papilio machaon* (Swallowtail) egg on buds of *Foeniculum vulgare* (fennel), 1 August 2012. *Christodoulos Makris*

Fig. 4.5. *Papilio machaon* (Swallowtail) eggs on *Citrus limon* (lemon), 27 October 2015. Swallowtail eggs are usually deposited singly, but laid here in a cluster of four – presumably laid by the same female. Courtesy of *Mike Lloyd*

Fig. 4.6. Unlike eggs of *Pieris rapae* (Small White) that are laid singly, those of *P. brassicae* (Large White) are laid in large clusters, 17 December 2015. *Christodoulos Makris*

Fig. 4.7. A much magnified *Pontia chloridice* (Small Bath White) egg on the hostplant *Cleome iberica*, 6 July 2013. *Christodoulos Makris*

Fig. 4.8. Egg of *Chazara briseis larnacana* (The Hermit), 3 October 2007. Courtesy of *Martin Gascoigne-Pees*

Fig. 4.9. Egg of *Glaucopsyche paphos* (Paphos Blue), 17 May 2014. Courtesy of *Marios Aristophanous*

change in appearance, while in others, the young larva can be seen hibernating within the transparent shell.

Butterflies at all stages, especially in the immature stages, have many predators; accordingly, hundreds of eggs may be laid in order to ensure the survival of a few to attain adult stage and produce later generations. Accordingly, butterflies form an essential role in the food chain.

Important predators of ova include ants, parasitoids and even larvae of the same species. Hermann *et al.* (2019) reported on a study of predation of *Danaus plexippus* Linnaeus, 1758 (Monarch) ova and neonate larvae, where no fewer than nine Orders (and 47 species) of milkweed-visiting arthropods were implicated in predation.

Larva (Caterpillar)

On emergence, the young larva may make the eggshell its first meal (the chorion or shell is rich in protein), or quickly feeds on the tender parts of the hostplant, including buds and flowers. In many species, the larva develops rapidly and may be fully grown after a few weeks. During this time, each of the five or so larval stages (instars) ends in the shedding (ecdysis) of the larval skin, which although highly flexible cannot cope with the massive change in body size from hatching to point of pupation, which may see a bodyweight increase up to 3000 times in some species (Reavey, 1993: 249; see also Callier and Nijhout, 2011). Larval appearance alters during each of the instars, sometimes significantly so, as with *Papilio machaon* that changes from mainly black in the first instar to a bright green with red and black markings (Fig. 4.10) prior to pupation. As with *P. machaon*, aposematic (warning) coloration is also seen in the larvae of *Pontia chloridice* (Fig. 4.12). In contrast, larvae from other species, e.g. *Glaucopsyche paphos* (Fig. 4.13), *Charaxes jasius* (Linnaeus, 1767) (Fig. 4.15), *Libythea celtis* (Fig. 4.16) and *Hipparchia cypriensis*

(Fig. 4.17), retain a camouflaged appearance throughout the development period. When fully grown, the larva seeks a pupation site that needs to ensure a secure and safe position during this fundamentally immobile stage of the life cycle.

Fig. 4.10. The aposematic *Papilio machaon* (Swallowtail) larva on *Foeniculum vulgare* (fennel), 8 May 2015. *Eddie John*

Fig. 4.11. Larval mass of *Pieris brassicae* (Large White) on one of many hostplants, the endemic *Brassica hilarionis*, 17 November 2017. *Christodoulos Makris*

Fig. 4.12. The aposematic *Pontia chloridice* (Small Bath White) larva on the hostplant *Cleome ornithopodioides*, 9 August 2011. *Christodoulos Makris*

Fig. 4.13. The cryptic final instar larva of *Glaucopsyche paphos* (Paphos Blue), attended by an ant of the species *Tapinoma simrothi*, 17 May 2014. Courtesy of *Marios Aristophanous*

Fig. 4.14. Larva of *Danaus chrysippus* (Plain Tiger) on *Cynanchum acutum*, the hostplant in Cyprus, 16 November 2014. *Christodoulos Makris*

Fig. 4.15. A cryptic *Charaxes jasius* (Two-tailed Pasha) reared larva on *Arbutus unedo* hostplant. *Eddie John*

Fig. 4.16. The cryptic *Libythea celtis* (Nettle-tree Butterfly) larva on the hostplant *Celtis australis*, 20 May 2007. *Eddie John*

Fig. 4.17. *Hipparchia cypriensis* (Cyprus Grayling), reared larva. Courtesy of *Martin Gascoigne-Pees*

Fig. 4.18. Larva (pre-pupa) of reared *Ypthima asterope* (African Ringlet) suspended in preparation for pupation. Courtesy of *Martin Gascoigne-Pees*

Larvae are susceptible to bacteria and viruses, and attack from numerous predators and parasitoids. Birds are major predators, particularly when they are feeding their young (Fig. 4.20). Arthropod predators, including spiders and insects such as mantids and beetles, also take a toll but are important in controlling numbers. Parasitoids, especially from the order Hymenoptera, such as *Cotesia glomerata* (Fig. 4.19) and *Hyposoter* spp. (Figs 4.21 and 4.22), play a major role, with specific species attacking larvae at various stages of development (Makris, 2003: 65; John *et al.*, 2008a, 2020), but for a detailed account on parasitoids of European butterflies, see Shaw *et al.* (2009: 130). Mortality, concealment, evasion and defence are comprehensively explained in Dennis (2020: 168).

Fig. 4.19. *Pieris brassicae* (Large White) larva from which many parasitic wasp larvae of *Cotesia glomerata* (Hymenoptera, Braconidae) have emerged as the pierid larva was preparing to pupate. The parasitoids have killed the larva in the process, before forming their own cocoons. 22 May 2014. Courtesy of *David and Rosalyn Sparrow*

Fig. 4.20. *Oenanthe cypriaca* (Cyprus Wheatear) gathering food for its brood, including a larva of *Pontia daplidice* (Bath White), Tsada, 19 May 2014. *Christodoulos Makris*

Fig. 4.21. *Hyposoter* sp. (Hymenoptera: Ichneumonidae, Campopleginae) are known to attack various spp. of pierid butterflies in Cyprus. The single large wasp emerges through a hole created in the side of the larva, leaving a typical 'mummy-like' skin, as in the above remains of a *Pontia chloridice* larva, where *Hyposoter ebenitor* was implicated. Courtesy of *Richard Lyszkowski*

Fig. 4.22. The *Hyposoter* sp. illustrated here emerged from a similarly mummified larva of *Pontia glauconome* (Desert White) in Turkey. Courtesy of *Erdem Seven*

Pupa (Chrysalis)

'Chrysalis' derives from the Greek χρυσός (*chrysos*) = gold. Depending on the species, the chrysalis may be attached to the larval hostplant or to some other support by means of cremaster and girdle, for example *Papilio machaon* (Swallowtail) (Figs 4.23 and 4.24) and *Pontia chloridice* (Small Bath White) (Fig. 4.25), or suspended as in *Vanessa cardui* (Painted Lady) (Fig. 4.26) and *Ypthima asterope* (African Ringlet) (Fig. 4.29). Other pupae are formed without support at the base of the hostplant or even created within a loosely structured silken cocoon beneath stones (Figs 4.27, 4.28 and 4.30). Mobility is restricted to a wriggling movement of the abdominal segments (a weak defensive act in order to deter potential predators) and in some species the pupa may also produce sound. As with earlier stages of the life cycle, pupae vary in structure and coloration and often are recognizable to species or at least to genus. Careful examination of the pupal case will reveal outlines of adult structure, especially the wings, which appear in miniature form and become especially apparent shortly before eclosion (see Chapter 12, Fig. 12.27). Many species have evolved to spend long periods in this stage, particularly overwintering species that may enter pupal diapause in autumn, timing their emergence in spring to coincide with the availability of fresh growth of their hostplants. In some desert species, where environmental conditions might not favour emergence, pupae may remain in diapause for years (Benyamini, 2008; John *et al.*, 2020).

As with other early stages, pupae are also prone to predation, especially before the pupal cuticle has hardened. Cryptic coloration is the most common form of defence, helping to camouflage the pupa against its background. In some species, e.g. *P. machaon* (Swallowtail), pupal coloration may be green or brown (Figs 4.23 and 4.24), depending on substrate

Fig. 4.23. *Papilio machaon* (Swallowtail) pupa (green form) attached by girdle and cremaster, 3 November 2012. *Christodoulos Makris*

Fig. 4.24. *Papilio machaon* (Swallowtail) pupa (brown form) attached by girdle and cremaster, 3 November 2012. *Christodoulos Makris*

texture and various environmental factors (Lindstedt *et al.*, 2019; Dennis, 2020: 289). The markings and coloration of some pupae, for example those of *Pontia chloridice* (Fig. 4.25), mimic a bird dropping. Parasitoids such as *Pteromalus puparum* (Linnaeus, 1758) (Hymenoptera: Chalcidoidea,

Fig. 4.25. *Pontia chloridice* (Small Bath White) pupa attached by girdle and cremaster to the hostplant *Cleome ornithopodioides*. Courtesy of *Martin Gascoigne-Pees*

Fig. 4.26. *Vanessa cardui* (Painted Lady) pupa attached by its cremaster hooked into a pad of silk formed below the leaf, 28 December 2012. *Christodoulos Makris*

Fig. 4.27. *Glaucopsyche paphos* (Paphos Blue) reared pupa, formed just below the soil surface or beneath leaves, 20 May 2014. Courtesy of *Marios Aristophanous*

Fig. 4.28. *Aricia agestis* (Brown Argus), the (reared) pupa is formed at the base of the hostplant. *Eddie John*

Fig. 4.29. *Ypthima asterope* (African Ringlet) pupa (reared) suspended by its cremaster, 7 December 2007. Courtesy of *Martin Gascoigne-Pees*

Fig. 4.30. *Hipparchia cypriensis* (Cyprus Grayling) pupa (reared). In this species, the pupa is formed on the ground below the hostplant. Courtesy of *Martin Gascoigne-Pees*

Pteromalidae), a gregarious endoparasitoid wasp, are widely distributed in the region and known to attack species such as *Papilio machaon* and *Pieris brassicae* (John *et al.*, 2021a).

Imago (Adult Butterfly)

During the period of pupation, larval tissues are liquefied and, sometimes in a little over a week, transformed into the adult butterfly. Nearing emergence, the wing markings often become visible through the pupal case, indicating that eclosion is imminent. Once fully extracted from the chrysalis, the butterfly expands its wings by means of pumping haemolymph through the wing veins. In an hour or two, when the wings have fully expanded and had time to harden, the butterfly is ready to take to the wing, but even then it is very vulnerable to predation from other invertebrates, e.g. spiders, mantids and dragonflies, as well as vertebrates such as birds, lizards, mantids and chameleons (Figs 4.31–4.34).

For further examples of predatory behaviour in Cyprus, please refer to Makris (2003: 64).

Fig. 4.31. Spiders of the Thomisidae family (Crab spiders) are frequent predators, attacking both large and small butterflies, 9 May 2015. *Eddie John*

Fig. 4.32. *Zizeeria karsandra* (Dark Grass Blue), attacked by a spider of the family Thomisidae, 15 August 2009. Courtesy of *Sami Tamson*

Fig. 4.33. *Machimus* sp. (Diptera: Asilidae) (Robber Fly), predating on *Lasiommata maera* (Large Wall Brown), 28 April 2016. *Christodoulos Makris*

Fig. 4.34. *Merops apiaster* (European Bee-eater) with a captured *Hipparchia syriaca* (Eastern Rock Grayling), 29 September 2007. Courtesy of *Dietmar Schuphan*

5 Habitats

Cyprus, with a total area of 9251 km^2, is an island of greatly contrasting ecological regions. From the Akamas in south-west, the land rises to the richly forested Troodos Mountains, with *Pinus brutia* (Calabrian Pine) dominating from sea level to 1400 m. *Pinus nigra* ssp. *pallasiana* (Black Pine) is found at higher elevations to Mount Olympos at 1950 m. On the northern and southern slopes, *Quercus infectoria* ssp. *veneris* (Aleppo Oak) is present, with *Quercus alnifolia* (Golden Oak) growing from 400 m to 1700 m. On the northern side of the Troodos, the land falls away to the Mesaoria Plain that stretches the width of the island from Morfou Bay to Ammochostos Bay, averaging 100 m in elevation. The plain separating the Troodos Mountain range from the northern Pentadaktylos (Keryneia) Mountain range is largely flat and constitutes the extensively cultivated agricultural heartland of the island.

The Pentadaktylos range extends across much of the northern part of the island and unlike the largely igneous Troodos consists mainly of limestone, producing a somewhat different flora, including forests of *Cupressus sempervirens* (Mediterranean Cypress). *Arbutus andrachne* (Eastern Strawberry Tree) is common in both mountainous regions of Cyprus with extensive stands, especially in the region of Kantara in the east of the Pentadaktylos range. Kyparissovounos (1024 m) represents the highest peak of the Pentadaktylos and at its eastern point, entering Karpasia, *Juniperus phoenicea* (Phoenician Juniper) dominates, as in coastal plains and areas of the Akamas Peninsula, Kormakitis in the far north-west and Kavo Gkreko at the south-east tip of the island. In the valleys, riparian forests, principally with *Platanus orientalis* (Oriental Plane), *Alnus orientalis* (Oriental Alder) and galleries with *Salix alba* (White Willow), *Populus alba* (White Popular) and *Nerium oleander* (Oleander) may be found, providing shade or dappled sunlight, especially

DOI: 10.1079/9781800621275.0005

in lowland areas, for butterfly species such as *Pararge aegeria* (Linnaeus, 1758) (Speckled Wood).

The two natural salt lakes at Larnaka and Akrotiri, the permanent wetland of, for example, Oroklini and temporary wetland areas such as Sygkrasi and Kouklia, together with watercourses and areas abutting dams, provide habitats for a variety of species of butterfly during periods when the island's vegetation begins to desiccate in the heat of summer. Lycaenids in particular, thrive in such areas. It may surprise that hundreds of wetland sites, both permanent and temporary, were identified in an island-wide survey carried out by Markogianni *et al.* (2014), and the importance of the preservation of 'riverine vegetation habitats and moist microhabitats for butterfly species conservation' was highlighted by Tzirkalli *et al.* (2019).

The following photographs (Figs 5.1–5.31) provide an impression of the varied topography of Cyprus; for detailed information on habitats and on the geology of the island, the reader is directed to Makris (2003: 18–26), Wagstaffe (2016: 12–23) and Vogiatzakis *et al.* (2016: 24–39).

Fig. 5.1. Facing east from north-western Cyprus (526 m) with the Mesaoria Plain in the far distance separating the Pentadaktylos Mountains (L) and Troodos Mountains (R), 21 May 2013. *Eddie John*

Fig. 5.2 Tripylos peak, Troodos Mountains, 22 October 2007. Courtesy of *Martin Gascoigne-Pees*

Fig. 5.3. Mountain slope with *Cedrus brevifolia* and *Quercus alnifolia* in Paphos Forest, habitat of *Favonius quercus*, 29 May 2007. *Christodoulos Makris*

Fig. 5.4. Paphos Forest, 1 October 2015. *Christodoulos Makris*

Fig. 5.5. Rocky valley just below Troodos peak with *Juniperus foetidissima* and *Genista fasselata*. 6 June 2009. *Christodoulos Makris*

Fig. 5.6. Rocky areas at Madari peak, habitat of *Pseudochazara anthelea*. 9 October 2015. *Christodoulos Makris*

Fig. 5.7. Forest track (760 m) above St Hilarion, Pentadaktylos (*Arbutus andrachne* in centre), habitat of a variety of species, including *Charaxes jasius*, 23 May 2013. *Eddie John*

Fig. 5.8. Kantara Mountains (400 m), Pentadaktylos range, with *Arbutus andrachne*, 25 May 2013. *Eddie John*

Fig. 5.9. Flowery slopes at St Hilarion (550 m) Pentadaktylos, 11 May 2011. Courtesy of *David and Rosalyn Sparrow*

Fig. 5.10. Grassland habitat (180 m) near Mousere, Troodos foothills, 14 May 2007. *Eddie John*

Fig. 5.11. Grassy areas and road margin with the endemic thistle *Onopordum cyprium* near Agia Varvara, habitat of *Chazara briseis larnacana*. 10 May 2020. *Christodoulos Makris*

Fig. 5.12. Grassy valley near Anarita, habitat of *Ypthima asterope*. 8 June 2015. *Christodoulos Makris*

Fig. 5.13. Akaki River (456 m), NW of Machairas Forest, with oleander in flower, 29 May 2013. *Eddie John*

Fig 5.14. Diarizos River (460 m), Arminou Dam, 1 June 2013. *Eddie John*

Fig. 5.15. Phrygana with *Sarcopoterium spinosum*, habitat of *Apharitis acamas*, near Vretsia, 11 June 2009. *Christodoulos Makris*

Fig. 5.16. Phrygana with *Sarcopoterium spinosum*, hillside habitat of *Apharitis acamas*, near Vretsia, 20 October 2021. *Christodoulos Makris*

Fig. 5.17. Riparian vegetation at Diarizos River (460 m), Kelephos Bridge, 1 June, 2013. *Eddie John*

Fig. 5.18. Dry bed of Xeros River (220 m) below Lefka Reservoir, 21 May 2013. *Eddie John*

Fig. 5.19. Dry riverbed with reeds, habitat of *Gegenes pumilio* and *Pelopidas thrax*. Diarizos River near Kouklia, 4 May 2021. *Christodoulos Makris*

Fig. 5.20. Cultivated fields with isolated shrubs of *Ziziphus lotus* near Peristerona, habitat of *Tarucus balkanicus*, 20 October 2021. *Christodoulos Makris*

Fig. 5.21. Kioneli Dam (190 m) Mesaoria Plain, habitat of several species of blue, with the Pentadaktylos range in the background, 23 May 2013. *Eddie John*

Fig. 5.22. Temporary wetland at Sygkrasi Dam (50 m), east of Lefkonoiko, on 7 May 2015 before drying out completely during the summer. *Eddie John*

Fig. 5.23. Temporary wetland at Kouklia (20 m), 27 May 2013. *Eddie John*

Fig. 5.24. Oroklini Marsh, habitat of various lycaenids, including *Tarucus balkanicus*, 20 June 2006. *Eddie John*

Fig. 5.25. Storm approaching Cape Kormakitis, 11 May 2015. In better spring weather, migrant *Vanessa cardui* have been observed heading out to sea from this location. *Eddie John*

Fig. 5.26. Coastal locations provide valuable habitats for a wide variety of butterflies, including *Chilades trochylus*. Cape Apóstolos Andreas, with Kleides Islands in the distance, 3 May 2016. *Eddie John*

Fig. 5.27. *Limonium sinuatum* (winged sea lavender) and various spp. of thistle at Agios Filon, Karpasia, 5 May 2016. *Eddie John*

Fig. 5.28. Coastal maquis north of Eptakomi, Karpasia, 18 October 2017. *Eddie John*

Fig. 5.29. Wild flowers in abandoned vineyards near Agios Nikolaos, 1 May 2009. *Christodoulos Makris*

Fig. 5.30. Garigue with *Genista fasselata* near Paramali, habitat of *Glaucopsyche paphos*, 11 March 2014. *Christodoulos Makris*

Fig. 5.31. Hills and cultivated fields near Marki, 11 March 2020. *Christodoulos Makris*

6 Butterfly Behaviour

Entire books have been devoted to this subject, so to discuss in detail the many fascinating and sometimes inexplicable aspects of butterfly behaviour would require much more space than is available here – in what is essentially a field guide – and we make no apology for the inevitable 'light touch' to some of the behavioural traits commented upon below. In the preface to his recent magnum opus, Dennis (2020) encapsulated the subject beautifully: 'Butterfly biology is not physics and there are rarely neat solutions to explain traits in life history and behaviour … '.

While courtship and mating, territoriality, hilltopping and thermoregulation, among other behavioural activities, are mentioned occasionally under the various species entries, here we discuss migration, aestivation, overwintering habits and feeding behaviour with particular relevance to the Cyprus butterfly fauna. Brief mention is also made of interspecific/intergeneric mating.

Butterfly Migration and Dispersal

Keen observers, especially those living in or visiting Cyprus at times of peak bird migration, will be familiar with this fascinating and regular seasonal event in the avian world. Worldwide, some 568 species of butterfly are also considered migratory, with the greatest proportion of migrants belonging to the family Pieridae, followed by those of the families Papilionidae, Nymphalidae, Lycaenidae and Hesperiidae (Chowdhury *et al.*, 2021). Migration of butterflies has been documented for centuries; Williams *et al.* (1942), for example, referred to an account from Germany of a possible migration of *P. brassicae* in the year 1100 (see also Williams, 1930: 12). Numerous species of butterfly in the Levant also migrate, but

DOI: 10.1079/9781800621275.0006

because of their much shorter lifespan – for many, measured in days or weeks – the passage of each long-distance individual migrant is unidirectional. In some species, a return migration is undertaken by the progeny of those migrants, e.g. *Vanessa cardui* (Painted Lady) and *Vanessa atalanta* (Linnaeus, 1758) (Red Admiral). In America, *Danaus plexippus* (Monarch) is a notable exception, as this travels thousands of kilometres from breeding areas in northern regions of America to overwintering sites in Mexico, with the same individuals setting off north in spring from winter hibernation roosts, breeding en route to the north.

Note that uphill dispersal by *Hipparchia cypriensis* (Cyprus Grayling) during late spring/early summer, with return autumn movement, is undertaken for an entirely different reason and is discussed further under 'Aestivation' (below) and in the entry for this species in Chapter 12, as well as in John and Parker (2002). Elsewhere in the Levant, four butterfly species, only one of which is known from Cyprus (*Issoria lathonia*, Queen of Spain Fritillary), are known to 'migrate' to lower levels in order to hibernate as imagines (Larsen, 1976).

The term 'dispersive migration' as used by Schneider (1962) might conveniently be applied to species that appear gradually to extend their range. In recent years, for example, areas of Turkey and Syria have experienced the arrival and establishment of *Papilio demoleus* (Lime Swallowtail), a species that appeared in Cyprus for the first time in 2021 (see species entry).

Unlike birds, timing, extent and frequency of butterfly migrations is less predictable. In most years, migratory numbers of butterflies are low and fail to attract attention; in other years, such as in 2001, 2014 and spectacularly in 2019, many millions of butterflies, particularly of *V. cardui* (Painted Lady), may pass through the island in spring (John, 2001a; John *et al.*, 2015a). Sporadic butterfly migration, as occurs across Cyprus, is often triggered by an irruption of large numbers resulting from favourable breeding conditions in southern mainland areas of the Levant, requiring mass movement in order to seek out suitable breeding areas for a subsequent generation. Writing on *V. cardui*, Larsen (1976) referred also to there being 'much intraregional migration with the Mediterranean and its fringes; much of this appears to be of an irregular and unpredictable nature, almost nomadic'. Pittaway (1985), too, referred to the nomadic nature in Saudi Arabia of species such as *V. cardui* and *Belenois* (*Anaphaeis*) *aurota* Fabricius, 1793 (Caper White), the latter discussed in Chapter 13.

Insect migration, especially of the nearly cosmopolitan *V. cardui* (Fig. 6.1), has been well studied in recent years (e.g. Stefanescu *et al.*, 2012; Stefanescu *et al.*, 2016; Talavera *et al.*, 2018). However, this rarely attracts public attention unless the migrating swarm becomes highly noticeable or results in human inconvenience, as when butterflies become casualties when colliding with vehicle windscreens.

The island's proximity to mainland countries of the eastern Mediterranean, perhaps in concert with a warming climate, has attracted unusual migratory species and the current checklist of 56 species includes

Fig. 6.1. One of *ca.* 500 *Vanessa cardui* (Painted Lady) 'refuelling' on cultivated *Pelargonium crispum* (Geraniaceae) in Karpasia during a brief pause in the 2014 migration, 23 April 2014. *Eddie John*

a number of extremely rare migrants to the island. The infrequency with which some appear perhaps warrants a description of 'stray' or 'vagrant', rather than being considered as the passage of large numbers of migrants. However, the possibility should not be overlooked that these species might arrive more frequently than is assumed, but in numbers that do not draw attention – they may be overlooked in a mixed migration, for example, where a species such as *V. cardui* is predominant. Furthermore, the main body of a migration might not pass directly over Cyprus, with perhaps just the fringe crossing over some of the island's peninsulas, thereby greatly reducing the chances of detection. This phenomenon may have happened in the autumn of 1986, when small numbers of *Catopsilia florella* (African Migrant) were seen in south-eastern Cyprus where, at three locations between Ayia Napa and Frenaros, 11 specimens were collected by Pavlos Neofytou (Makris, 2003: 130). Unfortunately, there is no known record of the extent of this immigration, or that of an earlier event in the 1960s, or of a more recent appearance in 2019, when a single female of *C. florella* was photographed in Karpasia in the extreme north-east of the island (John *et al.*, 2019b), when very large numbers of *V. cardui* were migrating.

Even fewer numbers of other rare migrants are worth noting. In August 1936, Prof. Bytinski-Salz collected a single worn male *Hypolimnas misippus* in Limassol, the only known record of this species in Cyprus. In July 1940, Themistokles Shiakides collected four specimens of *Azanus jesous* (Guérin-Méneville, 1849) (African Babul Blue) in Nicosia (Makris, 2003: 180), almost certainly in the same habitat as that used by *Luthrodes galba* (Lederer, 1855) (Small Desert Blue), with which the species flies in other countries. Seventy-five years later, in broadly the same area of Nicosia, a singleton of *A. jesous* (see Chapter 12, Fig. 12.172) was photographed by Hasan Bağlar in July 2015. *A. jesous* and *L. galba* are similar in size and appearance, requiring an experienced observer to detect the differences between the two species. This remark also applies to the discovery of *Deudorix livia* (Klug, 1834) (Pomegranate Hairstreak) by Gabriel Georghiou in September 1994 (Makris, 2003: 146), the single female taken strongly resembling *Lampides boeticus* (Linnaeus, 1767) (Long-tailed Blue).

Three other very recent discoveries are worthy of mention.

In June 2010, Daniel and Hilary Haines photographed a 'grayling-type' butterfly in the Kantara Mountains, which was subsequently identified with an acceptable degree of certainty as *Chazara persephone* (Great Steppe Grayling) (see Chapter 12, Fig. 12.355), last known in Cyprus about 100 years ago. Only a single specimen was seen, but whether this represented a vagrant or a species maintaining a tenuous presence is not known (John *et al.*, 2011).

In 2016, Yiannakis Makriyianni photographed *Issoria lathonia* (Queen of Spain Fritillary) (see Chapter 12, Fig. 12.330), nectaring on *Clinopodium vulgare* at Pano Amiantos on 26 May, and formally reported here for the first time. This, too, might have been a vagrant (see species entry).

Meanwhile, the rapid range expansion of *Papilio demoleus* (Lime Swallowtail) in nearby Mediterranean Syria and south into Lebanon has been followed with interest, as arrival in Cyprus and Israel had been anticipated (John *et al.*, 2021a). This highly attractive butterfly was initially photographed in Cyprus by Feriha Bağlar and shortly afterwards by her son, Hasan Bağlar (see Chapter 12, Fig. 12.15), in early August 2021; it is considered that these were progeny of migrants that had arrived undetected a month or so earlier (John *et al.*, 2021b; John *et al.*, 2022a).

That so few migrants are known to reach Cyprus may be masked by the fact that a number of well-known migrant species, for example *L. boeticus*, *Leptotes pirithous* (Linnaeus, 1767) (Lang's Short-tailed Blue), *V. atalanta*, *V. cardui* and *Pelopidas thrax* (Hübner, [1821]) (Millet Skipper), also breed on the island; these are rightly regarded as established resident species, as well as maintaining the status of 'migrant'. Furthermore, unless spectacularly large immigrations occur, it is unlikely that such movement will be apparent other than to an experienced observer able to distinguish migratory activity from normal flight behaviour. 'Thin' migrations may involve just a few specimens passing through an area every minute, but with each heading purposefully in the same direction and usually at the same height – see Williams *et al.* (1942) and Williams (1958) for a classification of migrant flight patterns.

In the great *V. cardui* migration years of 2001 and 2019, tens or even hundreds of millions passed across Cyprus from the south-east, with daily waves of immigrants arriving before dawn and continuing until early afternoon across an estimated island-wide 200 km front, heading generally N/NW through the island. In 2001, peak numbers occurred over a four-day period, with numbers dwindling as the migration continued into a second week. Across Cyprus, the migrants were often seen following topographical features (roads, valleys and dry riverbeds) with some, having reached the northern coastline, tracking the coast from east to west before leaving at Cape Kormakitis, the most northerly point in the west of the island. From there, the migrants headed out to sea in the direction of Turkey, where they were later observed arriving in off the sea (John, 2001a; see also Williams, 1970). Of course, 'corridors' (see Chapter 14) such as the landscape features described are not a prerequisite for large-scale migrations (see Shreeve and Dennis, 2009), as migrants are required to cross the

Mediterranean before and after leaving Cyprus, and may form a migratory population far exceeding the width of the island, as in 2019. However, there seems little doubt that within a region, landscape features assist dispersal of some species (e.g. John *et al.*, 2020; Başbay *et al.*, 2020b). Such was the intensity of the migration in 2001 that many hundreds, or even thousands, of migrating butterflies were killed as they flew across or along lanes of traffic.

In 2019, however, even the 2001 migration was dwarfed by an extraordinarily huge migration of *V. cardui* across large swathes of the eastern Mediterranean. Numerous reports indicated that the 'swarm' of migrants extended from Israel, north through Lebanon, Syria and Turkey, and continued north-westwards across Cyprus into western Turkey and further westwards to the Greek islands and mainland Greece from Athens to the Thessaloníki area (numerous observers, including from Lebanon, Syria, Turkey, Cyprus, Rhodes, Crete, Greece, 2019, personal communications to first author). From 21 March, experienced observers in Cyprus frequently referred to 'clouds' of migrants, with one noting 'the movement was tending to be west towards the coast and could, at times, be described as a blizzard and was truly not countable' (David Whaley, Cyprus, 2019, personal communication to first author). The initial migration reduced after a week, but was followed by successive waves of migrant *V. cardui* throughout April and into the first week of May. Over a period of about five weeks, the first author received reports from areas of the eastern Mediterranean ranging (in the east) from Israel, Lebanon and south-central areas of Turkey, to Rhodes, Athens and the Halkidhikí Peninsula (Greece) in the west.

Return migration by the progeny of *V. cardui* and *V. atalanta* largely accounts for the reappearance in Cyprus of these species in autumn. At a time when larval activity for both species is at a peak in more northerly latitudes, such activity is unsustainable in Cyprus due to the desiccation of hostplants in a parched landscape. Conversely, when larval hostplants are dying off in northern Europe, autumn rains in Cyprus and in areas to the south, such as the Negev in southern Israel, trigger hostplant growth, occasionally inducing a period of intensive egg-laying by returning migrants, encouraging successive broods over the winter and a considerable build-up of numbers.

Undoubtedly, one of the most spectacular migrant species, in appearance rather than substantial numbers, is *Danaus chrysippus* (Plain Tiger). This species is infrequently reported in most years, but appears in larger numbers in others, as in autumn 1998 when it could be found with some regularity along shorelines throughout the island (John, 2000; Makris, 2003: 208; John *et al.*, 2019a).

Other species of butterfly (and moth) migrate to or from the island, with numbers varying considerably year by year, so that in some years influx or departure may be so minimal as not to arouse interest, and requiring an experienced eye to appreciate the significance of such movement (John *et al.*, 2008b). Migrations may include mixed butterfly species, along with various day-flying moths, particularly *Macroglossum stellatarum* (Linnaeus) (Hummingbird Hawkmoth), *Heliothis peltigera*

Fig. 6.2. A migrant *Hyles livornica* (Striped Hawkmoth) nectaring on *Convolvulus arvensis*, Karpasia, 23 April 2014. *Eddie John*

Table 6.1. Butterfly species recently documented as migrants to Cyprus.

Frequent migrants	Occasional migrants	References
	Papilio demoleus[1]	John *et al*., 2021b
Danaus chrysippus		Makris, 2003: 208; John and Vane-Wright, 2019; John *et al*., 2019a
Vanessa cardui		John, 2001a; Makris, 2003: 226; John *et al*., 2015a
Vanessa atalanta		John, 2001a; Makris, 2003: 230
	Pieris rapae	John *et al*., 2008b; McArthur and John, 2019
	Catopsilia florella	Makris, 2003: 130; John *et al*., 2019b

[1]Very recent arrival (2021); may establish and/or appear as an occasional or frequent migrant.

(Denis & Schiffermüller) (Bordered Straw) and the spectacular *Hyles livornica* (Esper, 1780) (Striped Hawkmoth) (Fig. 6.2), as well as other insects (John *et al*., 2015a, 2019b).

Numerous publications are available for those wishing to read more on the fascinating subject of migration, e.g. Williams (1930, 1958); Baker (1968, 1978); Johnson (1969); Shields (1974); Petrakis & Legakis (2005); Dennis (2020: 71).

Table 6.1 lists those butterfly species in Cyprus for which evidence of migration has recently been both observed and documented, while

Table 11.1 ('Residency, broods and main habitats of butterflies in Cyprus', see Chapter 11) indicates the migratory capability of butterfly species historically reported for Cyprus.

Aestivation

Although some species, e.g. *Gegenes pumilio* (Hoffmansegg, 1804) (Pigmy Skipper), are highly tolerant of hot summer temperatures, others cannot survive in overly harsh conditions; aestivation (or summer diapause) describes the habit of some species to spend the summer in a reduced state of activity, in situations where they can escape hot, dry conditions. The cooler temperatures, shade and access to moisture found in mountainous areas provide a haven for species such as *Hipparchia cypriensis* (Cyprus Grayling) and *Hipparchia syriaca cypriaca* (Staudinger, 1879) (Eastern Rock Grayling). This seems a curious way for these species to have evolved, from which it might be deduced that they are not well adapted to survive in the Cyprus climate, but the fact that both are very common species points to a successful adaptation to the climatic conditions of the island.

Upon emergence in spring, *H. cypriensis* will mate early before both sexes disperse into the hills, where they remain only partially active until the cooler weather and rains of autumn induce a return to lower elevations (John and Parker, 2002; Parker and John, 2003). During the period of aestivation, delayed ovarian maturation ensures that the females are able to postpone oviposition until growth of the hostplants (species in the family Poaceae) is stimulated by rain in October and November. Although fewer males appear to survive aestivation, their persistence until the autumn is somewhat puzzling – if mating takes place in spring, what is the purpose of their survival until autumn? Possibly, their presence at aestivation sites helps reduce predation of gravid females due to the dilution effect (but must surely have minimal benefit, due to the small number of males present), or their role is to mate with those few females that did not mate in spring. Alternatively, some may simply live longer! While very few autumn matings of *H. cypriensis* have been reported (see species entry), post-aestivation survival of males evidently provides a further opportunity to seek out virgin females.

Large aggregations of aestivating groups of *H. syriaca* and *H. cypriensis* often occur in mountain refuges, where a disturbed butterfly is likely to provoke others into brief flight. However, for the most part they remain at rest on trunks of pine trees or deep in the shade of bushes from which they may emerge periodically, individually or in groups, to seek out moisture at streams or puddles. Other species, such *Argynnis pandora* (Cardinal), normally present at these higher elevations, may enter a short period of aestivation, perhaps accounting for the presence in autumn of occasional (and apparently fresh) specimens of a species usually regarded as univoltine (but see species entry).

The uphill movement by satyrines to suitable aestivation sites, sometimes involving kilometres of travel into the mountains from lower elevations, and the return in autumn to lower breeding areas is best described as seasonal dispersal or seasonal return migration rather than true migration.

Overwintering and Hibernation

Although short, winters in Cyprus can occasionally be cold, with snow a common sight at the highest elevations of the Troodos Mountains. Nevertheless, it is a time of year when several non-hibernating species may still be seen on the wing on sunny days, particularly in milder coastal areas (see Chapter 8). Most species, however, overwinter in the larval or pupal stage. Further information may be found under the entries for each of the species.

Intergeneric/Interspecific Mating

Although the authors are unaware of documented reports of intergeneric or interspecific mating relating to butterflies in Cyprus, this has been reported elsewhere involving Papilionidae, Pieridae (for example the *Pontia daplidice/edusa* complex) and especially Nymphalidae (e.g. Russell, 2013; Baidya *et al.*, 2018), sometimes resulting in fertile hybrid offspring in the case of interspecific mating (Descimon and Mallet, 2009: 219). Such activity appears less commonly reported in Lycaenidae, but has been documented by Bhakare and Smetacek (2010). Figures 6.3 and 6.4 appear to illustrate an example of apparent intergeneric courtship between

Fig. 6.3. Apparent intergeneric courtship involving *Leptotes pirithous* ♀ (above) and *Tarucus balkanicus* ♂ (below), 27 July 2011. *Courtesy of Jean Massry*

Fig. 6.4. *Leptotes pirithous* ♀ (above) and *Tarucus balkanicus* ♂, 27 July 2011. *Courtesy of Jean Massry*

a female *Leptotes pirithous* and a male *Tarucus balkanicus*; a sequence of (six) photographs sent to the first author seemed to show the female taking the initiative. However, no pairing was achieved before the butterflies flew off separately. Coutsis and Pamperis (2012) also reported on a homosexual 'pairing' between males of *Hyponephele lycaon* (Rottemburg, 1775) and *Maniola jurtina*.

Mate Location

Various methods are used in order to enhance mating opportunities. 'Hilltopping', first described by Shields (1967), is a mate location activity associated with a number of species, but especially so (in Cyprus) with *Papilio machaon* (Swallowtail), where males assemble at summits, ridges or hilltops and await the arrival of virgin females. Females spend little time in the area and once mated return to lower levels, creating the impression that only the males engage in hilltopping (Henning, 1990). In Cyprus, such activity is also associated with, for example, *Charaxes jasius* (Two-tailed Pasha) and *Lasiommata maera* (Linnaeus, 1758) (Large Wall Brown). Males of other species may patrol along tracks or field margins in search of a female; *Anthocharis cardamines* (Linnaeus, 1758) (Orange Tip) is a typical example. *Pararge aegeria* (Speckled Wood) establishes a territory into which a female might fly and may aggressively intercept intruders into its chosen territory; many male Hesperiidae and Lycaenidae will also select a favoured perch and await the arrival of a female (Dennis and Shreeve, 1988). *Lasiommata megera* (Linnaeus, 1767) (Wall Brown) may switch from perching to patrolling when on a hilltop (Dennis and Shreeve, 1988; Rutowski, 1991) and, as stated by Shields (1967), hilltopping 'almost certainly represents an independent and additional activity to perching and patrolling'.

'Ravining', a term coined by Tennent (1995), involves the patrolling of a territory along ravines or gullies, usually in semi-desert areas, and although this has yet to be documented for Cyprus, the behaviour of *Gegenes pumilio* (Pigmy Skipper), which often establishes territories along hot, dusty, sloping tracks, most closely resembles this activity. For detailed accounts of wider aspects of mate-locating behaviour in relation to habitat, see Dennis and Shreeve (1988), Rutowski (1991) and Dennis (2020: 233).

Adult Nutrition – Nectaring and Puddling

The presence of nectar plants varies according to season, with polyvoltine species adapting to seasonal nectaring opportunities. Figures 6.5–6.17 show a selection of important nectar plants favoured by some butterfly species in Cyprus; the majority of elevations and flowering periods are from Christofides (2017).

Fig. 6.5. *Chazara briseis* (The Hermit) nectaring on *Limonium sinuatum*, winged sea lavender (Plumbaginaceae), a common plant found on coasts (March–June) and widely used for early-season nectaring by numerous species, 8 May 2016. *Eddie John*

Fig. 6.6. *Glaucopsyche paphos* (Paphos Blue) nectaring on *Phagnalon rupestre* (Asteraceae), sea level to 800 m (March–May), 5 May 2015. *Eddie John*

Fig. 6.7. *Limenitis reducta* (White Admiral) nectaring on *Cistus creticus*, Cretan rockrose (Cistaceae), sea level to 1400 m (March–May), 11 May 2012. Courtesy of *Alison McArthur*

Fig. 6.8. *Glaucopsyche paphos* (Paphos Blue) nectaring on *Helichrysum stoechas*, gold flower (Asteraceae), sea level to 800 m (March–May), 29 April 2015. *Christodoulos Makris*

Fig. 6.9. *Pontia daplidice* (Bath White) nectaring on *Helichrysum italicum*, curry plant (Asteraceae), 300–1500 m (May–September), 16 July 2020. *Christodoulos Makris*

Fig. 6.10. *Echium angustifolium*, narrow-leaved bugloss (Boraginaceae) found from sea level to 1100 m (March–July), a popular nectar plant, here visited by *Papilio machaon* (Swallowtail), 16 April 2010. *Christodoulos Makris*

Fig. 6.11. *Chazara briseis larnacana* (The Hermit) nectaring on *Onopordum cyprium*, Cyprian donkey thistle (Asteraceae), sea level to 1400 m, (April–July), 10 May 2020. *Christodoulos Makris*

Fig. 6.12. *Gonepteryx cleopatra* (Cleopatra) nectaring on *Ptilostemon chamaepeuce* (Asteraceae), 300–1500 m (May–July), 20 June 2019. *Christodoulos Makris*

Fig. 6.13. *Hipparchia syriaca cypriaca* (Eastern Rock Grayling) on *Echinops spinosissimus*, globe thistle (Asteraceae), sea level to 1200 m (July–August); note the phoretic mite infestation around the head and basal wing areas, 23 July 2012. *Christodoulos Makris*

Fig. 6.14. *Luthrodes galba* (Small Desert Blue) nectaring on the frequently used *Rubus sanctus*, blackberry (Rosaceae), sea level to 1500 m (June–October), 25 July 2015. Courtesy of *Nick Morgan*

Fig. 6.15. *Pontia daplidice* (Bath White) nectaring on *Heliotropium hirsutissimum* (Boraginaceae), another valuable nectar plant found from sea level to 1400 m (May–November), 9 September 2011. Courtesy of *Colin Baker*

Fig. 6.16. *Colias croceus* (Clouded Yellow) nectaring on *Pterocephalus multiflorus*, 50–1800 m (June–November), 11 June 2013. *Christodoulos Makris*

While most species of butterfly derive their energy from nectar, others take nectar sparingly or not at all. *Charaxes jasius* (Two-tailed Pasha) is well known as a species that is attracted to highly noxious substances such as faeces (including dog) and carrion, but is also one of a number of species to imbibe sap from damaged trees or to feed on fermenting fruit (Figs 6.22, 6.23) and to puddle at liquids ('mud-puddling') for nutrients and minerals (Figs 6.19, 6.20). Other species, such as *Favonius quercus*

Fig. 6.17. *Gonepteryx cleopatra* (Cleopatra) nectaring on *Dittrichia viscosa*, woody fleabane (Asteraceae), sea level to 1400 m (August–November), 19 October 2017. Thirteen other species of butterfly and two species of moth were photographed on this valuable late source of nectar in October 2017. *Eddie John*

(Linnaeus, 1758) (Purple Hairstreak) also mud-puddle but honeydew from aphid secretions is a main source of nutrient for this species. See also Erhardt and Mevi-Schütz (2009) and especially Dennis (2020: 223) for a more detailed discourse on adult food resources in butterflies.

Although more commonly observed in tropical areas (Tea *et al.*, 2020), a phenomenon rarely reported in the Levant is that of seawater puddling; this has occasionally been reported in Cyprus for various satyrines and *Thymelicus acteon* (Rottemburg, 1775) (John and Tennent, 2012), as has that of mud-puddling by post-aestivation *Maniola cypricola* and *Chazara briseis* (Fig. 6.21) (John and Dennis, 2019). In Syria, marine puddling has recently been observed for *Papilio demoleus* (Fig. 6.18) (John *et al.*, 2021b).

Fig. 6.18. *Papilio demoleus* (Lime Swallowtail) marine puddling in Syria, 12 August 2021. The dead fish had just been discarded by a fisherman (see entry under *P. demoleus*). Courtesy of *Mudar Salimeh*

Fig. 6.19. An aggregation of 10 *Pieris rapae* (Small White) mud-puddling, 10 May 2012. Courtesy of *David and Rosalyn Sparrow*

Fig. 6.20. Three of a larger group of *Glaucopsyche paphos* (Paphos Blue) mud-puddling, 14 May 2007. *Eddie John*

Fig. 6.21. *Chazara briseis* (The Hermit) ♀ mud-puddling post aestivation, 16 October 2017. *Eddie John*

Fig. 6.22. *Charaxes jasius* (Two-tailed Pasha) probing at tree sap, 24 October 2015. As many as four individuals were present – a rare sight in Cyprus. Courtesy of *Mike Lloyd*

Fig. 6.23. *Charaxes jasius* (Two-tailed Pasha) feeding on rotting fruit, 3 August 2010. Courtesy of *Michael Field*

7 'Forms', Aberrations and Dwarfism

Aberrations, or an unusual 'form' will occasionally be seen, as with *Colias croceus* f. 'erateformis' (John *et al.*, 2006) (Fig. 7.13) and *Danaus chrysippus* ff. 'alcippus' and 'alcippoides' (John and Vane-Wright, 2019; John *et al.*, 2019a) (Fig. 7.3). The white form of *Colias croceus* f. 'helice' (Fig. 7.1) appears to be less common in Cyprus than in some other countries, representing < 5% of the *C. croceus* populations on the island (Makris, 2003: 122). Collections housing hundreds of specimens were often assembled in the past by lepidopterists keen to apply 'form' names based on small morphological variations. Perhaps an extreme example is provided by Turner (1920), who, writing on *'Anthocharis crameri'* (= *Euchloe ausonia*), stated, 'This common Mediterranean species, to which between thirty and forty varietal and aberrational names have been applied and more suggested, is a very difficult one to deal with.' Seasonal variations in *E. ausonia* (and other species) are recognized, and are influenced by external factors, as illustrated by differences in first and second broods (Makris, 2003: 300). However, it is important to acknowledge that variation 'is a fundamental characteristic of all animals' (Smart, 1991: 66); minor differences within a population are commonplace.

Examples of four aberrant specimens are shown alongside typically marked specimens in Figs 7.5–7.12. The appearance of *P. machaon* in Fig. 7.5 might have resulted from a failure of the yellow wing scales to develop properly. The partial exposure of the wing membrane beneath produces an effect in which the wing appears white and partially transparent (Dick Vane-Wright, Kent, 2015, personal communication to first author). See also *Pontia chloridice* in Chapter 12.

DOI: 10.1079/9781800621275.0007

Fig. 7.1. *Colias croceus* (Clouded Yellow) f. 'helice', 27 May 2013. *Eddie John*

Fig. 7.2. *Colias croceus* (Clouded Yellow) displaying the usual yellow coloration, 5 May 2015. *Eddie John*

Fig. 7.3. *Danaus chrysippus* (Plain Tiger) f. 'alcippus', 7 December 2018. Courtesy of *Matt Smith*

Fig. 7.4. *Danaus chrysippus* (Plain Tiger) normal form, 28 November 2016. *Eddie John*

Fig. 7.5. Example of a 'black and white' *Papilio machaon* (Swallowtail) exhibiting possible scale defect, 4 July 2015. Courtesy of *Andy Grant*

Fig. 7.6. Typical coloration of *Papilio machaon* (Swallowtail), 19 October 2017. *Eddie John*

Fig. 7.7. *Zizeeria karsandra* (Dark Grass Blue) lacking spots on the hindwing underside, 17 March 2015. Courtesy of *Mike Lloyd*

Fig. 7.8. Typically marked *Zizeeria karsandra* (Dark Grass Blue), 23 April 2014. *Eddie John*

Fig. 7.9. *Glaucopsyche paphos* (Paphos Blue) showing marked distortion of ocelli. Courtesy of *Bob Lambert*

Fig. 7.10. Typically marked *Glaucopsyche paphos* (Paphos Blue), 4 May 2016. *Eddie John*

Fig. 7.11. A possible *Hipparchia syriaca* (Eastern Rock Grayling), with an unusually strongly marked hindwing submarginal band and lacking yellow coloration, 7 October 2020. Courtesy of *Ray Atkinson*

Fig. 7.12. A typically marked specimen of *Hipparchia syriaca* (Eastern Rock Grayling), 21 June 2012. *Christodoulos Makris*

Dwarfism

Lack of foodplant caused, for example, by desiccation of hostplants in hot summer conditions, may induce (force) some final instar larvae into premature pupation before the normal larval size is attained. Hence, 'the size of adult insects varies in response to both temperature and diet' (Solonkin *et al.*, 2021). Dwarf butterflies may not be particularly common, but have been recognized for well over a century and occasionally were the cause of confusion among early lepidopterists. For example, Gibbs (quoted in Turner, 1920) remarked on a dwarf specimen thought to have been *P. rapae*: 'It then dawned upon me it was quite a small *brassicae*, which it proved to be.' The multiple-brooded *Colias croceus* (Fig 7.13) and *Vanessa cardui* are among species that occasionally produce undersized specimens in summer. John (2000) remarked on a dwarf form of *L. maera* and, in late April/early May 2016, following a dry winter, dwarf specimens of *Pieris rapae*, *Pontia daplidice*, *Colias croceus* and *Polyommatus icarus* were occasionally seen.

Fig. 7.13. This plate illustrates examples both of an unusual form of *Colias croceus* (f. 'erateformis') and of a dwarf female specimen (lower row). An even smaller dwarf form appears in Makris (2003: 52). *Christodoulos Makris*

8 Phenology

The Cyprus climate, with warm days possible at low elevations even in winter, allows some species to remain on the wing for much of the year and in some years encourages earlier than usual emergence of spring-flying species such as *Allancastria cerisyi cypria* Stichel, 1907 (Eastern Festoon) and *Euchloe ausonia* (Eastern Dappled White).

Species Recorded in Every Week/Month

Aggregated data over more than 20 years of recording show seven species to have been noted in all weeks of the year and a further ten species recorded at some point in every month (Table 8.1), but it should be emphasized that these data represent aggregated records and are not typical of a single year.

Number of Species Recorded per Month

Twenty-one butterfly species have been recorded on the wing at some point during January, with numbers rising slowly during February, before quickly increasing with the arrival of spring to a peak of 50 species in May. Note from Fig. 8.1 that at the height of summer there is a noticeable decline in the number of species recorded (44 in July and August), thereafter tailing off gradually from the end of September to the end of the year. Note also that this phenogram and the more detailed information shown in Table 8.2 are not truly representative of a 'typical' butterfly year because it includes exceptional sightings (unusually early or late) observed outside the normal flight period, such emergences most probably influenced by particularly favourable weather conditions. In a typical year, approximately three or four fewer species at the start or end of the year could be anticipated. Rare migrants have been included in the chart.

DOI: 10.1079/9781800621275.0008

Table 8.1. Butterfly species recorded in every week/month in Cyprus (aggregated CBRS data).

Species	Recorded in every week	Recorded in every month
Papilio machaon (Swallowtail)		*
Pieris brassicae (Large White)	*	*
Pieris rapae (Small White)	*	*
Pontia daplidice (Bath White)	*	*
Colias croceus (Clouded Yellow)	*	*
Gonepteryx cleopatra (Cleopatra)		*
Lycaena phlaeas (Small Copper)		*
Lampides boeticus (Long-tailed Blue)	*	*
Leptotes pirithous (Lang's Short-tailed Blue)	*	*
Zizeeria karsandra (Dark Grass Blue)		*
Vanessa atalanta (Red Admiral)		*
Vanessa cardui (Painted Lady)	*	*
Pararge aegeria (Speckled Wood)		*
Lasiommata maera (Large Wall Brown)		*
Lasiommata megera (Wall Brown)		*
Carcharodus alceae (Mallow Skipper)		*
Gegenes pumilio (Pigmy Skipper)		*

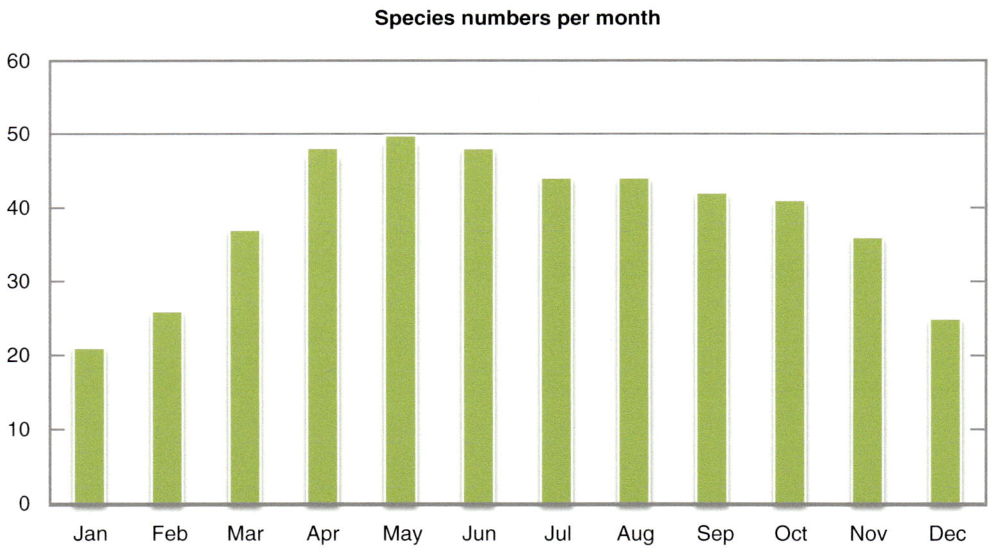

Fig. 8.1. Temporal activity of butterfly species in Cyprus, with a slight peak shown in May.

Phenogram

Table 8.2 (overleaf) shows the main flight periods and includes months in which early-emerging, or late-flying, specimens may be on the wing, subject to seasonal variations. Months in which very rare migrants have been observed are denoted by 'M', whereas *Nymphalis polychloros* (considered extinct) is shown as 'E'. Polyvoltine species such as *C. croceus* and *V. cardui* are present throughout the year, but desiccation of hostplants and probable emigration reduces numbers in high summer. Data have been aggregated from a vast number of personal observations by the authors of this volume and from direct contributions to the CBRS and published records. In total, 279 individuals have contributed to the recording scheme run by the first author since 2000, with some contributors reporting consistently for many years (see 'Acknowledgements').

Table 8.2. Butterfly phenology in Cyprus.

Species		J	F	M	A	M	J	J	A	S	O	N	D
Papilio machaon	Swallowtail	±	+	+	+	+	+	+	+	+	+	±	±
Papilio demoleus	Lime Swallowtail				±	+	?	?	+	+	+	+	?
Allancastria cerisyi	Eastern Festoon	±	±	+	+	+	±						
Aporia crataegi	Black-veined White				+	+	±						
Pieris brassicae	Large White	±	+	+	+	+	+	+	+	+	+	+	±
Pieris rapae	Small White	+	+	+	+	+	+	+	+	+	+	+	+
Pontia daplidice	Bath White	±	+	+	+	+	+	+	+	+	+	+	±
Pontia chloridice	Small Bath White			±	+	+	±	±	±	+	+	±	
Euchloe ausonia	Eastern Dappled White	±	±	+	+	±	±						
Anthocharis cardamines	Orange Tip	±	±	+	+	±							
Catopsilia florella	African Migrant				M						M	M	
Colias croceus	Clouded Yellow	±	+	+	+	+	+	+	±	+	+	+	±
Gonepteryx cleopatra	Cleopatra	±	±	+	+	+	+	±	±	±	±	±	±
Apharitis acamas	Levantine Leopard					±	+	+	+	+	±	±	
Lycaena phlaeas	Small Copper	±	±	+	+	+	+	+	+	+	±	±	±
Lycaena thersamon	Lesser Fiery Copper		±	±	+	+	+	+	+	+	+	+	±
Deudorix livia	Pomegranate Hairstreak									M			
Favonius quercus	Purple Hairstreak					±	+	+	+	±			
Lampides boeticus	Long-tailed Blue	±	±	±	±	+	+	+	+	+	+	+	±
Leptotes pirithous	Lang's Short-tailed Blue	±	±	±	+	+	+	+	+	+	+	+	±
Azanus jesous	African Babul Blue							M					
Tarucus balkanicus	Little Tiger Blue			±	±	+	+	+	+	+	±	±	±
Zizeeria karsandra	Dark Grass Blue	±	±	±	±	+	+	+	+	+	+	±	±
Celastrina argiolus	Holly Blue		±	±	±	+	+	+	+	±	±		
Pseudophilotes vicrama	Eastern Baton Blue			+	+	+	±		±				
Glaucopsyche paphos	Paphos Blue	±	±	+	+	+	±	±					
Luthrodes galba	Small Desert Blue		±	±	±	±	+	+	+	+	+	±	
Freyeria trochylus	Grass Jewel		±	+	+	±	+	+	+	+	±	±	
Aricia agestis	Brown Argus			±	+	+	+	+	+	+	±		
Polyommatus icarus	Common Blue		±	+	+	+	+	+	+	+	+	+	±
Libythea celtis	Nettle-tree Butterfly				±	+	+	+	+	±	±		
Danaus chrysippus	Plain Tiger			+	+	±	±	±	±	±	+	+	±

Charaxes jasius	Two-tailed Pasha				±	+	+	±	+	+	+	±	±
Hypolimnas misippus	False Plain Tiger / Diadem								M				
Limenitis reducta	Southern White Admiral				±	+	+	+	+	+	±	±	
Vanessa atalanta	Red Admiral	+	+	+	+	+	±	±	±	±	+	+	+
Vanessa cardui	Painted Lady	+	+	+	+	+	+	±	±	+	+	+	+
Nymphalis polychloros	Large Tortoiseshell					E	E	E	E				
Argynnis pandora	Cardinal				±	±	+	+	+	+	±	±	±
Issoria lathonia	Queen of Spain Fritillary					?							
Hipparchia syriaca	Eastern Rock Grayling				±	±	+	+	+	+	±	±	
Hipparchia cypriensis	Cyprus Grayling			±	+	+	+	+	+	+	+	±	
Chazara briseis	The Hermit				±	+	+	+	+	+	+	±	
Chazara persephone	Great Steppe Grayling						?						
Pseudochazara anthelea	White-banded Grayling				±	±	+	+	+	±			
Maniola cypricola	Cyprus Meadow Brown			±	+	+	+	+	+	+	±	±	±
Hyponephele lupina	Oriental Meadow Brown				±	+	+	+	+	+	±	±	
Pararge aegeria	Speckled Wood	±	+	+	+	+	+	+	+	+	+	±	±
Lasiommata megera	Wall Brown	±	+	+	+	+	+	+	±	±	±	±	±
Lasiommata maera	Large Wall Brown	±	+	+	+	+	+	+	±	±	±	±	±
Kirinia roxelana	Lattice Brown			±	±	±	+	+	+	±	±		
Ypthima asterope	African Ringlet			±	+	+	±	±	±		+	±	±
Carcharodus alceae	Mallow Skipper	±	+	+	+	+	+	+	+	+	+	+	±
Thymelicus acteon	Lulworth Skipper			±	+	+	±	±					
Gegenes pumilio	Pigmy Skipper	±	±	±	+	+	+	±	±	+	+	+	±
Pelopidas thrax	Millet Skipper			±	±	+	+	+	+	+	+	±	
Potential species per month:		21	26	37	48	50	48	44	44	42	41	37	25

Key:
+ = Months denoting peak flight activity;
± = Months in which the species may be seen in smaller numbers;
E = Considered extinct (and excluded from count);
M = Very rare migrant, showing month(s) observed (*P. demoleus* appears to have overwintered successfully in 2021/2022 and may become established).
NB: *Deudorix livia* and *Azanus jesous* might have been introduced accidentally;
? = status uncertain (see species entries for details).

9 Biodiversity and Endemics

Species Richness

Although less species-rich than might be expected from the widely considered view of Cyprus being an island of considerable biodiversity (Myers *et al.*, 2000; Hadjisterkotis, 2001) and located in a 'hotspot' region (Werner and Buszko, 2005), the butterfly fauna is considered depauperate (Wiltshire, 1948; Olivier, 1993: 209, 221) due to isolation, despite its proximity to mainland countries, especially to the north and east. Dennis and Hardy (2018: 9) stated, 'Offshore island faunas are typically characterized by two patterns: (i) depleted faunas and (ii) disharmonic faunas. Disharmony again describes two basic features: (i) island faunas comprise non-random assemblages of the source fauna, effectively filtered by dispersal and migration; and (ii) novel species not found on the mainland' (see also Dennis *et al.*, 2008 and Vodă *et al.*, 2016).

Indeed, the island is host to seven endemic species or subspecies of butterfly and to a number of species not known in other countries of the Mediterranean basin or in Europe. Dennis and Schmitt (2009: 250) remarked on the importance of the Mediterranean zone, as it 'contains many of Europe's endemic species, and most of the European butterfly species' evolutionary history.'

Endemic Species and Subspecies

The endemic species status of *Glaucopsyche paphos* (Paphos Blue) has been unchallenged for many years, but the increasing use of molecular techniques with particular reference to *Maniola* spp. has hinted 'at one single incompletely differentiated species complex', seemingly challenging

DOI: 10.1079/9781800621275.0009

the endemic status of *Maniola cypricola* (Cyprus Meadow Brown). Genetic data obtained by Kreuzinger *et al.* (2015) indicated that *M. cypricola*, along with six other *Maniola* spp. (including other island endemic 'Meadow Browns' from Sardinia and Chios), suggested variation, rather than speciation, leading the authors to conclude that the genus should be considered as one 'superspecies', namely *M. jurtina*, which is commonly found throughout Europe. However, at the time of writing, this conclusion is not wholly accepted and verification of these results is required before the name *M. cypricola* is consigned to the wastebin (see species entry). *Hipparchia cypriensis* (Cyprus Grayling) is a relatively recent elevation from hitherto subspecific *H. pellucida cypriensis*, and the raising to species rank seems justified, yet not universally recognized (see species entry later).

Four endemic subspecies are also currently recognized, namely: *Allancastria cerisyi cypria* (Eastern Festoon), *Hipparchia syriaca cypriaca* (Eastern Rock Grayling), *Chazara briseis larnacana* (The Hermit) and *Hyponephele lupina cypriaca* (Oriental Meadow Brown). Other historically listed subspecies have not withstood scrutiny and are now considered synonymous with the nominotypical subspecies; for example, *Pseudochazara anthelea acamanthis* (White-banded Grayling), considered by Olivier (1993: 196) to be a Cyprus endemic subspecies, was later synonymized with *P. anthelea anthelea* in Olivier (1996). The subspecific status of *Apharitis acamas cypriaca* Riley, 1925 (Levantine Leopard) is uncertain and may be invalid, hence its exclusion from Table 9.1 (see the individual species entries for further details).

Table 9.1. The endemic butterfly species and subspecies of Cyprus, with references to original descriptions or further supporting work.

	Endemic species	Endemic subspecies
Papilionidae		*Allancastria cerisyi cypria*[1]
Lycaenidae	*Glaucopsyche paphos*[2]	
Nymphalidae (Satyrinae)	*Hipparchia cypriensis*[3]	*Hipparchia syriaca cypriaca*[4]
	Maniola cypricola[5]	*Chazara briseis larnacana*[6]
		Hyponephele lupina cypriaca[7]

[1]Stichel (1907); Nazari and Carbonell (2018);
[2]Chapman (1920);
[3]Kudrna, (1986: 184); Jutzeler *et al.* (2019);
[4]Olivier (1993: 196, 222);
[5]Graves (1928); Thomson (1987: 12);
[6]Olivier (1993: 196, 221);
[7]Riley (1921); Eckweiler and Bozano (2011: 49); Lukhtanov and Pazhenkova (2021).

10 The Cyprus Butterfly List

Taxonomy

Butterfly taxonomic names used in this book largely follow the revision of the Taxonomy Commission of Butterfly Conservation Europe (Verovnik and Wiemers, 2010; Wiemers *et al.*, 2018). However, as mentioned earlier, there is no consensus on the scientific naming of all European species. Where such divergence of opinion exists, for example between Verovnik and Wiemers (2010), Kudrna *et al.* (2011), Wiemers *et al.* (2018) and Kudrna (2019), a suitable note appears in the text for the relevant species.

Breakdown by Family

The least well-represented family is the Papilionidae, with just three species, whereas Nymphalidae (into which the former 'family' of Satyridae has been subsumed) has 22 species (Fig. 10.1). It is anticipated that, in the near future, *Gegenes nostrodamus* will boost the Hesperiidae to five species.

List of Species

For the sake of completeness, all species known to have been recorded in Cyprus over the past 100 years or so have been included in the main listing (Table 10.1). Some, namely *Catopsilia florella* (African Migrant), *Deudorix livia* (Pomegranate Hairstreak), *Azanus jesous* (African Babul Blue), *Hypolimnas misippus* (False Plain Tiger), *Issoria lathonia* (Queen of Spain Fritillary) and *Chazara persephone* (Great Steppe Grayling), have been recorded very infrequently and *Papilio demoleus* (Lime Swallowtail) is a very recent but seemingly successful addition in 2021. The migratory

DOI: 10.1079/9781800621275.0010

Nymphalis polychloros (Large Tortoiseshell), once present in small numbers up to 1931 (see Makris, 2003: 224), may reappear in Cyprus but this seems unlikely.

Potential species are listed separately (see Chapter 13) and are considered as possible species for Cyprus based on their presence in nearby countries, known migratory capability, range expansion, or movement associated with climate change.

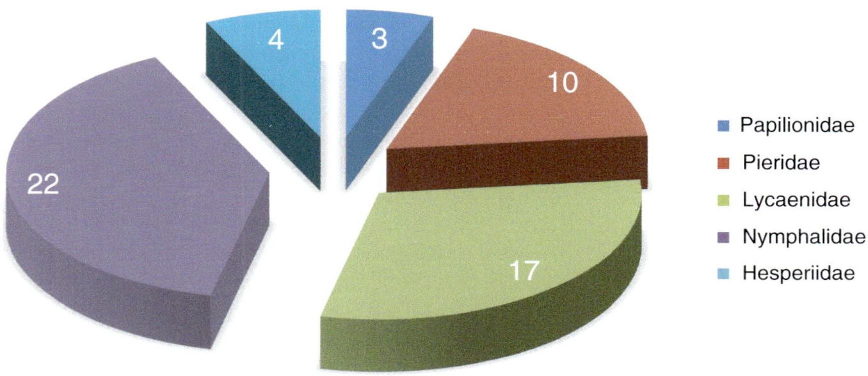

Fig. 10.1. Representation of species within the five butterfly families in Cyprus.

Table 10.1. List of butterfly species recorded in Cyprus by family, subfamily and species.

Family	Subfamily	Scientific name	Vernacular name
Papilionidae	Papilioninae	*Papilio machaon* Linnaeus, 1758	Swallowtail
		Papilio demoleus Linnaeus, 1758	Lime Swallowtail
	Parnassiinae	*Allancastria cerisyi cypria* Stichel, 1907	Eastern Festoon
Pieridae	Pierinae	*Aporia crataegi* (Linnaeus, 1758)	Black-veined White
		Pieris brassicae (Linnaeus, 1758)	Large White
		Pieris rapae (Linnaeus, 1758)	Small White
		Pontia daplidice (Linnaeus, 1758)	Bath White
		Pontia chloridice (Hübner, [1813])	Small Bath White
		Euchloe ausonia (Hübner, [1804])	Eastern Dappled White
		Anthocharis cardamines (Linnaeus, 1758)	Orange Tip
	Coliadinae	*Catopsilia florella* (Fabricius, 1775)	African Migrant
		Colias croceus (Geoffroy, 1785)	Clouded Yellow
		Gonepteryx cleopatra (Linnaeus, 1767)	Cleopatra
Lycaenidae	Aphnaeinae	*Apharitis acamas* (Klug, 1834)	Levantine Leopard
	Lycaeninae	*Lycaena phlaeas* (Linnaeus, [1760])	Small Copper
		Lycaena thersamon (Esper, 1784)	Lesser Fiery Copper
	Theclinae	*Deudorix livia* (Klug, 1834)	Pomegranate Hairstreak
		Favonius quercus (Linnaeus, 1758)	Purple Hairstreak

Continued

Table 10.1. Continued.

Family	Subfamily	Scientific name	Vernacular name
	Polyommatinae	*Lampides boeticus* (Linnaeus, 1767)	Long-tailed Blue
		Leptotes pirithous (Linnaeus, 1767)	Lang's Short-tailed Blue
		Azanus jesous (Guérin-Méneville, 1849)	African Babul Blue
		Tarucus balkanicus (Freyer, 1844)	Little Tiger Blue
		Zizeeria karsandra (Moore, 1865)	Dark Grass Blue
		Celastrina argiolus (Linnaeus, 1758)	Holly Blue
		Pseudophilotes vicrama (Moore, 1865)	Eastern Baton Blue
		Glaucopsyche paphos Chapman, 1920	Paphos Blue
		Luthrodes galba (Lederer, 1855)	Small Desert Blue
		Freyeria trochylus (Freyer, 1844)	Grass Jewel
		Aricia agestis ([Denis & Schiffermüller], 1775)	Brown Argus
		Polyommatus icarus (Rottemburg, 1775)	Common Blue
Nymphalidae	Libytheinae	*Libythea celtis* (Laicharting, 1782)	Nettle-tree Butterfly
	Danainae	*Danaus chrysippus* (Linnaeus, 1758)	Plain Tiger
	Charaxinae	*Charaxes jasius* (Linnaeus, 1767)	Two-tailed Pasha
	Limenitidinae	*Limenitis reducta* Staudinger, 1901	Southern White Admiral
	Nymphalinae	*Hypolimnas misippus* (Linnaeus, 1764)	False Plain Tiger/Diadem
		Nymphalis polychloros[1] (Linnaeus, 1758)	Large Tortoiseshell
		Vanessa atalanta (Linnaeus, 1758)	Red Admiral
		Vanessa cardui (Linnaeus, 1758)	Painted Lady
	Heliconiinae	*Argynnis pandora* ([Denis & Schiffermüller], 1775)	Cardinal
		Issoria lathonia (Linnaeus, 1758)	Queen of Spain Fritillary
	Satyrinae	*Hipparchia syriaca cypriaca* (Staudinger, 1879)	Eastern Rock Grayling
		Hipparchia cypriensis (Holik, 1949)	Cyprus Grayling
		Chazara briseis larnacana (Oberthür, 1909)	The Hermit
		Chazara persephone (Hübner, [1805])	Great Steppe Grayling
		Pseudochazara anthelea (Hübner, [1824])	White-banded Grayling
		Maniola cypricola (Graves, 1928)	Cyprus Meadow Brown
		Hyponephele lupina cypriaca Riley, 1921	Oriental Meadow Brown
		Pararge aegeria (Linnaeus, 1758)	Speckled Wood
		Lasiommata megera (Linnaeus, 1767)	Wall Brown
		Lasiommata maera (Linnaeus, 1758)	Large Wall Brown
		Kirinia roxelana (Cramer, 1777)	Lattice Brown
		Ypthima asterope (Klug, 1832)	African Ringlet
Hesperiidae	Pyrginae	*Carcharodus alceae* (Esper, 1780)	Mallow Skipper
	Hesperiinae	*Thymelicus acteon* (Rottemburg, 1775)	Lulworth Skipper
		Gegenes pumilio (Hoffmansegg, 1804)	Pigmy Skipper
		Pelopidas thrax (Hübner, [1821])	Millet Skipper

[1]*Nymphalis polychloros* is now considered extinct in Cyprus – see species entry.

11 Residency, Broods and Habitats of Butterflies in Cyprus

Table 11.1 (overleaf) summarizes the current state of knowledge concerning the 56 species of butterfly listed for Cyprus. The column entitled 'Principal Habitats' should be regarded as a general guide, rather than one that defines precisely where a species is likely to be encountered.

DOI: 10.1079/9781800621275.0011

Table 11.1. Residency, broods and main habitats of butterflies in Cyprus.

	Residency status			No. of broods			Principal habitats								
Species	R	M	(A)	1	2	≥3	F	GM	R	C	MP	W	AE	WF	U
Papilio machaon	•					•	•	•	•	•		•	•	•	•
Papilio demoleus	•?	•[4]					Data deficient in Cyprus								
Allancastria cerisyi	•			•			•	•	•	•			•	•	
Aporia crataegi	•	•[1]		•			•	•	•						
Pieris brassicae	•	•[1]				•	•	•	•	•	•	•	•	•	•
Pieris rapae	•	•[2]				•	•	•	•	•	•	•	•	•	•
Pontia daplidice	•	•[1]				•	•	•	•	•	•	•	•	•	•
Pontia chloridice	•					•	•		•				•		
Euchloe ausonia	•				•		•	•	•	•			•	•	•
Anthocharis cardamines	•			•			•	•	•				•	•	•
Catopsilia florella		•[3]					Data deficient in Cyprus								
Colias croceus	•	•[1]				•	•	•	•	•	•	•	•	•	•
Gonepteryx cleopatra	•			•			•	•	•	•					
Apharitis acamas	•					•	•	•							
Lycaena phlaeas	•					•	•	•	•	•	•	•	•	•	•
Lycaena thersamon	•					•		•		•		•	•	•	•
Deudorix livia		•[3]	•[3]				Data deficient in Cyprus								
Favonius quercus	•			•			•	•					•		
Lampides boeticus	•	•[1]				•	•	•	•	•			•	•	•
Leptotes pirithous	•	•[1]				•	•	•	•	•			•	•	•
Azanus jesous		•[3]	•[3]				Data deficient in Cyprus								
Tarucus balkanicus	•					•				•	•	•			
Zizeeria karsandra	•					•				•	•	•	•	•	•
Celastrina argiolus	•					•	•			•					
Pseudophilotes vicrama	•				•			•	•	•					

Glaucopsyche paphos	•		•	?		•	•	•	•					
Luthrodes galba	•				•				•	•	•	•	•	•
Freyeria trochylus	•				•	•	•	•	•			•		
Aricia agestis	•				•	•	•					•		
Polyommatus icarus	•				•	•	•	•	•	•	•	•	•	•
Libythea celtis	•		•			•								
Danaus chrysippus		•[2]	-	-	-				•		•	•	•	•
Charaxes jasius	•			•		•								
Limenitis reducta	•				•	•								
Hypolimnas misippus					Data deficient in Cyprus									
Nymphalis polychloros	**?**				Considered extinct in Cyprus									
Vanessa atalanta	•	•[2]			•	•	•	•	•			•	•	•
Vanessa cardui	•	•[2]			•	•	•	•	•		•	•	•	•
Argynnis pandora	•	•[1]	•			•		•						
Issoria lathonia					Data deficient in Cyprus									
Hipparchia syriaca	•		•			•	•					•		
Hipparchia cypriensis	•		•			•	•	•	•			•		
Chazara briseis	•		•			•	•	•	•			•		
Chazara persephone					Data deficient in Cyprus									
Pseudochazara anthelea	•		•			•	•	•						
Maniola cypricola	•		•			•	•	•	•	•	•	•	•	•
Hyponephele lupina	•		•			•	•	•	•				•	
Pararge aegeria	•				•	•	•		•			•		•
Lasiommata megera	•				•	•	•	•	•			•		•
Lasiommata maera	•				•	•	•	•	•			•		•
Kirinia roxelana	•		•			•						•		•
Ypthima asterope	•				•			•	•					
Carcharodus alceae	•				•	•	•	•	•	•	•	•	•	•
Thymelicus acteon	•		•			•	•	•	•			•	•	•

Continued

Table 11.1. Continued.

	Residency status			No. of broods			Principal habitats								
Species	R	M	(A)	1	2	≥3	F	GM	R	C	MP	W	AE	WF	U
Gegenes pumilio	•					•		•	•	•			•	•	
Pelopidas thrax	•	•[1]				•		•	•	•		•	•		

Key:
R: Resident
M (A): Migrant.
[1]Established migrant in the wider region, although there is little or no evidence to suggest that the resident Cyprus population is supplemented regularly.
[2]Observed migrant in Cyprus.
[3]Rare migrant/possible adventive, breeding unconfirmed.
[4]Recently arrived migrant; breeding confirmed in urban areas, and April 2022 observations indicate successful overwintering, possibly leading to establishment.
F: Forests
GM: Garigue and Maquis
R: Rocky areas
C: Coastal areas
MP: Mesaoria Plain
W: Wetland
AE: Agricultural ecosystems
WF: Waste or fallow ground
U: Urban areas

12 Presentation of Species

Explanatory Notes

Each of the butterfly families is introduced with the family name, followed by the scientific species' name, the naming authority and year in which described (e.g. Linnaeus, 1758), the English common name, the subfamily to which the species belongs and the Type Locality (TL) of the species, i.e. the geographical location where the name-bearing taxon was captured.

Photographs

Wherever possible, photographs (not to scale) show male and female uppersides and undersides. Attention is occasionally drawn to key identification characteristics.

Presence in the eastern Mediterranean (the Levant)

Where relevant, and in order to provide a degree of context, reference is made to the species' presence in neighbouring countries. As stated by Dapporto and Dennis (2009) in respect of large Mediterranean islands, 'An important observation for this study is that the biogeography and conservation status of organisms on islands cannot be assessed adequately without examining their distributions in neighbouring source pools.' However, this is not to suggest that the Cyprus butterfly fauna is supplemented regularly or extensively by migrations from mainland countries.

The area defining the Levant is considered to comprise Turkey south of the Taurus Mountains (Antalya and Hatay Provinces), Mediterranean Syria, Lebanon, Cyprus, Israel, Jordan, the Sinai Peninsula, and the northernmost

DOI: 10.1079/9781800621275.0012

parts of Upper Egypt between the Suez Canal and the Nile Delta, and the coastal areas west to Mersa Matruh (after Larsen, 2015). However, this book concentrates on the butterflies of Cyprus and should not be considered a comprehensive account of species in the Levant, where a far greater number of butterfly species is present (see, e.g., Benyamini and John, 2020).

Within this section are included some of the historical butterfly scientific (and occasionally colloquial) names of relevance to Cyprus, as used by earlier authors. Species' names and style are frequently at variance with today's usage but cited entries are exactly as published at the time. Often, the entries also provide an indication of when a species was first formally recorded for Cyprus. However, it is not our intention to provide a comprehensive listing of synonyms (often running to 20 or more for various species); in this regard, Hesselbarth *et al.* (1995), Ackery *et al.* (1999) or GBIF (2021) may help further. Attention is also drawn to current nomenclatural disharmony among lepidopterists (see also 'forms' in Chapter 7).

Distribution in Cyprus

An indication is given of the distribution of each species, both elevational and geographical. Note that where, for example, 'sea level to mountain top' is stated, some species may be present only at higher elevations during summer.

Distribution maps

Map grids are based on the Universal Transverse Mercator (UTM) coordinate system and record species to within 5 km^2 UTM squares. Complete and partial squares are shown, in total numbering 445, 5 km^2 UTM squares. For presentation purposes, these are shown grouped within grid lines representing 10 km^2 UTM squares.

Key to Distribution Map entries

Coloured discs represent records from:

- ● 2010 to date
- ● 2000–2009
- ● 1990–1999
- ○ Pre-1990

What do the maps tell us?

- Some species – *Papilio machaon* and *Pieris rapae* for example – may be found extensively throughout the island; others, such as the more sedentary *Luthrodes galba* and *Tarucus balkanicus*, have a limited distribution mirroring that of their respective hostplants.

- The maps also illustrate that some species, such as *Freyeria trochylus*, may be found at all altitudes, whereas *Luthrodes galba* (occasionally erroneously recorded at high altitude) is strictly found at lower elevations in association with its larval hostplant *Prosopis farcta* and is present from sea level to 200 m.
- Despite efforts by the authors and colleagues to survey as much of the island as possible, there is inevitably a degree of bias in the distribution of some, but not necessarily all, species. The Mesaoria Plain, for example, is an inhospitable area stretching across the island and is rarely visited by those reporting to the Cyprus Butterfly Study Group (CBSG). However, the recording implications are minimized, as this area is not home to many species and to none exclusively. For the opposite reason, the west of the island, the Troodos and Pentadaktylos ranges, are favoured areas to visit because of attractive scenery and greater variety of species present.
- Note that reduction of scale has occasionally failed to show the presence of fragments of coastal squares, e.g. VD35NE and WD44NE. These *do* exist as small areas of land.
- The presence of a dot in the centre of a 10 km^2 UTM square signifies that we have been unable to pinpoint the precise 5 km^2 square in which the species was recorded.
- Such is the ubiquitous nature of some species, for example *Pieris rapae* (Small White), *Colias croceus* (Clouded Yellow) and *Vanessa cardui* (Painted Lady), that these will eventually be recorded in all of the island's 5 km^2 UTM squares. Others, as described above, are so closely tied to their larval hostplants that these occupy areas of Cyprus only where ecological conditions are favourable for the growth of the hostplants.
- The maps do not attempt to indicate a species' abundance.

The distribution maps are continually being improved and updated – they represent 'work in progress' by the first author over 20+ years – and should not be seen as defining precisely the complete distribution of any species within Cyprus.

With permission, distribution maps have been created using a modified version of that originally prepared by the Department of Forests, Cyprus, for Makris (2003: 20), as shown in Fig. 12.1, which includes elevational gradients.

Phenology

As elsewhere in the wider region, the number of broods of certain polyvoltine species in Cyprus will be dependent on foodplant availability. The text and the phenogram provided earlier indicate the main flight periods; however, the timing of early season emergences is influenced by fluctuations in the island's weather in springtime.

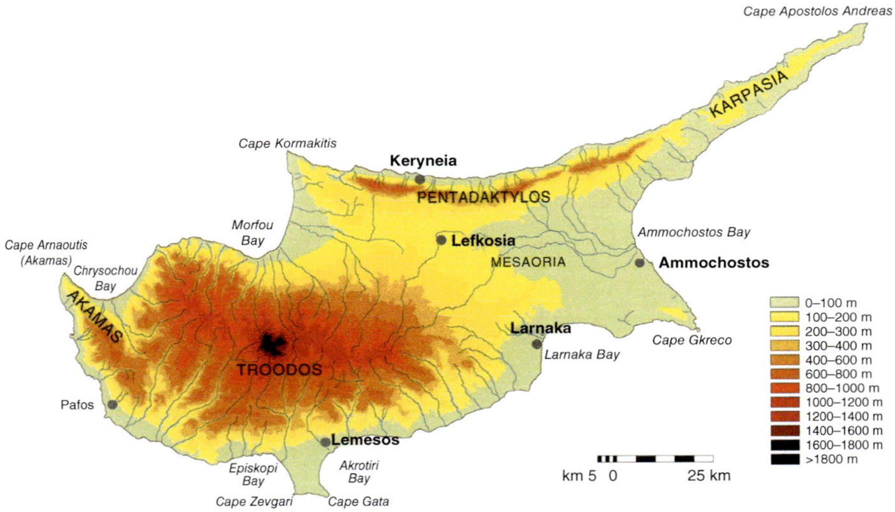

Fig. 12.1. Topographic map of Cyprus, courtesy of the Dept of Forests, Cyprus.

Larval hostplants

In general, those listed refer to known larval hostplants in Cyprus, rather than to the Levant in its entirety. With specific reference to Cyprus, details of confirmed larval feeding *to maturity* (not simply adult egg-laying) on hostplants other than those listed would be appreciated by the authors.

Ecology and behaviour

Behavioural traits, for example hilltopping, mud-puddling and migrational tendencies, are noted in this section.

Description/identification aids

Butterflies are not shown to scale but an indication of size is given here, along with details of flight characteristics (where relevant) and descriptions of wing colour and markings. *By means of italicized text*, attention is occasionally drawn to key diagnostic features, for example chequering of wing fringes, presence of ocelli, etc.

Similar species

While most species can be identified without difficulty, reference to the preceding section along with guidance given here should enable accurate determination.

Conservation status in Cyprus

The status of butterfly species in Cyprus has been assessed by the authors; where relevant, mention is made of the assessment of each species as listed in, for example, Van Swaay *et al.* (2010a) and Numa *et al.* (2016). Categories of relevance to Cyprus are:

NT = Near Threatened
LC = Least Concern
DD = Data Deficient.

Wing areas and venation

Throughout the descriptions of species, mention may be made of specific areas of the wings or particular wing veining in assisting with identification (see Figs 12.2. and 12.3 overleaf).

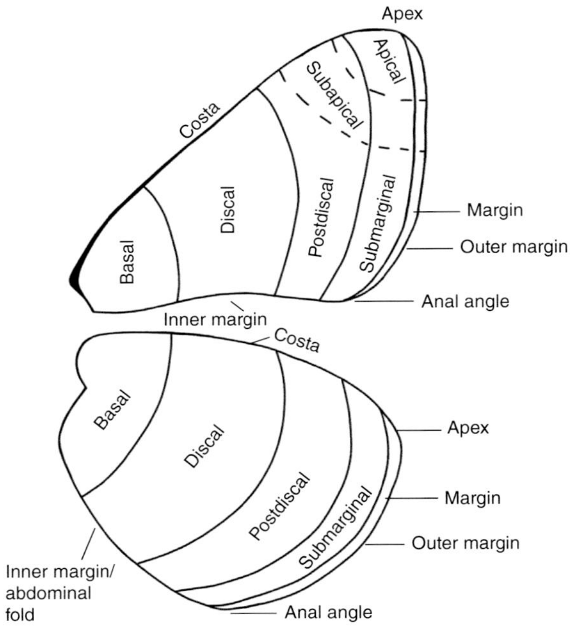

Fig. 12.2. Schematic representation of wing areas of *Aporia crataegi*.

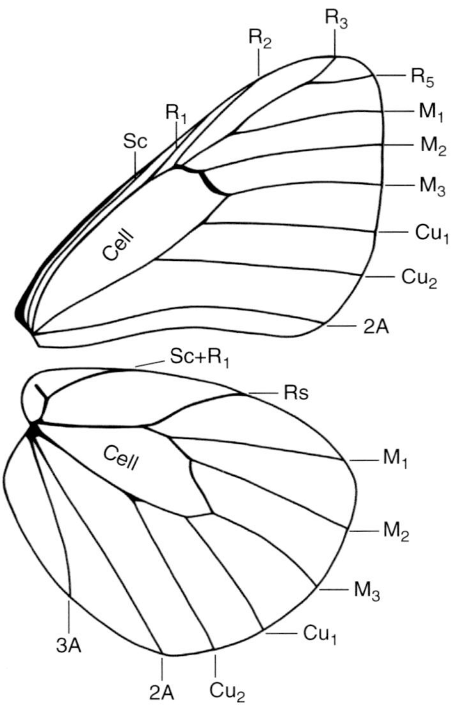

Fig. 12.3. Schematic representation of wing venation of *Aporia crataegi* (Sc = subcostal; R = radius; Rs = radius sector; M = median; Cu = cubitus). Drawings: *Bethan Wallace*

Presentation of Species

Families:
PAPILIONIDAE
PIERIDAE
LYCAENIDAE
NYMPHALIDAE
HESPERIIDAE

Family: PAPILIONIDAE

Papilio machaon Linnaeus, 1758	Swallowtail	Papilioninae
Papilio demoleus Linnaeus, 1758	Lime/Citrus Swallowtail	
Allancastria cerisyi cypria Stichel 1907	Eastern Festoon	Parnassiinae

Papilio machaon Linnaeus, 1758 **Papilionidae** (Papilioninae)

Swallowtail

TL: Sweden.

Figs 12.4. and 12.5. *Papilio machaon* (Swallowtail), 13 November 2012 (left) and 13 July 2013 (right), showing natural variation in the extent and depth of yellow scaling and coloration in the basal areas and body. *Christodoulos Makris*

Fig. 12.6. *Papilio machaon* (Swallowtail) nectaring on *Limonium sinuatum*, 5 May 2015. *Eddie John*

Fig. 12.7. *Papilio machaon* (Swallowtail) nectaring on *Dittrichia viscosa*, 19 October 2017. *Eddie John*

Fig. 12.8. *Papilio machaon* (Swallowtail), 20 June 2015. *Christodoulos Makris*

Fig. 12.9. *Papilio machaon* (Swallowtail) pairing, 18 February 2013. *Christodoulos Makris*

Presence in the eastern Mediterranean

Widespread in countries of the Levant, where it is represented throughout the region (including Cyprus), by the ssp. *Papilio machaon syriacus* Verity, 1908 (Benyamini and John, 2020: 28).

For Cyprus, Turner (1920) referred to 'race asiatica Mén. = sphyrus auct.' and Rebel (1939) to 'Papilio machaon sphyrus Hb.', a ssp. name now considered invalid for Cyprus. See also Hesselbarth *et al.* (1995: 293), on which page '*Papilio machaon cypriae* Verity, 1905', '*Papilio machaon* gen. aest. var. *sphyroides cypria* Verity, R. [1905]' and '*Papilio machaon* forma *cypricus* Eller, K., 1939', among others, are synonymized with ssp. *Papilio m. syriacus*. A dwarf form, 'ab. cypria', described by Verity (1905) and included in Turner (1920), also appears invalid (see Chapter 7).

Distribution in Cyprus

Altitudinal: from sea level to the tops of mountains, sparingly above 1800 m (Makris, 2003: 287), but more common at lower elevations.
Geographical: common and widespread throughout the island, using a variety of habitats.

Phenology

Trivoltine to polyvoltine, February to November. In warmer lowland and coastal areas may be seen from January into early December (Cyprus Butterfly Recording Scheme (CBRS) records). Overwinters as a pupa.

Larval hostplants

Polyphagous: principally *Foeniculum vulgare* (fennel) (Fig. 12.11), but also many other Umbelliferae (Apiaceae), for example *Ammi majus* (bishop's weed), *Daucus carota* (wild carrot), *Opopanax hispidus* and increasingly frequently reported on *Petroselinum crispum* (garden parsley) and occasionally *Crithmum maritimum* (rock samphire – Mike Hardman, Cyprus,

2012, personal communication to first author). Also Rutaceae: *Citrus limon* (Fig. 12.12: eggs were laid on the lemon tree blossom at the end of October), *C. aurantium* (sour orange), *Ruta chalepensis* (fringed rue), *Haplophyllum buxbaumii*.

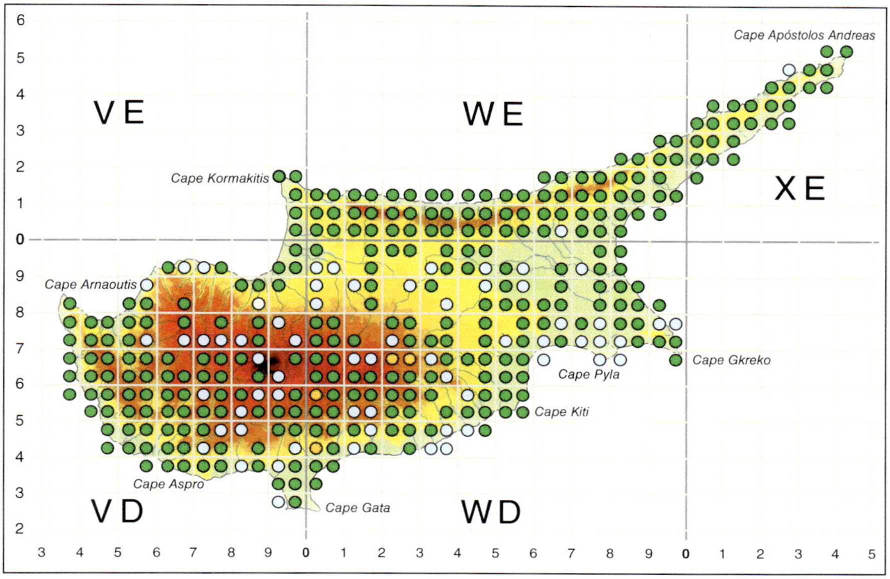

Fig. 12.10. *Papilio machaon* distribution map showing 5 km^2 observations within 10 km^2 UTM squares. © *Eddie John*

Fig. 12.11. *Foeniculum vulgare*, 11 August 2012. Courtesy of *Yiannis Christofides*

Fig. 12.12. *Papilio machaon* (Swallowtail) final instar larva on *Citrus limon*, 3 December 2014. Courtesy of *Mike Lloyd*

Ecology and behaviour

The use of a wide range of hostplants attracts *P. machaon* into varied habitats, including irrigation ditches, field margins, verges, mountain tracks and agricultural areas, especially where fennel is present. *P. machaon* is often attracted into gardens in summer, where cultivated parsley will readily be used for egg-laying. Males are frequent hilltoppers; at Cape Kormakitis on 27 April 2014, three were seen to select a rise just a few metres higher than the surrounding area. The first author also observed an example of thermoregulation in late October 2017, when a nectaring specimen immediately responded to a cloud-induced, temporary fall in temperature by flying to a sun-warmed stone, allowing body temperature to be maintained (see also Dennis, 2020: 211 for an extended discussion on thermoregulation). Nectar sources include *Limonium sinuatum* (Plumbaginaceae), *Centaurea aegialophila* (Asteraceae), *Dittrichia viscosa* (Asteraceae), *Heliotropium hirsutissimum* (Boraginaceae) and *Vitex agnus-castus* (Lamiaceae).

Description/identification aids

Wingspan: very large – up to 80 mm, making *P. machaon* one of the island's largest tailed butterflies. Appears black and yellow on the wing, with a vigorous flight that becomes *fluttery* as the butterfly hovers when nectaring or ovipositing. Sexes are similar, **uppersides** with a light yellow ground colour, darker areas in the forewing basal and submarginal areas and heavy black veining. Hindwing uppersides have a wide dark submarginal/postdiscal band 'dusted' with blue scaling. Near the anal angle is a large orange-red ocellus and also a characteristic tail. **Undersides** are lighter overall, with orange-red postdiscal scaling and a wide submarginal band irrorated with blue scaling. *P. machaon* is one of only two Swallowtails known for Cyprus, but the recently arrived *P. demoleus* (see next species) is markedly different in appearance.

Similar species

Papilio alexanor Esper, 1800 (Southern Swallowtail) and *P. saharae* Oberthür, 1879 (Saharan Swallowtail) are present elsewhere in the Levant, but are unrecorded in Cyprus, as is *Iphiclides podalirius* Linnaeus, 1758 (Scarce Swallowtail), but see Chapter 13.

Conservation status in Cyprus

Not threatened. Listed by Van Swaay *et al.* (2010a) and Numa *et al.* (2016) as 'LC'.

Papilio demoleus Linnaeus, 1758 **Papilionidae** (Papilioninae)

Lime/Citrus Swallowtail

TL: Guangzhou [Canton], SE China.

Fig. 12.13. *Papilio demoleus* (Lime Swallowtail) ♂, nectaring on *Zinnia* sp. in a domestic garden, 25 October 2021. *Christodoulos Makris*

Fig. 12.14. *Papilio demoleus* (Lime Swallowtail) ♀, 12 November 2021. *Christodoulos Makris*

Fig. 12.15. *Papilio demoleus* (Lime Swallowtail) nectaring on *Carathanthus roseus,* Lysi, Cyprus, 12 August 2021. Courtesy of *Hasan Bağlar*

Fig. 12.16. *Papilio demoleus* (Lime Swallowtail), 16 November 2021. *Christodoulos Makris*

Presence in the eastern Mediterranean

A potential pest of citrus (particularly of seedlings), this Indo-Australian species has appeared in recent years in Syria and eastern Turkey, resulting from range expansion, perhaps in association with climate change (Benyamini *et al.*, 2007). Since its arrival in the eastern Mediterranean regions, where *P. demoleus* was first reported in Hatay Province, Turkey in the autumn of 2018 (Atay and Tatlı, 2019), this papilionid has spread rapidly throughout Syria's coastal provinces of Latakia and Tartus (Başbay *et al.*, 2020a,b). *P. demoleus* was reported from Lebanon for the first time in 2020 (appearing again in 2021), followed by Cyprus in 2021 (John *et al.*, 2021a,b, 2022a,b) and Israel later in the same year (Hecht, 2021). *P. demoleus* appears not (yet) to have progressed westwards along the Mediterranean coast from Hatay Province in Turkey, but its dispersal into the provinces of Adana and Mersin is expected.

In the western Palaearctic, Azerbaijan is probably the most northerly country from which *P. demoleus* has been reported: the first observation in that country was at Länkäran in 2013 (Snegovaya and Petrov, 2019) and in 2018, Yusifova and Rákosy (2019) encountered the species in much the same area. It is considered that presence there was attributable to infestation of early stages of *P. demoleus* on imported *Citrus* (Alireza Naderi, Tehran, 2020, personal communication to first author).

Molecular analysis of *P. demoleus* from Cyprus has shown identical results with those from specimens taken earlier in Mediterranean Syria, confirming the taxon in both countries to be nominotypical subspecies *P. demoleus demoleus* (John *et al.*, 2021a, 2022a). The results support the earlier expectation of the species' arrival in Cyprus by means of migration and of the hypothesis of range expansion in the region (Başbay *et al.*, 2020a,b; John *et al.*, 2021b).

Distribution in Cyprus

Altitudinal: in Cyprus, provisionally known from coast to *ca.* 250 m; sea level to intermediate elevations in nearby mainland countries.

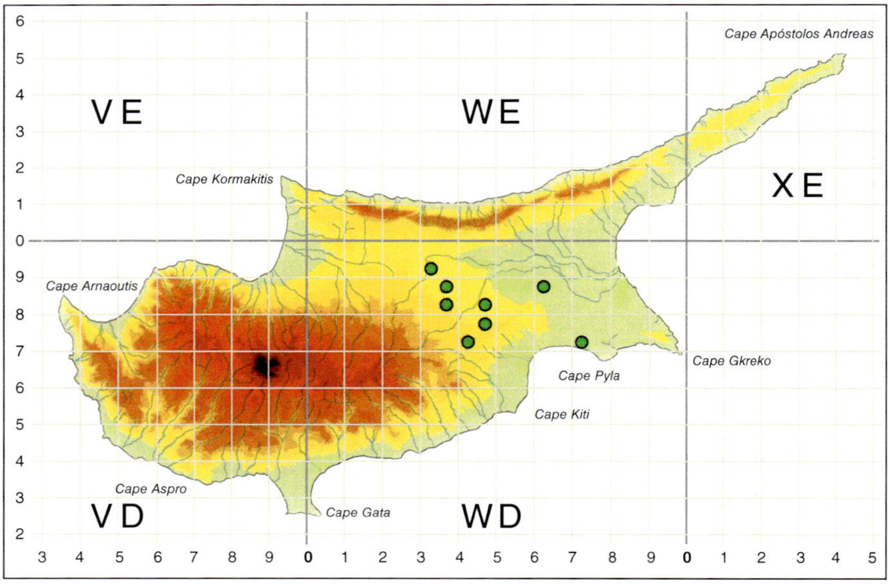

Fig. 12.17. *Papilio demoleus* distribution map showing 5 km² observations within 10 km² UTM squares. © *Eddie John*

Geographical: as this migrant has been known in Cyprus only since August 2021, the extent of dispersal has yet to be determined; further spread is expected. Presence of *Citrus* is a prerequisite.

Phenology

Polyvoltine, with an anticipated five generations in the Levant (Başbay *et al.*, 2020b). Although overwintering success has been reported in Mediterranean Turkey and Syria, there was some doubt about this being achieved in Cyprus, not least because the winter of 2021/2022 was unusually cold. However, during the proofing stages of this book, we were notified of an unexpectedly early observation of *P. demoleus* in flight at Geri (south-east of Nicosia) on 16 April 2022 in an area where the species had been observed on several occasions in November of the previous year (George Konstantinou, Geri, 2022, personal communication to first author). If sufficient numbers have overwintered successfully, the species can be expected to remain on the wing until November.

Larval hostplants

Polyphagous on numerous varieties of *Citrus* spp. (Rutaceae), with *Citrus limon* (Fig. 12.19) and *C. aurantium* often listed as hosts (e.g. Benyamini and John, 2020: 43). Yunus and Munir (1972) showed acceptance by *P. demoleus* of 18 species or varieties of *Citrus*, but with the maximum number of eggs laid on *C. limon*. To date, in Cyprus, larval activity has been noted on *C. reticulata* (mandarin), *C.* x *sinensis* (blood orange) and *C. limon* (lemon).

Ecology and behaviour

Westwards trans-Mediterranean migration was expected to account for the future arrival of *P. demoleus* in Cyprus, as the species is known as a migrant (Larsen, 1984a: 88; Larsen, 1988; Wiemers, 2007) and for which short sea crossings pose no significant barrier (Braby, 2016). Discovery of the species on 11 August 2021 at Lysi in the arid Mesaoria Plain of Cyprus, where citrus is grown only domestically, indicates that the accidental introduction of early stages through importation of young plants, while possible, seems less likely, and molecular results now lend support to this view (John *et al.*, 2021a, 2022a).

In early September, *P. demoleus* was observed at separate locations, a week or so apart, at Lympia, some 23 km to the south-west of Lysi. These observations were followed by reports in mid-October of the presence of adults or larvae on or around citrus at Athienou, Nicosia, Lysi, Ormideia and Geri (John *et al.*, 2022a,b). Observations continued into November, with the last confirmed report of 2021 received from Geri on 5 November. Adult longevity is considered to be *ca.* three weeks, with the life cycle occupying about a month under optimal conditions. Therefore, with nearly three months having elapsed from the first sighting at Lysi on 11 August to the final observation on 5 November 2021, this has allowed the completion of a probable two broods during the autumn. A captive eclosion (the reared female specimen from Athienou shown in Figs 12.14 and 12.16) occurred on 11 November 2021.

Experience elsewhere in the Levant has shown that *P. demoleus*, a species that was unknown in Mediterranean Syria until 28 June 2019 (Başbay *et al.*, 2020a), has quickly colonized coastal areas of Syria and has dispersed into Lebanon (2020) and Israel (2021). Its presence is by no means confined to citrus orchards; in parts of Turkey and Syria *P. demoleus* has become a common sight in towns and cities, where it thrives on domestically grown citrus as, to date, has been the experience in Cyprus. Widespread dispersal throughout Cyprus is highly likely if, as in nearby mainland regions, winter survival is achieved.

P. demoleus is an avid mud-puddler and, among four other papilionids and one pierid, was classified by Tea *et al.* (2020) as a frequent marine puddler in Sulawesi. An individual photographed at coastal Syria (see Chapter 6, Fig. 6.18) is the only known example of such behaviour by this species in the Levant. Initially, the butterfly had alighted on one of several dead fish recently discarded by a fisherman, but in the absence of putrefaction, the fish appeared to offer little attraction. At Lysi, frequent nectaring was observed on *Carathanthus roseus*, syn. *Vinca rosea*, *Bougainvillea spectabilis* and *Lantana camara* (John *et al.*, 2021b); at Nicosia, *Plumbago auriculata* attracted; and at Athienou, *P. demoleus* was seen nectaring on *Zinnia* sp. (John *et al.*, 2022a). At all locations, *P. demoleus* was seen to be very active around *Citrus*.

Exceptionally for this book, all early stages are illustrated, as the species is a very recent arrival to the eastern Mediterranean and, indeed (for the nominotypical ssp.), to a country of the European Union. First and

Fig. 12.18. *Papilio demoleus* (Lime Swallowtail) egg on *Citrus limon*, Athienou, 25 October 2021. *Christodoulos Makris*

Fig. 12.19. Flowering *Citrus limon* with *Papilio demoleus* (Lime Swallowtail) first instar larva (upper centre and arrowed in inset, right), Athienou, 25 October 2021. *Christodoulos Makris*

second instar larvae bear a resemblance to those of *P. machaon*, although with a light brown body colour, rather than black, turning darker brown in the third and fourth instars. Larvae typically show a distinctive white abdominal 'saddle' with occasional diagonal lateral markings on a green body, but may be more plainly marked in the fifth (final) instar (Figs 12.24 and 12.28). As is evident from the illustrations taken in October 2021 (Figs 12.18–12.28), eggs and larvae at various stages of development were present concurrently in the domestic garden of Giorgos Koronides (Athienou). Typical early stages of *P. demoleus* are also illustrated in Igarashi (1979: Plate 121).

Figs 12.20–12.24. First, second, third, fourth and a lightly marked fifth instar *P. demoleus* (Lime Swallowtail) larvae on various *Citrus* sp., Athienou, 26 October 2021. *Christodoulos Makris*

Figs 12.25–12.27. *P. demoleus* (Lime Swallowtail) pre-pupa and pupa (green form), Athienou, 28 and 29 October 2021, and pre-eclosion, 10 November 2021. *Christodoulos Makris*

Fig. 12.28. Final instar *P. demoleus* (Lime Swallowtail) larva (showing typical markings) on *Citrus* x *sinensis*, Athienou, 17 October 2021.

Description/identification aids

Wingspan: very large – up to 100 mm, the ♀♀ slightly the larger of the sexes. Males and females are very similar in appearance, with numerous white markings on a black ground colour and with a large black hindwing costal spot, heavily irrorated with blue scaling. However, differences in the appearance of the tornal spot on the hindwing **upperside** help separate the sexes: in the male, this is almost entirely orange, with a black inner spot capped with a light dusting of blue scaling, whereas in the female the orange is less extensive and the black mark more pronounced (Jahnavi *et al.*, 2018; Islam *et al.*, 2019). **Undersides** are similar in both sexes.

Similar species

Papilio demodocus Esper, [1798] is very similar in appearance but although expanding its range along the Saudi Arabian coast of the Red Sea, there have yet to be confirmed records of its appearance in the far south of Sinai (the southernmost point of the Levant).

Conservation status in Cyprus

Not applicable.

Allancastria cerisyi cypria Stichel, 1907 — Papilionidae (Parnassiinae)

Eastern Festoon

TL: Cyprus.

Fig. 12.29. *Allancastria cerisyi cypria* (Eastern Festoon) ♂ nectaring on *Allium neapolitanum*, 10 March 2015. Courtesy of *Clive Burrows*

Fig. 12.30. *Allancastria cerisyi cypria* (Eastern Festoon) ♂, 23 April 2012. *Christodoulos Makris*

Fig. 12.31. *Allancastria cerisyi cypria* (Eastern Festoon) ♀, 7 April 2016. *Christodoulos Makris*

Fig. 12.32. *Allancastria cerisyi cypria* (Eastern Festoon) ♀, 10 March 2019. Courtesy of *Matt Smith*

Fig. 12.33. *Allancastria cerisyi cypria* (Eastern Festoon) ♂, 23 April 2012. *Christodoulos Makris*

Presence in the eastern Mediterranean

The species is present throughout the region, represented mainly by the nominotypical ssp. *Allancastria cerisyi cerisyi*, but many subspecific names have appeared in the literature (see Nazari and Carbonell, 2018). There is also inconsistency in the use of the generic name to describe the Cyprus subspecies, with some authors favouring *Allancastria*, others *Zerynthia* (e.g. Tshikolovets, 2011: 88; Kudrna 2019: 253). Earlier, Nazari *et al.* (2007) had demonstrated that around 16 MYA, '*Zerynthia* in Europe and *Allancastria* in the Zagros Mts would have diverged from a common ancestor'. Nazari and Sperling (2007) and Nazari and Carbonell (2018) listed the Cyprus ssp. as *Allancastria cerisyi cypria* Stichel, 1907 in contrast to Hesselbarth *et al.* (1995: 248) who synonymized *Zerynthia* (*Allancastria*) *cerisyi cypria* with the nominotypical subspecies, citing Koçak (1981). Of four subspecies in Europe and the Middle East, Nazari and Carbonell (2018) further stated that '*A. c. cerisyi, A. c. ferdinandi, A. c. martini* and *A. c. cypria* are the only subspecies that are well isolated from one another and are systematically identifiable'. However, more molecular sampling is required to confirm the status of *A. c. cerisyi* as a valid ssp. (Vazrick Nazari, Ottawa, 2013, personal communication to first author). Nazari further stated, 'The DNA evidence shows that *Allancastria* and *Zerynthia* indeed share a common ancestor sometime around 20–30 million years ago. The genetic distance between these two genera is comparable to that between *Archon* and *Luehdorfia*.'

The species was initially listed as '*Thais Cerysii* God.', by Lederer (1855), '*Thais cerysii. God. 'The Courtezan.*', by Bucknill (1911), and 'Thais (Zerynthia Godt. [race cypria Stich. (1907)]', by Turner (1920). Stichel (1907) had earlier described this as 'Thais cerisyi cypria nov. subsp.', a description adopted by Rebel (1939) and Wiltshire (1949).

Distribution in Cyprus

Altitudinal: all elevations, although less common at the highest.
Geographical: a locally common subspecies, recorded over much of the island, but unrecorded from the Mesaoria Plain and the south-east of the island.

Phenology

Univoltine, generally March to May. Such is the erratic timing of emergence of this single-brooded spring species that individuals may be seen in February or even late January in mild winters, with stragglers on the wing into June. Overwinters as a pupa.

Larval hostplants

Aristolochiaceae: *Aristolochia sempervirens* (birthwort family) (Figs 12.35 and 12.36). *A. parvifolia* is also found in Cyprus, but there are no records of this *Aristolochia* species being used.

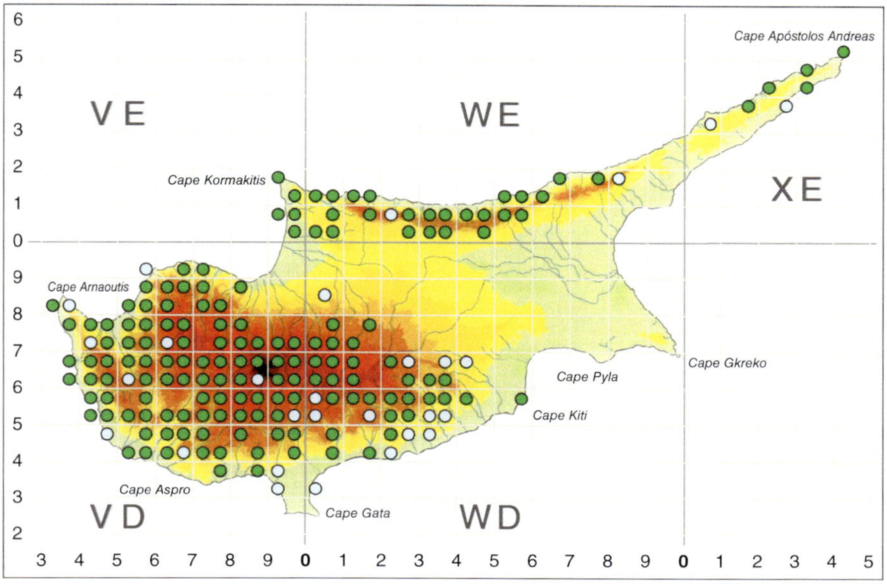

Fig. 12.34. *Allancastria cerisyi* distribution map showing 5 km^2 observations within 10 km^2 UTM squares. © *Eddie John*

Fig. 12.35. *Aristolochia sempervirens*, Diarizos Valley, 17 October 2007. *Eddie John*

Fig. 12.36. *A. cerisyi* (Eastern Festoon) larvae on the hostplant, *Aristolochia sempervirens*, Vasa Kellakiou, 19 April 2017. *Christodoulos Makris*

Ecology and behaviour

Frequents diverse habitats, including dry, rocky, bushy areas, but also lush valleys, riverbanks and ravines supporting the hostplant. Most commonly found in the mountains, particularly in areas with running water that provides conditions supporting growth of its hostplant (Makris, 2003: 88). Males have been observed to fly to the tops of escarpments/cliffs, in a manner resembling hilltopping activity.

Description/identification aids

Wingspan: large – up to 65 mm. *Exhibits sexual dimorphism*, ♂ being lighter and much less heavily marked with black than the darker ♀. Male forewing **upperside** has a *light yellow-white ground colour* and strong black markings along the forewing costa (weakening as they extend into the postdiscal and submarginal areas) and apex. ♂ hindwing upperside has diffuse irregular black discal markings ending in a red mark near the costa, diffuse black and faint-red submarginal markings leading to a small red ocellus near the anal angle, which has a short tail and on the inner margins of which are two rudimentary tails. ♀ upperside is a *deeper yellow*, similarly marked, but with *more extensive areas of black* on fore- and hindwings and noticeable *blue scaling along the submarginal area and with red markings more intense*. The red spotting is characteristic and not found in any other Cyprus species. **Undersides** generally mirror the uppers, other than on the forewing apex, where the black is more diffuse.

Similar species

None in Cyprus.

Conservation status in Cyprus

Not threatened locally or in the Mediterranean (Numa *et al.*, 2016), although *A. cerisyi* is considered a species of European conservation concern, where it is listed as 'NT' (Van Swaay *et al.*, 2010a).

Family: PIERIDAE

Aporia crataegi (Linnaeus, 1758)	Black-veined White	Pierinae
Pieris brassicae (Linnaeus, 1758)	Large White	
Pieris rapae (Linnaeus, 1758)	Small White	
Pontia daplidice (Linnaeus, 1758)	Bath White	
Pontia chloridice (Hübner, [1813])	Small Bath White	
Euchloe ausonia (Hübner, [1804])	Eastern Dappled White	
Anthocharis cardamines (Linnaeus, 1758)	Orange Tip	
Catopsilia florella (Fabricius, 1775)	African Migrant	Coliadinae
Colias croceus (Geoffroy, 1785)	Clouded Yellow	
Gonepteryx cleopatra (Linnaeus, 1767)	Cleopatra	

Aporia crataegi (Linnaeus, 1758) Pieridae (Pierinae)

Black-veined White

TL: Sweden.

Fig. 12.37. *Aporia crataegi* (Black-veined White) ♂ nectaring on *Cistus creticus*, 4 June 2017. *Christodoulos Makris*

Fig. 12.38. *Aporia crataegi* (Black-veined White) ♀ nectaring on *Medicago × varia*, 4 June 2012. Courtesy of *David and Rosalyn Sparrow*

Fig. 12.39. *Aporia crataegi* (Black-veined White) pairing, the larger ♀ above, on the host plant *Crataegus azarolus*, Lofou, 21 May 2011. *Christodoulos Makris*

Presence in the eastern Mediterranean

Ranges throughout the Levant and is represented in southern Turkey and Cyprus by nominotypical *Aporia crataegi crataegi*.

This species was unrecorded in Cyprus by Lederer's collector Franz Zach, and hence not listed in Lederer (1855). Bucknill (1911) stated that Guillemard (1888) had listed *A. crataegi* among just three species of butterfly observed during his 1887 ornithological visit, but Turner (1920) notes that neither Bucknill nor Wilson (see Chapter 2) had encountered the species, an indication of its rarity then, as now.

Distribution in Cyprus

Altitudinal: usually found above 500 m, appearing to be most numerous in areas of the Troodos Mountains between 800–1000 m, but is known to 1800+ m.
Geographical: mainly restricted to the Troodos Mountains and foothills, where it is usually a rare species; occasional sightings at lower elevations are possibly the result of vagrancy. *A. crataegi* remains unrecorded from the entire Pentadaktylos range, despite the presence there of its hostplant. If there is an altitudinal shift in distribution to higher elevations brought about by climate change, as appears to be happening in central Spain (Merrill *et al.*, 2008), *A. crataegi* seems unlikely to colonize the lower northern range in the future.

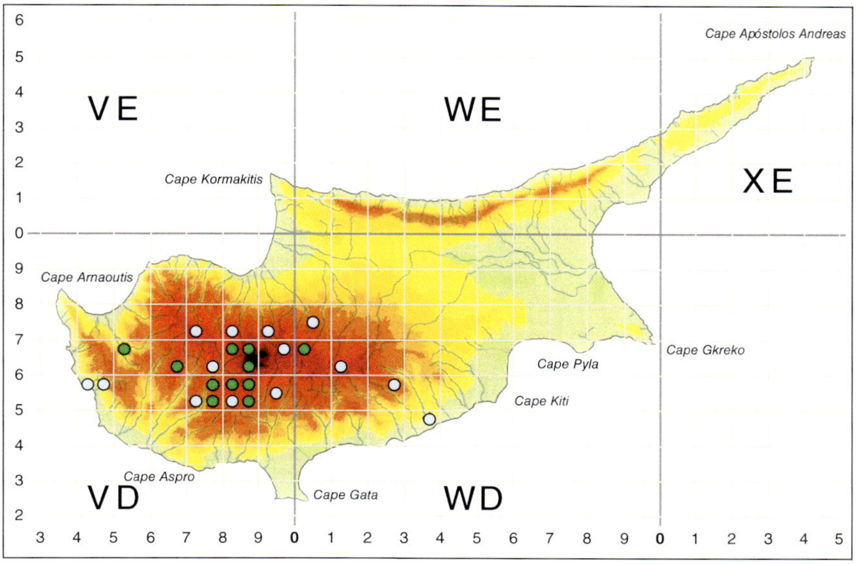

Fig. 12.40. *Aporia crataegi* distribution map showing 5 km^2 observations within 10 km^2 UTM squares. © *Eddie John*

Phenology

Univoltine, April to June. Overwinters communally in a larval hibernaculum.

Larval hostplants

Rosaceae: *Crataegus azarolus* (Mediterranean medlar) (Makris, 2003: 94; Fig. 12.41). Various other hostplants are known elsewhere, e.g. *Amygdalus communis* (almond), and there is some evidence that this may be used in Cyprus (Peter Russell, West Sussex, 2016, personal communication to first author). In Syria, *A. crataegi* has been described as an important pest species on almond (Shlalo and Kassis, 2016).

Ecology and behaviour

A. crataegi is mainly found in mountainous regions at intermediate to higher elevations, where it is a reluctant flyer in cloudy conditions, quickly settling when the temperature falls. Moreover, it has been noted that the species is reported to spend less time in flight than other pierids; in Slovenia, Jugovic and Kržič (2019) calculated male flight activity at 58% and female at just 28%. Nevertheless, *A. crataegi* is highly mobile and a known migrant species capable of crossing expanses of water (e.g. Bolotov *et al.*, 2021), but there is no confirmation of such activity in Cyprus, although Larsen (1977) noted range expansion in Lebanon, which he attributed to 'dispersal ... of a passive nature', rather than one linked to active migration. The species is an avid mud-puddler, observed engaging in this activity in large assemblages in other countries (e.g. Benyamini and John, 2020: 65). Jugovic and Kržič (2019) observed that most commonly, batches of 30–50 eggs are laid, and upon emergence larvae are gregarious, living within a silken web formed on the hostplant (see Makris 2003: 97), from which refuge they leave in order to feed. The web becomes their hibernaculum in which the larvae overwinter, resuming feeding in spring upon leaf regrowth of the deciduous host.

Description/identification aids

Wingspan: large – up to 65 mm. *A. crataegi* differs from all other pierids in Cyprus by the presence of extensive and *pronounced black veining* on the upper- and undersides of both fore- and hindwings. Sexes are similar, but ♀♀ have brown veining near the costal margin of the forewing **uppersides** and undersides. Forewings of mated ♀♀ become somewhat translucent due to scale loss during the act of coupling. Explaining this further, Thomas and Lewington (1991: 201) stated, 'the female grasping the male tightly between her powerful wings. These rub backwards and forwards against her partner's, until many of the scales are missing'. The mating arrangement is supported by an illustration that compares well with a photograph in Benyamini and John (2020: 65), although that in Makris (2003: 95) shows the female to be enclosed within the wings of the *male*.

Fig. 12.41. *Crataegus azarolus*, 1 May 2015. *Eddie John*

This suggests that mating might commence with the latter positioning, but if disturbed, the reverse arrangement necessarily applies, as the female carries the male *in cop.*, the only pierid female known to do so (Wiklund and Forsberg, 1991).

Similar species

None in Cyprus.

Conservation status in Cyprus

The status of this rather rare species in Cyprus is difficult to determine, but it is concerning that it has not recently been seen at many of its previously known locations; fires are a known hazard affecting local habitats. The species is now considered vulnerable in Cyprus, although listed by Van Swaay *et al.* (2010a) and Numa *et al.* (2016) as 'LC' for Europe and the Mediterranean, respectively.

***Pieris brassicae* (Linnaeus, 1758)** **Pieridae (Pierinae)**

Large White

TL: Sweden.

Fig. 12.42. *Pieris brassicae* (Large White), ♂ nectaring on *Leontodon tuberosus*, 11 February 2014. *Christodoulos Makris*

Fig. 12.43. *Pieris brassicae* (Large White) ♀ nectaring on *Dittrichia viscosa*, 1 November 2015. *Christodoulos Makris*

Fig. 12.44. *Pieris brassicae* (Large White) ♂ nectaring on *Echium angustifolium*, 3 May 2016. *Eddie John*

Fig. 12.45. *Pieris brassicae* (Large White) ♀ nectaring on *Salvia fruticosa*, 26 June 2011. *Christodoulos Makris*

Presence in the eastern Mediterranean

Present throughout the Levant as the nominotypical subspecies *Pieris brassicae brassicae* (Feltwell and Vane-Wright, 1982). The validity of the eastern Mediterranean ssp. *Pieris brassicae catoleuca* Röber, 1896 is questionable. Feltwell and Vane-Wright (1982) referred to *catoleuca* as a 'Syrian subrace', while noting that Larsen (1974: 88), who had regarded this as a subspecies, 'at one and the same time ... admits the weakness of the separation'. Hesselbarth *et al.* (1995: 392) synonymized this with the nominotypical ssp. throughout the region, including Cyprus.

In Turner (1920), *Pieris brassicae* ab. 'cypria' Verity, R. [1908] was used to describe a dwarf race, encouraging Graham-Smith and Graham-Smith (1930) to write, 'Specimens of normal dimensions are, however, obtained from parts of the island'. Small (dwarf) specimens of *P. brassicae* and other species *are* known to occur in Cyprus (see Chapter 7). Reflecting practice at the time, Turner (1920) discussed various 'form' names at some length, including '*chariclea* Steph.' for the spring emergence.

Distribution in Cyprus

Altitudinal: found from sea level to the highest elevations.
Geographical: a very common species, present throughout the island.

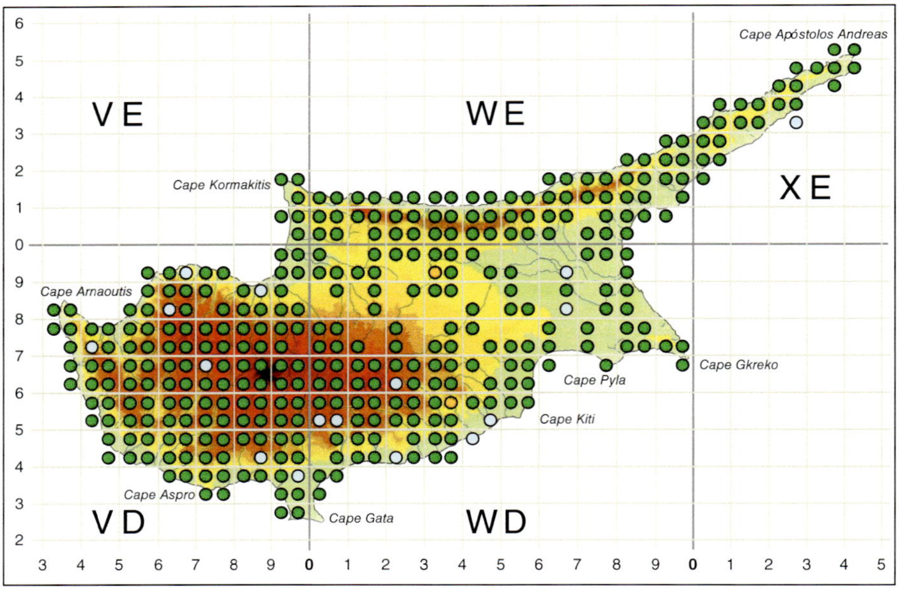

Fig. 12.46. *Pieris brassicae* distribution map showing 5 km^2 observations within 10 km^2 UTM squares. © *Eddie John*

Phenology

Polyvoltine. Generally February to December, but observed on the wing in all weeks of the year in successive overlapping generations. Overwinters as a pupa.

Larval hostplants

A wide range of cultivated and wild Brassicaceae is used, including: *Eruca sativa* (rocket) (Fig. 12.47), *Hirschfeldia incana* (Fig. 12.64), *Rapistrum rugosum*, *Sinapis alba* (white mustard) (Fig. 12.56), *S. arvensis* (charlock), *Sisymbrium* spp.; also Capparaceae: *Capparis spinosa* (caper) (Fig. 12.48); Tropaeolaceae: *Tropaeolum majus* (garden nasturtium).

Ecology and behaviour

Found in all types of habitat, particularly agricultural and urban areas. In summer, the species appears to retreat from hotter coastal areas, but may still be found on the wing in the mountains. On 18 May 2003, in hot afternoon temperatures (30°C), four individuals were disturbed as they sheltered in the shade of trees at Larnaka salt lake (first author, personal observation). Spieth *et al.* (2011) reported on aestivation of this species in the Iberian Peninsula, but this has not been noted in studies of populations elsewhere, including the eastern Mediterranean (Israel). Aestivation is not thought to occur in Cyprus, the species remaining active throughout the summer at higher, cooler elevations, although the absence in lowland areas might point to pupal aestivation, as reported for the species in southern areas of the Iberian Peninsula (Spieth and Schwarzer, 2001). Although known as a migrant species in the wider region (Larsen, 1976), there is no evidence of such activity in Cyprus.

Fig. 12.47. *Eruca sativa* (rocket), 1 April 2010. Courtesy of *Yiannis Christofides*

Fig. 12.48. *Capparis spinosa* (caper), 1 July 2006. *Eddie John*

Description/identification aids

Wingspan: large – usually up to 60 mm, with occasional ♀♀ to 70 mm; individuals of the summer brood are noticeably larger than those of earlier or later broods. Appears off-white on the wing. Both sexes have a large black apical/submarginal patch on the forewing **upperside**, *extending over half way down the outer margin* (*cf. Pieris rapae*) and visible on the forewing underside; the female has two large black forewing spots, absent in the male. Winter and spring specimens have a light suffusion of black scaling and the hindwing **underside** has a more pronounced yellow ground colour.

Similar species

Although there is usually an obvious difference in size between the *Pieris* spp. in Cyprus, smaller specimens of *P. brassicae* may be mistaken for *P. rapae* and larger specimens of *P. rapae* for *P. brassicae*.

Conservation status in Cyprus

Not threatened. Listed by Van Swaay *et al.* (2010a) and Numa *et al.* (2016) as 'LC'.

Pieris rapae (Linnaeus, 1758)

Pieridae (Pierinae)

Small White

TL: Sweden.

Fig. 12.49. *Pieris rapae* (Small White) ♂, 9 January 2015. *Christodoulos Makris*

Fig. 12.50. *Pieris rapae* (Small White) ♀ nectaring on *Cistus creticus*, 7 May 2011. Courtesy of *David and Rosalyn Sparrow*

Fig. 12.51. *Pieris rapae* (Small White) nectaring on *Heliotropium hirsutissimum*, 29 July 2021. Courtesy of *Clive Burrows*

Fig. 12.52. *Pieris rapae* (Small White) ♀ nectaring on *Narcissus obsoletus*, 9 November 2008. Courtesy of *Alison McArthur*

Fig. 12.53. *Pieris rapae* (Small White) pairing, ♂ above, ♀ below, 9 January 2015. *Christodoulos Makris*

Fig. 12.54. *Pieris rapae* (Small White) pairing, ♀ left, ♂ right, 9 January 2015. *Christodoulos Makris*

Presence in the eastern Mediterranean

Widely distributed throughout the Levant as nominotypical *Pieris rapae rapae*. The ssp. *Pieris rapae leucosoma* (Schawerda, 1905) had been described for the region, although Turner (1920) and Rebel (1939), considered this a race or form; its validity as a ssp. is questionable and it was synonymized with the nominotypical subspecies by Hesselbarth *et al.* (1995: 402). As with *P. brassicae*, Turner (1920) again discussed various 'forms' and Rebel (1939) referred to an occasional dwarf form.

Distribution in Cyprus

Altitudinal: present at all elevations, although in greater numbers in lowland areas.
Geographical: very commonly seen throughout the island and the most frequently recorded butterfly species in Cyprus; the few gaps remaining in the distribution map simply require visits to the areas in order to ensure full coverage.

Phenology

Polyvoltine. Recorded in all weeks of the year (CBRS data) in successive overlapping generations. Note that the pairing in Figs 12.53 and 12.54 was photographed on 9 January 2015, in a temperature of 5°C. All immature stages may also be found throughout a mild winter.

Larval hostplants

Principally Brassicaceae (including cultivated forms): *Hirschfeldia incana*, *Rapistrum rugosum*, *Sinapis alba* (white mustard) (Fig. 12.56), *S. arvensis* (charlock), *Raphanus raphanistrum*, *Sisymbrium* spp.; Capparaceae: *Capparis spinosa* (caper); Tropaeolaceae: *Tropaeolum majus* (garden nasturtium).

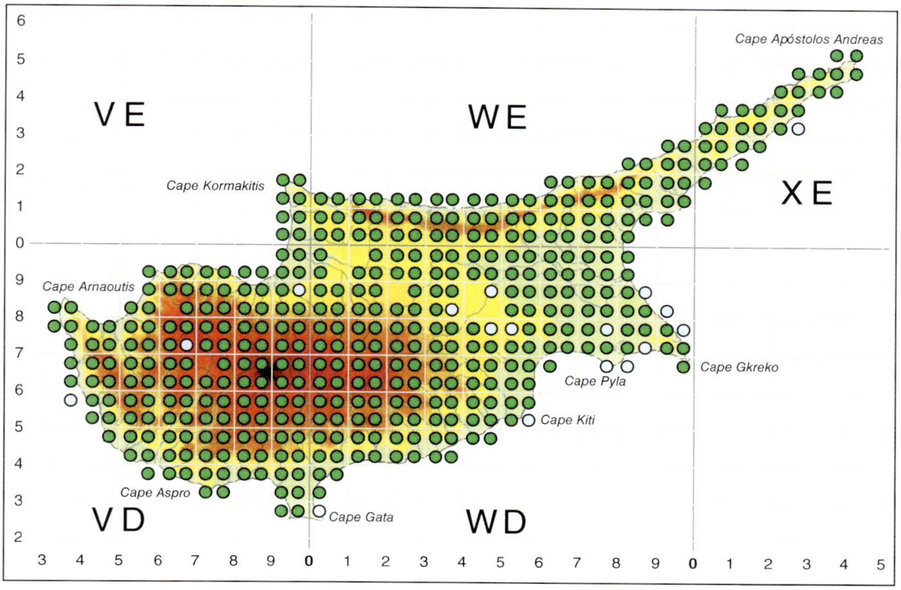

Fig. 12.55. *Pieris rapae* distribution map showing 5 km^2 observations within 10 km^2 UTM squares. © *Eddie John*

Fig. 12.56. *Sinapis alba*, 20 May 2013, one of several hostplants for *Pieris rapae*. *Eddie John*

Ecology and behaviour

Found in all habitat types, but especially common in agricultural areas where many species in the family Brassicaceae are found as weeds in areas under cultivation. Occasionally, small aggregations form at mud (see Chapter 6, Fig. 6.19). *P. rapae* is a migratory species, although such observations are rarely reported in Cyprus (see John *et al.*, 2008b; McArthur and John, 2019); certainly, immigration is not a requirement for the support of Cyprus populations. Nectar is sourced from many flower species.

Description/identification aids

Wingspan: medium-sized – usually up to 50 mm, occasionally slightly larger. Unlike similarly sized pierids in Cyprus, *P. rapae* is a *slow-flying*, seasonally variable white butterfly, the first generation of both sexes with diffuse black markings. In later broods, the forewing black apical mark is stronger, *rarely as black as in P. brassicae and the mark does not extend as far down the outer margin* (*cf. P. brassicae*). The male has a single black forewing spot (*cf.* ♂ *P.* brassicae), the ♀ two black spots.

Similar species

Pieris napi (Linnaeus 1758), present elsewhere in the Levant, is not found in Cyprus. *P. rapae* is unlike any other small white butterfly species found in Cyprus, but, as with *P. brassicae*, specimens may vary in size.

Conservation status in Cyprus

Not threatened. Listed by Van Swaay *et al.* (2010a) and Numa *et al.* (2016) as ‘LC’.

Pontia daplidice (Linnaeus, 1758) Pieridae (Pierinae)

Bath White

TL: Northwest Africa.

Fig. 12.57. *Pontia daplidice* (Bath White) ♂ nectaring on *Dittrichia* viscosa, 10 November 2021. *Christodoulos Makris*

Fig. 12.58. *Pontia daplidice* (Bath White) ♀, 9 May 2016. *Eddie John*

Fig. 12.59. *Pontia daplidice* (Bath White) ♂ nectaring on *Dittrichia viscosa*, 5 November 2011. *Christodoulos Makris*

Fig. 12.60. *Pontia daplidice* (Bath White) ♀ nectaring on *Limonium sinuatum*, 10 May 2015. *Eddie John*

Fig. 12.61. *Pontia daplidice* (Bath White) ♀ nectaring on *Centaurea hyalolepis*, 24 May 2013. *Eddie John*

Fig. 12.62. *Pontia daplidice* (Bath White) pairing, ♂ above, 16 April 2017. *Christodoulos Makris*

Presence in the eastern Mediterranean

The 'Bath White complex' (*P. daplidice* and *P. edusa*), two related and morphologically inseparable species, has been investigated in a number of countries, including those of the eastern Mediterranean. Recent DNA work (John *et al.*, 2013b) indicated that *P. daplidice* is present in Cyprus and Lebanon and confirmed that *P. edusa* is widespread in Turkey. Both taxa appear to meet in south-east Turkey, but their Asian contact zone where they may interbreed is not well known (Martin Wiemers, Vienna, 2013, personal communication to first author) and requires further research. The aforementioned molecular work established that 12 Cyprus specimens, taken over a 4-year period and from different parts of the island, were identified as *P. daplidice* (confirming the use of this name by Parker, 1983) and not *P. edusa*. Therefore, references to *P. edusa* in previous publications (e.g. John, 2000: 5, 23; Makris, 2003: 106) should be disregarded. However, it should be borne in mind that both species are known as vigorous migrants (Larsen, 1976).

Recorded by Lederer (1855) as '*Pieris Daplidice* L.' and by Bucknill (1911) as '*Leucochloe daplidice L. Bath White*.' Turner (1920) referred to the species as currently known but, in the manner adopted at the time, he discussed a dwarf form ('ab. miniscula Vrty'), taken on Troodos in July, as well as various other 'forms' or 'aberrations'.

Distribution in Cyprus

Altitudinal: most commonly encountered at low to intermediate elevations, but may occur anywhere across the island.
Geographical: a very common species, well distributed across the island and could be expected to appear anywhere.

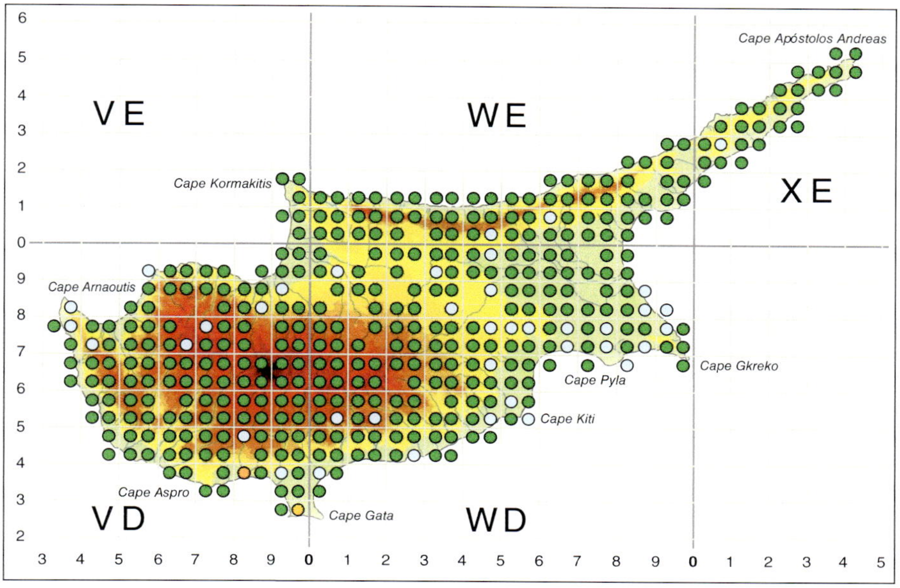

Fig. 12.63. *Pontia daplidice* distribution map showing 5 km^2 observations within 10 km^2 UTM squares. © *Eddie John*

Phenology

Polyvoltine. Observed on the wing in all weeks of the year (CBRS data), but generally February to November. Overwinters as a pupa.

Larval hostplants

Various cultivated and wild Brassicaceae: *Erucaria hispanica* (Spanish mustard), *Rapistrum rugosum*, *Hirschfeldia incana* (Fig. 12.64), *Sinapis alba* (white mustard) (Fig. 12.56), *S. arvensis* (charlock), *Sisymbrium* spp. (Fig. 12.65); also Resedaceae: *Reseda lutea* (wild mignonette); Cleomaceae: *Cleome ornithopodioides*, *C. iberica* (John *et al.*, 2008a, 2013b). Elsewhere in the Levant, Capparaceae: *Capparis spinosa* (caper) (Fig. 12.48), has been recorded as a hostplant (Benyamini and John, 2020: 106).

Ecology and behaviour

Found in all types of habitat, especially field margins, waste ground and agricultural land over which *P. daplidice* flies rapidly, often randomly in a zigzag manner. The species is a known migrant, but such activity has not been reported in Cyprus, although Larsen (1974: 92, 1975) documented major migrations of *P. daplidice* in Lebanon and described the pierid as 'a powerful opportunistic migrant' (Larsen, 1994). On 5 June 2015, the first author encountered a pair *in cop.*, the limp wings of the female indicating that mating had occurred very shortly after emergence from the pupa.

Fig. 12.64. *Hirschfeldia incana*, 18 May 2007. *Eddie John*

Fig. 12.65. *Sisymbrium orientale*, 19 February 2016. *Christodoulos Makris*

Description/identification aids

Wingspan: medium-sized – up to 50 mm. A *fast-flying* butterfly that is darker on the wing than *P. rapae*. Early broods are darker overall (especially undersides) appearing a 'muddy' green, contrasting with the yellow-green found in summer generations. Occasional summer specimens have been found where black markings have been much less intense. **Upperside** ground colour of both sexes is white, with black apical markings, larger in the ♀. A forewing black mark at the end of the cell is common to both sexes, but is larger in the ♀. **Undersides** of both sexes are similar, the ♀ having more pronounced black forewing markings. In both sexes, the black mark at the end of the cell *extends to the costa* (*cf. P. chloridice*), sometimes becoming diffuse at the margin; there is also a smaller black postdiscal mark near the forewing anal angle. Apical markings are yellow-green. In ♂ and ♀, hindwing undersides are heavily marked in green or yellow-green, with yellow veining; these markings becoming *wider and rounded* near the hindwing margin (*cf. P. chloridice*).

Similar species

See 'Description/identification aids' under *Pontia chloridice* and *Euchloe ausonia*.

Conservation status in Cyprus

Not threatened. Listed by Van Swaay *et al.* (2010a) and Numa *et al.* (2016) as 'LC'.

Pontia chloridice (Hübner, [1813]) Pieridae (Pierinae)

Small Bath White

TL: Uncertain: Südrussland? (Russia).

Fig. 12.66. *Pontia chloridice* (Small Bath White), first brood ♂ nectaring on *Matricaria recutita*, 4 May 2015. *Eddie John*

Fig. 12.67. *Pontia chloridice* (Small Bath White) ♀ on the hostplant *Cleome iberica*, 6 July 2013. *Christodoulos Makris*

Fig. 12.68. *Pontia chloridice* (Small Bath White) ♂ on *Rubus sanctus*, 8 August 2011. *Christodoulos Makris*

Fig. 12.69. *Pontia chloridice* (Small Bath White) ♂, 8 August 2011. *Christodoulos Makris*

Fig. 12.70. Upperside of an aberrant *Pontia chloridice* (Small Bath White), 1 April 1982. Coll. Gabriel Georghiou. *Christodoulos Makris*

Fig. 12.71. Underside of an aberrant *Pontia chloridice* (Small Bath White), 1 April 1982. Coll. Gabriel Georghiou. *Christodoulos Makris*

On 1 April 1982, an aberrant individual of *P. chloridice*, almost devoid of typical markings, was caught by Gabriel Georghiou at Lefka Dam. It is known that such aberrations may occur when pupae are exposed to extreme temperatures, but it is uncertain if the specimen shown in Figs 12.70 and 12.71 was subjected in this way and it appears to be unique in Cyprus.

Presence in the eastern Mediterranean

Sporadically present in Turkey, but unrecorded from other countries of the Levant except Cyprus, which represents the southernmost extent of the species' range in the region (Benyamini and John, 2020: 78).

P. chloridice was not listed for Cyprus by Lederer (1855) and appears to have been first recorded by a Mr Marsden on the island 'many years' prior to 1918, according to Bucknill and Wilson (in Turner, 1920). (See John and Coutsis, 2022, for further details.)

Distribution in Cyprus

Altitudinal: found at intermediate elevations between 400 m and 1500 m, but most frequently within the range 600–1000 m.
Geographical: an uncommon and very local species, not known from northern or eastern Cyprus. Unsuitable geology probably accounts for the absence of the hostplants from the northern ranges of limestones, chalks and marls, restricting *P. chloridice* to the igneous rocks of the Troodos range and foothills above *ca*. 400 m.

Phenology

Bivoltine to sparingly trivoltine, April to November. *P. chloridice* was suspected by Parker (1983) to breed in Cyprus and was confirmed to do so in 1997 (Makris, 2003: 110). In exceptional years, the summer brood can be locally common (second author, personal observation), although

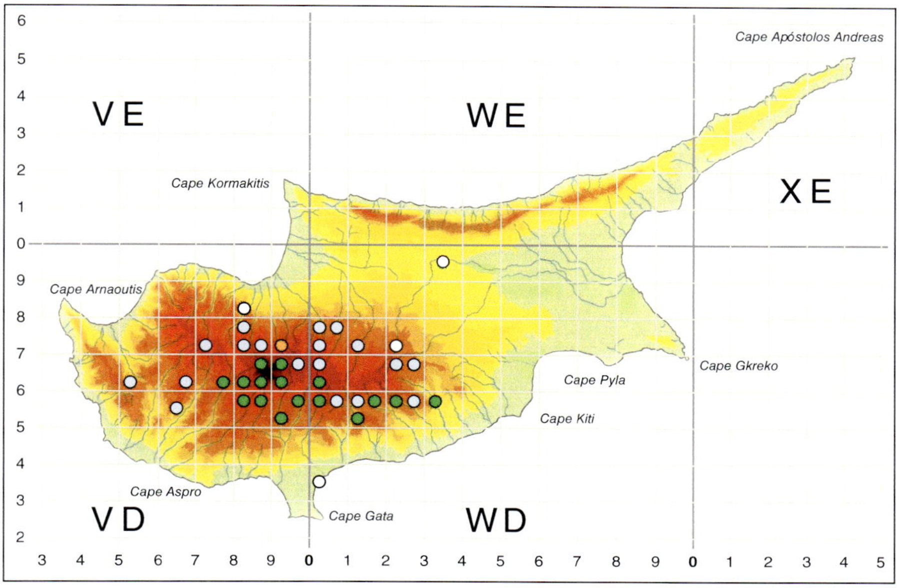

Fig. 12.72. *Pontia chloridice* distribution map showing 5 km² observations within 10 km² UTM squares. © *Eddie John*

in general very few observations are reported from the summer months, as in 2021, when there was just a single summer record (Elli Tzirkalli, Nicosia, 2021, personal communication to first author). However, the Covid-19 pandemic might have had an influence, as this has significantly reduced the overall number of records submitted to the CBRS in 2020 and 2021.

Eggs and larvae of all instars of *P. chloridice* have been found on the same plant in the autumn, indicative of a temporal spread of egg-laying activity by females of the same generation, rather than a 'blurring' of two late broods. In nearby Antalya (Turkey), Köstler (2001) expressed the view that four generations were possible there, although this was later questioned in Köstler (2002). In unfavourable years, the occasional absence of the species (and occasionally of its hostplants) from known locations hints at possible facultative pupal diapause, with some or all pupae from the spring generation remaining in diapause until the following year (John and Coutsis, 2022). Overwinters as a pupa.

Larval hostplants

Cleome ornithopodioides (Figs 12.73 and 12.74) and *C. iberica* (Cleomaceae) (Makris, 2003: 110; John *et al.*, 2008a, 2013a). In captivity, larvae accept Capparaceae: *Capparis spinosa* (caper) (Peter Russell, West Sussex, 2008, personal communication to first author). *C. ornithopodioides* is found from *ca.* 200 m to 1500 m, *C. iberica* from 800 m to 1800 m (second author, personal observation).

Fig. 12.73. Young plants of *Cleome ornithopodioides* germinating from disturbed gravel alongside a Troodos mountain track, 1 June 2013. *Eddie John*

Fig. 12.74. Seed pods and flowers of *Cleome ornithopodioides*, Kellaki area, 15 October 2007. *Eddie John*

Ecology and behaviour

Most often, found in small numbers on disturbed verges of hillside tracks above 400 m, in close association with its annual *Cleome* hostplants. In June 2012, two ♂♂ and one ♀ were found flying around *C. ornithopodioides* growing in a small, newly ploughed olive grove, devoid of other plants except *Heliotropium hirsutissimum* (John *et al.*, 2013a). An unusually large colony, comprising hundreds of *P. chloridice*, was reported from Xyliátos (500 m) in 1997, but was found to have disappeared from the biotope during the following two years (Makris, 2003: 291). *P. chloridice* was seen again there in September 2005, but spring storms in 2007 left the gravelly biotope buried in mud, and with no sign of hostplant or butterfly (John *et al.*, 2008a). Maintenance of roads and tracks in the hills after winter storms, with associated disturbance of verges, assists dispersal and germination of the hostplant's seeds, but the precise requirements for the germination of *Cleome* seeds (scarification and/or burial of seed, or exposure of seed to light) remain in question. Under African conditions, the related *Cleome gynandra* L. gave improved germination rates when seeds were buried compared with others that were exposed to light (Ochuodho and Modi, 2007). The answer to the question of the required optimum conditions for germination could help explain the disappearance of populations of *Cleome* (and *P. chloridice*) from certain areas, only to reappear in subsequent seasons. In October 2007, for example, numerous *C. ornithopodioides* were present at a biotope near Kelephos Bridge but 10 years later, with conditions seemingly still suitable, no plants were found during a visit at the same time of year. However, the fragile and unstable nature of some sites in which *C. ornithopodioides* grows is also likely to play a key role in the sporadic and unpredictable appearances of *P. chloridice* in previously known locations.

Although the species is known to overwinter in the pupal stage, doubt remains about the chosen site for pupation; annual hostplants are likely to be uprooted or destroyed during winter storms, leading to the hypothesis that pupation takes place on a stone or rock (Tolman, 1992; John *et al.*, 2008a).

Early accounts (e.g. Turner, 1920) suggested that immigrant *P. chloridice* might arrive in Cyprus from the mainland. However, while the species has been documented in the far north of Europe as one having migrant capability (e.g. Keynäs and Mikkola, 1970), there is no evidence of such activity in the Levant or in Greece. Besides, the likelihood of southerly migration from populations in Antalya Province, Turkey, to Cyprus seems remote, when the island represents the species' most southerly presence in the wider area (see also John and Coutsis, 2022).

Nectaring has been observed on various flowers, including *Leontodon tuberosus*, *Rubus sanctus*, *Matricaria recutita* and *Heliotropium hirsutissimum*.

Description/identification aids

Wingspan: medium-sized – up to 45 mm. **Upperside** ground colour of both sexes is white, with black apical markings, larger in the ♀, and closely resembling *P. daplidice*, but in both sexes the forewing upperside spot at the end of the cell is smaller. **Undersides** of both sexes are *less heavily marked* than *P. daplidice* and the ♂ forewing lacks the small black postdiscal mark near the anal angle. As on the upperside, the spot at the end of the cell is smaller than in *P. daplidice* and *does not reach the costa* (*cf. P. daplidice*). In both sexes, the yellow-green scaling surrounding the veins at the outer margin of the hindwing underside presents an elongated *toothed appearance, not club-shaped* as in *P. daplidice*. Adults of the second brood – the 'dry season' summer form – have lighter underside wing markings, lacking much of the dark green scaling of those that emerge in April.

Similar species

See 'Description/identification aids' under *P. daplidice* and *Euchloe ausonia.*

Conservation status in Cyprus

It is difficult to assess the local status of this species, due to its comparative rarity and apparent population fluctuations. Appropriate management of 'breeding sites' would appear to be an impracticable aim, as such areas are often transient in nature. Moreover, it would not be possible to compensate for lack of rainfall that must play a key role in hostplant germination. As the species is at the southernmost point of its range, climate change must be seen as a potential threat if global temperatures continue to rise. Categorized as 'LC' in Europe and in the Mediterranean as a whole (Van Swaay *et al.*, 2010a; Numa *et al.*, 2016), 'DD' in Cyprus, but considered vulnerable.

Euchloe ausonia (Hübner, [1804]) Pieridae (Pierinae)

Eastern Dappled White

TL: Italy.

Fig. 12.75. *Euchloe ausonia* (Eastern Dappled White) first brood ♂ on *Hirschfeldia incana*, 22 February 2015. *Christodoulos Makris*

Fig. 12.76. *Euchloe ausonia* (Eastern Dappled White) first brood ♂ nectaring on *Crupina crupinastrum*, 18 February 2013. Courtesy of *Alison McArthur*

Fig. 12.77. *Euchloe ausonia* (Eastern Dappled White) second brood ♀ nectaring on *Erucaria hispanica*, 20 April 2014. *Eddie John*

Fig. 12.78. *Euchloe ausonia* (Eastern Dappled White) second brood on *Erucaria hispanica*, 7 April 2014. *Christodoulos Makris*

Fig. 12.79. *Euchloe ausonia* (Eastern Dappled White) second brood pairing, ♀ left, ♂ right, 1 April 2015. *Christodoulos Makris*

Presence in the eastern Mediterranean

Two ssp. are present in the wider region: that found in Turkey and Cyprus is *Euchloe ausonia taurica* (Röber, 1907) TL: Taurus, while *E. a. melisande* Frühstorfer, 1908 has been given as the ssp. that occurs in southern regions of the Levant (Benyamini and John, 2020: 82). The species has been known by numerous names in the past: Lederer (1855) recorded *E. ausonia* as '*Anthocharis Ausonia* Esp.', while Turner (1920) referred to 'Anthocharis crameri Btlr. (belia auct.) [group ausonia Hb.; race taurica Röb.]'. Rebel (1939) used the name by which the species is known today.

Distribution in Cyprus

Altitudinal: rarely, if ever, flies at the highest elevations (i.e. above 1600 m) and is most commonly encountered at lower and intermediate levels.
Geographical: a widespread presence, but patchily recorded across the entire island (CBRS data); this may be a recording artefact reflecting the species' limited time on the wing.

Phenology

Bivoltine, February to May, *rarely later*. Overwinters as a pupa.

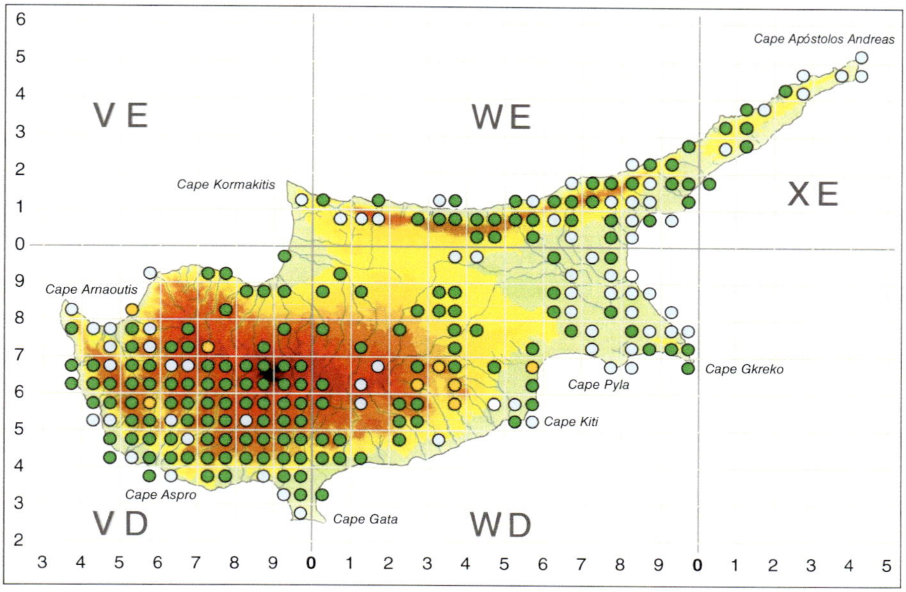

Fig. 12.80. *Euchloe ausonia* distribution map showing 5 km² observations within 10 km² UTM squares. © *Eddie John*

Larval hostplants

Various Brassicaceae. In Cyprus, the main hostplant is *Hirschfeldia incana* (Fig. 12.81); other reported hostplants are *Erucaria hispanica* (Figs 12.77 and 12.78), *Rapistrum rugosum*, *Sinapis alba* (white mustard) (Fig. 12.56) and *S. arvensis* (charlock). Dennis (1995) reported oviposition on *Brassica nigra* (black mustard) in Crete, but although present, this is a rare plant in Cyprus (Christofides, 2017: 102).

Ecology and behaviour

Found in a variety of habitats, flying close to the ground in agricultural areas, field margins and 'waste' ground, and is most commonly found in areas supporting *Hirschfeldia incana*. *E. ausonia* is not as frequently encountered as *Pontia daplidice* and has a *much* shorter flight period. The species has been reported hilltopping in Crete (Hardy, 1994), but this means of mate location has not been documented in Cyprus. On 11 April 2006, three male *E. ausonia* were observed aggressively attempting to separate a mating pair; the attempt was unsuccessful despite one of the males attaching to the wings of the hanging female (Nigel Cottle, Somerset, 2006, personal communication to first author). In parts of the Levant, pupal diapause over a period of several years has been reported (Benyamini and John, 2020: 82).

Fig. 12.81. *Euchloe ausonia* (Eastern Dappled White) ♂ on the principal hostplant *Hirschfeldia incana*, 22 February 2015. *Christodoulos Makris*

Description/identification aids

Wingspan: medium-sized – up to 50 mm. A fairly fast-flying butterfly, appearing somewhat similar on the wing to *P. daplidice*. **Upperside** ground colour of both sexes is white, with black forewing apical markings and a black mark at the end of the cell, similar to the two preceding species. Forewing **underside** is similar in both sexes, but *lacks the small black postdiscal mark near the anal angle* (*cf. P. daplidice*). The hindwing underside is yellow-green, with markings much less well-defined than in *P. daplidice* or *P. chloridice*. Underside wing-veining is *well suffused with yellow*, especially in the second brood, *which is also noticeably larger*. Note the phenology: any similarly marked butterfly seen after the end of May/early June is very unlikely to be *E. ausonia*, although early records show the species to have been taken on Troodos in June and July (Wilson, in Turner, 1920) and on Mt Olympos on 20 June 1973 (Bretherton, 1974).

Similar species

See 'Description/identification aids' under *P. daplidice*, *P. chloridice* and *Anthocharis cardamines* (♀).

Conservation status in Cyprus

Not threatened. Listed by Van Swaay *et al.* (2010a) and Numa *et al.* (2016) as 'LC'.

***Anthocharis cardamines* (Linnaeus, 1758)** **Pieridae** (Pierinae)

Orange Tip

TL: Europe.

Fig. 12.82. *Anthocharis cardamines* (Orange Tip) ♂, 9 March 2013. *Christodoulos Makris*

Fig. 12.83. *Anthocharis cardamines* (Orange Tip) ♂ nectaring on *Sinapis alba*, 9 April 2016. Courtesy of *David and Rosalyn Sparrow*

Fig. 12.84. *Anthocharis cardamines* (Orange Tip) ♂ nectaring on *Turritis laxa*, 17 April 2019. *Christodoulos Makris*

Fig. 12.85. *Anthocharis cardamines* (Orange Tip) ♀ nectaring on *Hirschfeldia incana*, 12 March 2016. *Christodoulos Makris*

Fig. 12.86. *Anthocharis cardamines* (Orange Tip) ♂ on the endemic *Onosma fruticosa*, 9 March 2013. *Christodoulos Makris*

Fig. 12.87. *Anthocharis cardamines* (Orange Tip) ♂ above, ♀ below, 22 March 2010. Courtesy of *Yiannis Christofides*

Presence in the eastern Mediterranean

Found throughout the eastern Mediterranean region as ssp. *Anthocharis cardamines phoenissa* von Kalchberg, 1894 TL: 'Haifa' (Israel) (Benyamini and John, 2020: 90). Hesselbarth *et al.* (1995: 354) referred to 'a broad zone of overlap and mixing [between *Anthocharis cardamines phoenissa* and the nominotypical subspecies] from SW to NE diagonally through Anatolia' (translated from the original German), with *ca.* 50% of individuals in populations on the western and northern edge of the hybridization zone showing characteristics of *cardamines*, while the remainder featured hybrid characteristics. See also 'Description/identification aids', below.

Gibbs (in Turner, 1920) referred to very small specimens and stated, 'They appear to prevail in the Mediterranean islands ...'. Turner (1920) referred to receiving specimens from Cyprus measuring 26.5 mm (♂♂) and 27 mm (♀♀), adding, 'This is the form ab. *minor* Ckrll'. Parker (1983) reported dwarf forms as being reasonably common, an observation explained by Hesselbarth *et al.* (1995: 354) who, when comparing the origins of the nominotypical ssp. compared with ssp. *A. c. phoenissa*, stated that 'The consistently smaller *phoenissa*, on the other hand, penetrated northwards from the Iranian / Syrian area'. In contrast, 'Postglacial *cardamines* spread from southeastern Europe and western Georgia into Anatolia' (both sentences translated from the original German). Examples of nominotypical *A. c. cardamines* and of the smaller ssp. *A. c. phoenissa* may be found in Hesselbarth *et al.* (1995: Vol 3, Plate 25).

Distribution in Cyprus

Altitudinal: all elevations, but flies mainly at lower to intermediate altitudes.

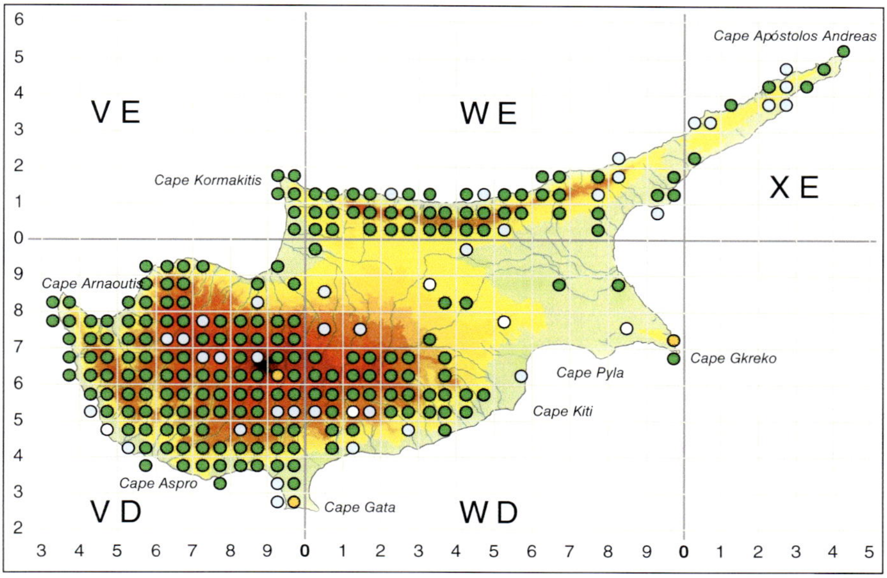

Fig. 12.88. *Anthocharis cardamines* distribution map showing 5 km² observations within 10 km² UTM squares. © *Eddie John*

Geographical: common in the western half of the island, sporadic in the north-east to the very tip of the Karpaz Peninsula, but with very few records from the Mesaoria Plain and the south-east (CBRS data).

Phenology

Univoltine, generally February to May, but may appear in mid–late January. Overwinters as a pupa. Extended pupal diapause over a second winter has been reported elsewhere (Benyamini and John, 2020: 90).

Larval hostplants

Brassicaceae: *Biscutella didyma* (Fig. 12.89), *Sinapis alba* (white mustard), *S. arvensis* (charlock), *Erucaria hispanica* (Figs 12.77 and 12.78), *Alliaria petiolata*, *Hirschfeldia incana* (Figs 12.64 and 12.81), *Sisymbrium irio*.

Ecology and behaviour

Mainly found in irrigated areas and river valleys, where flight is close to the ground. As with the nominotypical ssp., males are typical patrollers, flying along tracks and woodland margins in their search for females. Nectaring has been observed on a variety of flowers.

Fig. 12.89. *Biscutella didyma*, 21 April 2011. Courtesy of *Yiannis Christofides*

Description/identification aids

Wingspan: medium-sized – up to 45 mm. **Upperside** ground colour of both sexes is white, the male easily identified even in flight by its bright orange forewing apical/subapical patch, bordered by black at the apex. Occasionally there is a light suffusion of black scales along the inner border of the orange patch, a form known as f. 'umbrosa' Culot (see illustrations in Makris, 2003: 119 and Plate 7). In Cyprus, this suffusion (when present) seems consistently light, whereas in Turkey this varies from being barely recognizable in some individuals to being present as a strong band in others (Hesselbarth *et al.*, 1995: 354). A specimen from Lebanon illustrated in Larsen (1974: Plate 96), indicates that this 'band' may be very heavily marked – nearly forming a continuous black ribbon – and in which country, f. 'umbrosa' may outnumber typical specimens (Larsen, 1974: 99). In Israel, two photographs of *A. c. phoenissa* shown in Benyamini (2002a: 95) and three in Benyamini and John (2020: 90) depict specimens devoid of such scaling. **Underside** ground colour of both sexes is white, the forewing with a *small black spot* at the end of the cell. The forewing orange patch of the ♂ is reduced in size and bordered with a diffuse white or greenish-yellow at the apex, the latter markings less well defined in the ♀. Hindwing underside of both sexes is similar, with typical green markings dusted with yellow and with yellow veining.

Similar species

Male: no similar species is present in Cyprus. Female: similar to *E. ausonia*, but is much smaller and the *apical markings* on the forewing underside of *A. cardamines* are *far less extensive* and *much less well defined.*

Conservation status in Cyprus

Not threatened. Listed by Van Swaay *et al.* (2010a) and Numa *et al.* (2016) as 'LC'.

Catopsilia florella (Fabricius, 1775) **Pieridae** (Coliadinae)

African Migrant

TL: "Sierra Leon Africae".

Fig. 12.90. *Catopsilia florella* (African Migrant) ♀ nectaring on *Limonium sinuatum*, 8 April 2019, Karpasia. Courtesy of *Will Hawkes*

Fig. 12.91. *Catopsilia florella* (African Migrant) ♂ photographed in The Gambia. Courtesy of *David Penney*

Fig. 12.92. *Catopsilia florella* (African Migrant) ♀ photographed in The Gambia. Courtesy of *David Penney*

Fig. 12.93. *Catopsilia florella* (African Migrant), a light-coloured ♀ (captive specimen from Israel), 26 September 2013. *Christodoulos Makris*

Presence in the eastern Mediterranean

A migrant species into the Levant, where it is being seen with increasing frequency in southern countries, aided by the use of the species' hostplants as garden ornamentals and rising winter temperatures (Benyamini and John, 2020: 120). Larsen and Nakamura (1983) reported just two known records from the Levant, the most recent known account at the time being that by Larsen (1978). However, it is puzzling that Larsen should state in his 1978 paper that the first record from Lebanon 'was quoted by Lederer (1855, as *Callimorpha dryas*) ...', when Lederer's (1855) account lacks any mention of this name. Among moths collected by Zach, two species of *Callimorpha* (*C. hera*, syn. *Euplagia quadripunctaria* (Poda, 1761) and *C. dominula* (Linnaeus, 1758)), *were* listed by Lederer, but there is no mention of *C. florella* under its currently accepted name, or as one of its many earlier synonyms, which include *Callidryas florella* Fabricius. However unlikely, Larsen's seemingly inadvertent listing, therefore, might be explained by the similarity between the names *Callimorpha dryas* and genus *Callidryas*.

The species is known for Mediterranean Syria (Benyamini and John, 2020: 121), but not for Hatay Province, Turkey, immediately to the north (Atahan *et al.*, 2018). However, with three confirmed records for Cyprus, it is highly likely that other, small migrations from the mainland will have occurred but have passed undetected.

Distribution in Cyprus

Altitudinal: Insufficient data are available for Cyprus, but all known specimens were seen in low-lying areas at or near the coast.
Geographical: known in Cyprus only from localities in the far north-east of Karpasia and the south and south-east of Cyprus (see section on 'Migration' in Chapter 6 and text below).

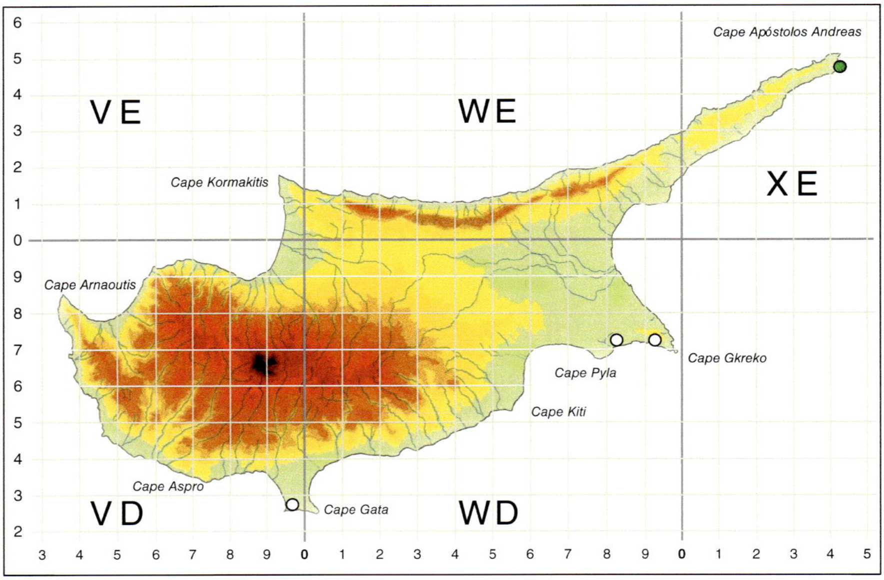

Fig. 12.94. *Catopsilia florella* distribution map showing 5 km^2 observations within 10 km^2 UTM squares. © *Eddie John*

Phenology

Polyvoltine, but migratory specimens are most likely to be seen early in the year.

Larval hostplants

Fabaceae, subf. Caesalpinioideae: *Cassia* and *Senna* spp. (Figs 12.95 and 12.96), non-native to the Mediterranean area, but planted as ornamental flowering shrubs. Eggs are most frequently laid on flower buds, on which they are easily seen (Matt Rowlings, Switzerland, 2019, personal communication to first author).

Ecology and behaviour

A vigorous, fast-flying migrant, *C. florella* appears to have been first taken in Cyprus in 1963 or 1964, probably at RAF Akrotiri (Jack Harrison, UK, 2019, personal communication to first author; John *et al.*, 2019b). *C. florella* appeared again in Cyprus in the autumn of 1986, when eleven specimens were taken by Pavlos Neofytou at three locations, at or close to the south-east coast over at least two days (Makris, 2003: 130; Aristos Aristophanous, Cyprus, 2019, personal communication to first author). The timing suggests that probably these were not part of a late migration, but rather the progeny of an earlier (unreported) migration that had successfully bred locally, perhaps having been part of a spring or early summer migration of *C. florella* observed in Israel earlier that year (John *et al.*, 2019b). The capture of 10 ♂♂ and only one ♀ further points to a recent local emergence of offspring, as males are usually the first to emerge and a sex ratio of 50/50 was noted during a huge migration of *C. florella* in Botswana (Larsen, 1992). None of any remaining Cyprus individuals

Fig. 12.95. *Senna didymobotrya*, 20 October 2007. *Eddie John*

Fig. 12.96. *Senna artemisioides*, 1 May 2015. *Eddie John*

survived the winter, as far as is known. The captured specimens were displayed at the Natural History Museum, Larnaka, but became vulnerable to attack by *Anthrenus* sp.; the current state of the specimens is not known.

In April 2019, a third (known) immigration was recorded, when a singleton (a ♀) was photographed (Fig. 12.90) near Cape Apostolos Andreas in Karpasia peninsula by Will Hawkes (John *et al.*, 2019b). It is possible that many more *C. florella* were involved in a migration along a broad front, as another female was photographed a week earlier at Harran, Şanlıurfa Province, south-eastern Turkey (Trakel, undated; Onat Başbay, 2019, Ankara, personal communication to first author). In recent years, both Turkey and Cyprus have records only for 1986 and 2019 (John *et al.*, 2019b), suggesting a common source for both migrations.

Description/identification aids

Wingspan: a large butterfly, up to 60 mm. Females are polymorphic (Figs 12.92 and 12.93). Ground-colour is usually light green/white, ♀♀ tending to become darker (yellow-green or yellow) in later generations. The ♂ **upperside** is largely without markings except for a *very small brown spot at the end of the cell* on the forewing, larger in the ♀. The ♀ also has *brown markings on the apex and outer margin* of the forewing (barely noticeable in the ♂). ♀ **undersides** of lighter-coloured forms are almost indistinguishable from those of the ♂ (see Figs 12.91 and 12.93), whereas yellow ♀♀ have three clear spots ringed in brown in the cell and a series of indistinct brown postdiscal marks and alternate veins on the hindwing marked in brown at the margin.

Similar species

The absence of a *pointed apical forewing and hindwing* easily separates this species from *G. cleopatra.*

Conservation status in Cyprus

Not applicable – infrequent migrant. Not assessed by Numa *et al.* (2016). ‘DD’ in Cyprus.

Colias croceus (Geoffroy, 1785) Pieridae (Coliadinae)

Clouded Yellow

TL: Paris.

Fig. 12.97. *Colias croceus* (Clouded Yellow) ♂ nectaring on *Echium angustifolium*. Courtesy of *Ray Atkinson*

Fig. 12.98. *Colias croceus* (Clouded Yellow), backlit ♂ nectaring on *Echium angustifolium*, 5 May 2016. *Eddie John*

Fig. 12.99. *Colias croceus* (Clouded Yellow) ♂ nectaring on *Limonium sinuatum*, 23 April 2008. Courtesy of *Alison McArthur*

Fig. 12.100. *Colias croceus* (Clouded Yellow) ♂ nectaring on *Crupina crupinastrum*, 21 March 2013. *Christodoulos Makris*

Fig. 12.101. *Colias croceus* (Clouded Yellow) f. 'helice', 26 April 2014. Courtesy of *Adrian Riley*

Fig. 12.102. *Colias croceus* (Clouded Yellow) f. 'helice', nectaring on *Pterocephalus multiflorus*, 11 July 2014. *Christodoulos Makris*

Fig. 12.103. *Colias croceus* (Clouded Yellow) pairing, ♂ above, 7 September 2016. *Christodoulos Makris*

Fig. 12.104. Top: *Colias croceus* (Clouded Yellow) f. 'erateformis' – a rare light yellow form of the Clouded Yellow. Note the dwarf appearance of the female in the second row from the top, with specimens of *C. croceus* of normal size and appearance shown in the bottom row. *Christodoulos Makris*

Presence in the eastern Mediterranean

C. croceus is very common throughout the Levant and is present as the nominotypical ssp.; there are 'no definable subspecies within its distribution area' (Olivier, 1990). Surprisingly, this was not encountered in Cyprus by Zach, and hence not listed for the island in Lederer (1855). While included as *Colias croceus* by Bucknill (1911), this was curiously listed, along with various forms, as 'Colias edusa Fab. 1787 (croceus Fourcr., 1786) [ab. helice ♀, Hb.; ab. helicina Obthr.; ab. aubuissoni Crdj.; ab. faillae Stef.; ab. obsoleta Tutt]', by Turner (1920), who had drawn heavily on Bucknill's experience. Somewhat similarly, Rebel (1939) listed the species as 'Colias croceus Fourcr. (*edusa* F.).' *Colias edusa* (Esper, [1805]) is now considered a separate species (see 'Description/identification aids').

Distribution in Cyprus

Altitudinal: found from sea level to the tops of mountains.
Geographical: a very common species well represented throughout the island. The few gaps in our distribution knowledge reflect recording effort rather than the species' absence from those areas.

Phenology

Polyvoltine, recorded in every week of the year (CBRS data), but with reduced numbers in winter and during the hottest summer months. Mating has been observed in late November and larval stages are likely to feed throughout mild winters.

Larval hostplants

Many species of Leguminosae/Fabaceae including: *Alhagi maurorum*, *Erophaca baetica* (Fig. 12.106), *Hippocrepis* spp., *Lotus* spp. (Fig. 12.159), *Medicago sativa* (lucerne/alfalfa), *Ononis spinosa* (spiny rest-harrow) (Fig. 12.107) and *Trifolium* spp. Ovipositing on *Erythrostemon* (*Caesalpinia*) *gilliesii* (bird of paradise), also of the family Fabaceae, was noted in 2013 (Mike Hardman, Cyprus, 2013, personal communication to first author).

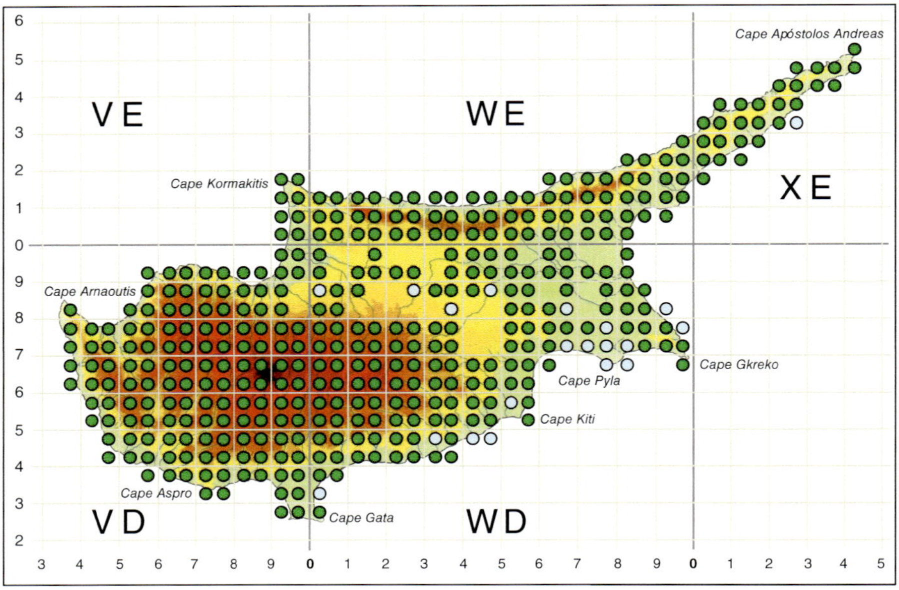

Fig. 12.105. *Colias croceus* distribution map showing 5 km² observations within 10 km² UTM squares. © *Eddie John*

Fig. 12.106. *Erophaca baetica*, 1 May 2015. *Eddie John*

Fig. 12.107. *Ononis spinosa*, 12 June 2012. *Christodoulos Makris*

Ecology and behaviour

Found in a wide range of habitats, but especially common in agricultural areas where lucerne/alfalfa is grown as a fodder crop. Numbers reduce in summer due to limited hostplant availability (desiccation) and possibly because of a northward migration, although there is little evidence to confirm this. Although mass migrations have never been documented in the Levant (Benyamini and John, 2020: 112), a noticeable movement of *C. croceus*, flying west to east on 20 April 2006 at a rate of 20–30 per hour over 3 hours at the north coast, west of Keryneia, was observed by Allen Holmes (Lancashire, 2006, personal communication to first author). The first author of this book has observed similar migratory behaviour, involving single numbers of *C. croceus* in view at any one time, concurring with the experience of Larsen (1976), who also noted that species of *Colias* 'tend to migrate singly'. While normal flight is rapid, that of migratory individuals is typically more direct and purposeful. Nectaring on a wide variety of flowers has been observed.

Description/identification aids

Wingspan: up to *ca.* 50 mm. The only medium-sized, deep yellow/yellow-orange butterfly found in Cyprus. **Upperside** ground colour of both sexes is as described, with a broad black submarginal band across fore- and hindwings, interspersed with orange/yellow markings in the ♀. Both sexes have a small round spot at the end of the forewing cell and an irregular orange spot at the end of the cell on the hindwing. The ♂ has a yellow bar of androconia on the hindwing costa at the inner margin. **Undersides** of both sexes are yellow or yellow-green with irregular submarginal forewing black spotting and a small round spot at the end of the forewing cell. The hindwing underside has two end-of-cell white spots (one very small) encircled with brown, and a series of small brown post-discal spots. To a degree, Cyprus populations are polymorphic, with several forms described by Turner (1920), but not to the extent of the Azores, where six forms have been described (Hutsebaut *et al.*, 2020). A white form, f. 'helice', occurs in less than 5% of the ♀ *C. croceus* population in Cyprus (Makris, 2003: 122), although on a single occasion *ca.* 50% of a large number of nectaring *C. croceus* were reported as f. 'helice' (Daniel Haines, London, personal communication to first author). In this form, the forewing upper- and underside ground colour is replaced by white, the hindwing underside by greenish-yellow.

A very rare light yellow form, previously erroneously recorded as *Colias erate* (Manil, 1990; John, 2000: 6, 26; Makris, 2003: 128) has since been shown to be *C. croceus* f. 'erateformis' Niculescu, 1976 (see Chapter 7, Fig. 7.13, and John *et al.*, 2006), although treated as f. 'cremonae' Verity, 1911 by Kudrna *et al.* (2011: 136, 2015: 182) and Kudrna (2019: 115). *C. croceus* f. 'erateformis' has also been recorded in Rhodes (Cuvelier and Mølgaard, 2012). Fewer than ten specimens of f. 'erateformis' have been reported in over 20 years of recording in Cyprus, although this was also known to Bucknill (1911), who noted individuals where 'the wings or even only a pair of the wings are lemon yellow'. The same author wrote, 'But many curious freaks occur and are not *very* rare here', and was quoted in Turner (1920) thus: 'many curious pale forms were taken, and ab. *helice*'.

Similar species

None in Cyprus.

Conservation status in Cyprus

Not threatened. Listed by Van Swaay *et al.* (2010a) and Numa *et al.* (2016) as 'LC'.

Gonepteryx cleopatra (Linnaeus, 1767) Pieridae (Coliadinae)

Cleopatra

TL: "Barbaria" (Algeria).

Fig. 12.108. *Gonepteryx cleopatra* (Cleopatra) ♂ nectaring on *Ptilostemon chamaepeuce*, 21 May 2013. *Christodoulos Makris*

Fig. 12.109. *Gonepteryx cleopatra* (Cleopatra) ♀ nectaring on *Ptilostemon chamaepeuce*, 5 June 2011. *Christodoulos Makris*

Fig. 12.110. *Gonepteryx cleopatra* (Cleopatra) ♀ nectaring on *Dittrichia viscosa*, 19 October 2017. *Eddie John*

Fig. 12.111. *Gonepteryx cleopatra* (Cleopatra) ♀ nectaring on *Dittrichia viscosa*, 19 October 2017. *Eddie John*

Fig. 12.112. *Gonepteryx cleopatra* (Cleopatra) ♂ mud-puddling, 11 August 2013. *Christodoulos Makris*

Presence in the eastern Mediterranean

Occasionally referred to as Cleopatra's Brimstone, this species is widely represented in the Levant, including Cyprus, by the ssp. *Gonepteryx cleopatra taurica* Staudinger, 1881; TL: Taurus (Olivier, 1993: 202; Hesselbarth *et al.*, 1995: 352; Benyamini and John, 2020: 116).

Listed as *Rhodocera Cleopatra* L. by Lederer (1855), the current name was applied by Bucknill (1911) and by Turner (1920), including the reference to race/ssp. *taurica*.

Distribution in Cyprus

Altitudinal: from coast to mountaintop, but most frequently found at lower to intermediate elevations.
Geographical: a common butterfly, generally well distributed throughout the island, but with few records from the Mesaoria Plain, due to the absence of the hostplant.

Phenology

Univoltine or partially bivoltine, although three, or even more, broods have been suggested for mainland Levant (Comay *et al.*, 2020; Benyamini and John, 2020: 116). Such is the adult longevity of this species that

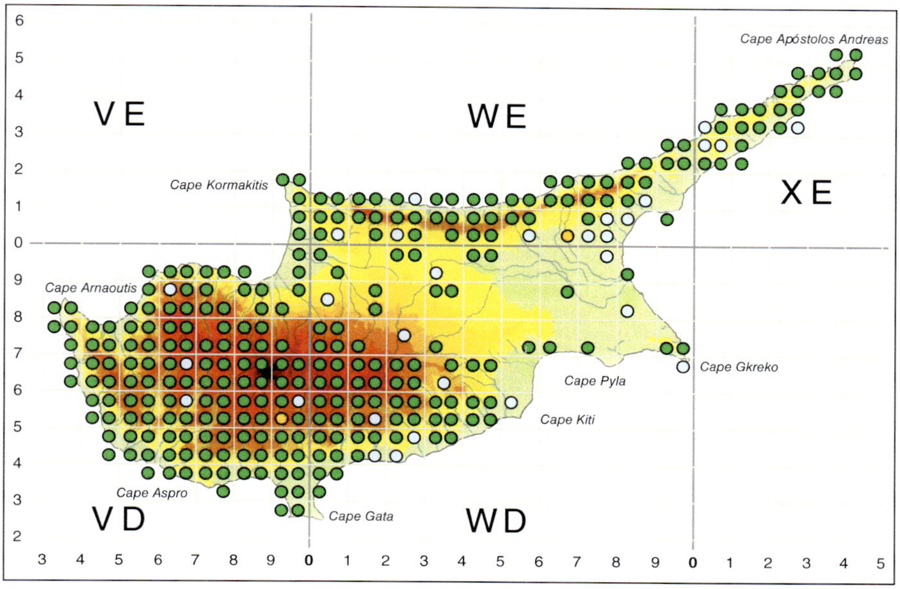

Fig. 12.113. *Gonepteryx cleopatra* distribution map showing 5 km² observations within 10 km² UTM squares. © *Eddie John*

overwintering butterflies seen in Cyprus in February (or even earlier) are those to have eclosed as first-generation adults in May (or later) of the previous year. Overwintered females oviposit throughout the spring, producing the new generation in May/June. The latter remain on the wing or in partial aestivation until the autumn, when they enter hibernation. The spread of egg-laying activity from the overwintered ♀♀ almost certainly results in some overlap between the previous year's and current year's generations. Rarely, a possible partial summer brood has been reported in Cyprus, where ovipositing and first instar larvae (which developed to maturity), were observed on 29 May 2002 and into June at two locations (Makris, 2003: 132). However, there is a possibility that the larvae seen in June might have resulted from delayed oviposition due to local conditions (the late Torben Larsen, personal communication to first author).

Turner (1920) quoted Wilson thus: 'A fairly common species of three broods; the first emerges [from *hibernation*] towards the end of February, the second and larger brood at the end of May, and a very small brood in October. A few are found in the plains generally in flower gardens, but the largest numbers occur in the mountains usually near small streams and among the bracken. Females are comparatively rare.' Turner (1920) concluded a lengthy discussion on voltinism by stating, 'Rober (Seitz), "Pal. Gr.-Schm.," very strongly doubts the existence of three broods or even two, but the long definite intervals during which no individuals are ever found [but see final sentence of this section] and the excellent condition of the specimens when captured seem to be against this opinion, and certainly in Cyprus the evidence of G.F.W. would seem almost conclusive. The October-caught specimens are probably a "precocious" emergence of the early spring brood.' Verity (1919), writing on 'race EUROPAEUS', went further, naming the second and third generations 'SECUNDA' and 'TERTIA', respectively.

When one considers the unblemished appearance of the females photographed in late October 2017 (Figs 12.110 and 12.111), it is tempting to consider that Wilson, Verity and Turner had a point. However, Simes (1920), who studied the behaviour of this pierid in many countries of the Mediterranean (not Cyprus), was vigorous in his defence of univoltinism for *G. cleopatra*, not least because he considered the hostplant leaves to be unacceptable to larvae by mid-summer.

One hundred years later, the matter remains unresolved to the satisfaction of all. However, it would appear from Wilson's remarks that he regarded the emergence in February of *overwintered G. cleopatra* as constituting the first of three broods, a point noted by Rebel (1939). Furthermore, adult inactivity in the form of partial aestivation during the summer months might contribute to the appearance of unblemished *univoltine* individuals temporarily emerging from aestivation late in the year to seek nectar prior to hibernation, a point also noted by Larsen (1974: 106) and Larsen and Nakamura (1983).

Having been recorded on the wing in 50 weeks of the year (CBRS, unpublished data), *G. cleopatra* will undoubtedly number among those species recorded in all weeks in Cyprus.

Larval hostplants

Rhamnaceae: mainly *Rhamnus alaternus* (Mediterranean buckthorn) (Fig. 12.114), rarely *Rhamnus lycioides* ssp. *graeca* (Fig. 12.115).

Fig. 12.114. *Rhamnus alaternus*, 23 February 2016. *Christodoulos Makris*

Fig. 12.115. *Rhamnus lycioides* ssp. *graeca*, 5 April 2012. Courtesy of *Yiannis Christofides*

Ecology and behaviour

Frequently encountered in flight along mountain valleys, in which *Rhamnus alaternus* may be encountered. Although not known as a migrant, *G. cleopatra* may disperse outside established breeding areas and has even been reported flying over the sea in small numbers along the northern coast (David Whaley and Judy Dawes, Cyprus, personal communication to first author), where the species appears to have a slightly greater presence. Larsen (1976) noted a tendency of the species 'to wander long distances in a random fashion after hibernation'.

For the purposes of nectaring, *G. cleopatra* is mainly attracted to flowers of *Ptilostemon chamaepeuce*, *Salvia fruticosa*, *Pterocephalus multiflorus*, *Dittrichia viscosa*, *Lonicera etrusca* (Etruscan honeysuckle), *Punica granatum* (pomegranate) and *Echium* spp. Attraction to garden pelargoniums (Parker, 1983) and other garden flowers has also been reported. *G. cleopatra* has been observed mud-puddling by the second author (Fig. 12.112).

Description/identification aids

Wingspan: up to 60 mm. The *pointed forewing apex and hindwing point at M3* are unique in Cyprus. *Sexes are dimorphic*, the ♂ an unmistakeable *bright yellow marked with a distinctive flush of orange*, the ♀ a pale green. Both sexes have a small brown forewing spot at the end of the cell and a larger brown spot at the end of the cell on the hindwing. Yellow females are found rarely, but *lack the forewing orange flush*, characteristic of the ♂.

Similar species

Gonepteryx rhamni (Linnaeus, 1758) (Brimstone) occurs in nearby mainland countries but has *not* been recorded in Cyprus; that reported by Guillemard (1888) during a four-month ornithological tour of the island, in which just two other species of butterfly received a mention, was highly likely to have been a misidentification. On a four-day visit to Cyprus in late October 1994, Larsen (1997) listed *G. rhamni* as one of only four species of butterfly seen; as with *Iphiclides podalirius* (Linnaeus, 1758) (see Chapter 13), this would appear to have been an error of recollection and certainly not one of misidentification at the time. See also *Catopsilia florella*, to which *G. cleopatra* is somewhat similar.

Conservation status in Cyprus

Not threatened. Listed by Van Swaay *et al.* (2010a) and Numa *et al.* (2016) as 'LC'.

Family: LYCAENIDAE

Apharitis acamas (Klug, 1834)	Levantine Leopard	Aphnaeinae
Lycaena phlaeas (Linnaeus, [1760])	Small Copper	Lycaeninae
Lycaena thersamon (Esper, 1784)	Lesser Fiery Copper	
Deudorix livia (Klug, 1834)	Pomegranate Hairstreak	Theclinae
Favonius quercus (Linnaeus, 1758)	Purple Hairstreak	
Lampides boeticus (Linnaeus, 1767)	Long-tailed Blue	Polyommatinae
Leptotes pirithous (Linnaeus, 1767)	Lang's Short-tailed Blue	
Azanus jesous (Guérin-Méneville, 1849)	African Babul Blue	
Tarucus balkanicus (Freyer, 1844)	Little Tiger Blue	
Zizeeria karsandra (Moore, 1865)	Dark Grass Blue	
Celastrina argiolus (Linnaeus, 1758)	Holly Blue	
Pseudophilotes vicrama (Moore, 1865)	Eastern Baton Blue	
Glaucopsyche paphos Chapman, 1920	Paphos Blue	
Luthrodes galba (Lederer, 1855)	Small Desert Blue	
Freyeria trochylus (Freyer, 1844)	Grass Jewel	
Aricia agestis ([D & S], 1775)	Brown Argus	
Polyommatus icarus (Rottemburg, 1775)	Common Blue	

Apharitis acamas (Klug, 1834) Lycaenidae (Aphnaeinae)

Levantine Leopard

TL: "in Syria et in Arabia felici".

Fig. 12.116. *Apharitis acamas* (Levantine Leopard) nectaring on *Carlina libanotica*, 22 August 2014. *Christodoulos Makris*

Fig. 12.117. *Apharitis acamas* (Levantine Leopard) ♀ perched on *Sarcopoterium spinosum*, 11 June 2009. *Christodoulos Makris*

Fig. 12.118. *Apharitis acamas* (Levantine Leopard) nectaring on coastal *Dittrichia viscosa*, 21 October 2017. *Eddie John*

Fig. 12.119. *Apharitis acamas* (Levantine Leopard) ♀, 30 July 2021. Courtesy of *Clive Burrows*

Presence in the eastern Mediterranean

Formerly included in the genus *Cigaritis*, this is found throughout the Levant, probably as nominotypical *Apharitis acamas acamas*. An endemic ssp. *Apharitis acamas cypriaca* Riley, 1925, TL: Episkopis, Cyprus (♂ and ♀ type specimens are in the BMNH) was described for the island (Riley, 1925) but validity is uncertain and the species would probably benefit from molecular analysis. Indeed, Larsen and Pittaway (1982) stated, 'Subsp. *cypriaca* Riley is so close to nominate *acamas* that its retention hardly seems worthwhile.' This Afro-eremic species is not known from any other Mediterranean island, Cyprus representing the westernmost extent of 'European' distribution.

On taxonomic grounds, Kudrna (2019: 33) considered *Apharitis* 'as a junior subjective synonym of *Cigaritis* Donzel, 1847, probably not worthy of recognition as a subgenus'. Larsen (2005: 182) stated a preference to maintain all three genera (to include *Spindasis*) as valid, adding 'though they are clearly closely related: *Cigaritis* as a Palaearctic derivate, *Apharitis* as an eremic derivate, and *Spindasis* as a Palaeotropical remnant of a joint ancestor – a pattern that must be the result of a long evolutionary process influenced by ecology'. The late Torben Larsen (personal communication to first author in 2007) further stated, 'I might even accept the inclusion of *Apharitis* in *Spindasis*, but *Cigaritis* is special.' Wiemers *et al.* (2018) referred to *Cigaritis acamas*.

Unsurprisingly, this species was not recorded in Cyprus by Zach (Lederer, 1855), but was listed by Bucknill (1911) as '*Cigaritis zohra. Jord. The Flare.*', with the comment 'has so far only been taken by Major Bolton in the Kyrenia Hills'. Turner (1920) listed the species as 'Cigaritis acamas Klug. = [Cigaritis zohra Donz.]', noting that Bucknill in the meantime had also succeeded in taking specimens in the same mountain range.

Distribution in Cyprus

Altitudinal: known from coastal locations, but mainly found at intermediate elevations to 800–900 m, with a very unusual single record on 4 July 2017 at 1160 m (David and Rosalyn Sparrow, Cyprus, 2017, personal communication to first author).
Geographical: a local and uncommon species found in small, disjunct colonies and principally known from central and western areas. Makris (2003: 142) showed a single historical record from northern Cyprus, since when extensive fieldwork in the north of the island has added records to a further eight 5 km^2 UTM squares (CBRS data).

Phenology

Bivoltine to trivoltine, late May to early November. Overwintering stage is probably as a pupa.

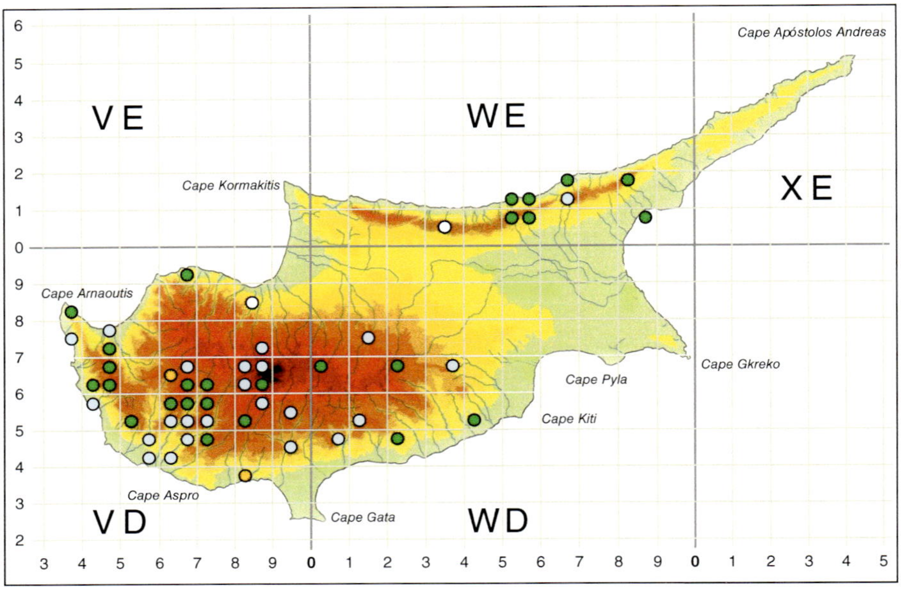

Fig. 12.120. *Apharitis acamas* distribution map showing 5 km^2 observations within 10 km^2 UTM squares. © *Eddie John*

Larval hostplants

Although strongly associated with *Sarcopoterium spinosum* (thorny burnet) (Rosaceae) (Fig. 12.121), and on which ovipositing has been observed, the young *A. acamas* larvae are quickly sought out by ants of the genus *Crematogaster* and 'groomed' to encourage the release of nutritious secretions. A mutualistic (obligate myrmecophilous) relationship ensures that, in turn, early stages of *A. acamas* larvae are fed with regurgitated food by the ants, while in their later instars preying on ant brood, thus also becoming parasitic (Sanetra and Fiedler, 1996). Larsen and Pittaway (1982) deduced that in some circumstances *A. acamas* larvae might live through all instars in ants' nests, and are truly aphytophagous and at no point herbivorous. While rare, this behaviour is not unique, as larvae of approximately 200 of the 5000 or so members of the Lycaenidae worldwide are thought not to feed on plants, but to be parasitic on ants (Pierce *et al.*, 2002).

Ecology and behaviour

Found in a variety of habitats, including coastal locations and open river valleys, but usually found on hillsides covered with low shrubs, especially those of *S. spinosum*. The late Torben Larsen (personal communication in 2002, to first author) wrote, 'Eggs are laid near colonies of appropriate ants. The caterpillars spend much of their time in ants' nests and have been bred in later instars inside ants' nests with no access to plants, obviously living on ants' brood or being fed regurgitated food by the ants. It is also likely that

Fig. 12.121. *Sarcopoterium spinosum*, 6 May 2011. *Eddie John*

depending on circumstances *A. acamas* might be only partly aphytophagous. Some *Spindasis*, so close to *Apharitis* that some [e.g. Heath, 1997] consider them congeneric, are definitely known to feed on plants.'

Flight is rapid, erratic and often difficult to follow against the shrubby habitat in which the species is usually found. However, once settled, *A. acamas* can be approached easily.

Although rarely seen to visit flowers, nectaring has been observed on *Cardopatium corymbosum*, *Chrozophora tinctoria*, *Echinops spinosissimus*, *Carlina libanotica* and *Dittrichia viscosa* (authors' personal observations). As is sometimes the experience in Cyprus, Larsen and Pittaway (1982) noted that in Oman, *A. acamas* is found in areas devoid of nectar sources or water, indicating that feeding may not be entirely necessary in the adult stage, due to the accumulation of fatty reserves at the immature stages.

Description/identification aids

Wingspan: up to 35 mm. **Upperside** ground colour of both sexes is light orange (not evident in flight), with diffuse brown scaling in the basal area of the forewing and in the basal and discal areas of the hindwing. Markings are similar in ♂ and ♀, with a complete submarginal brown band on fore- and hindwings and smaller broken brown bands crossing all wings. **Underside** ground colour of both sexes is off-white. Markings mirror the upperside in shape, but are of a lighter brown outlined in dark brown and with *a silver line* running more or less centrally along each of the bands. The hindwing anal angle has *two tails*, one long, the other short.

Similar species

None in Cyprus. *Apharitis myrmecophila* Dumont, 1922 and *Apharitis cilissa* Lederer, 1861 are known elsewhere in the Levant.

Conservation status in Cyprus

Not considered seriously threatened, although fire is a constant hazard in the species' typical habitat (see Chapter 14). Not assessed by Van Swaay *et al.* (2010a), or at the Mediterranean level by Numa *et al.* (2016).

Lycaena phlaeas (Linnaeus, [1760]) Lycaenidae (Lycaeninae)

Small Copper

TL: Västmanland, Sweden.

Fig. 12.122. *Lycaena phlaeas* (Small Copper) nectaring on *Dittrichia viscosa*, 23 October 2016. *Christodoulos Makris*

Fig. 12.123. *Lycaena phlaeas* (Small Copper) on the hostplant *Rumex cyprius*, 17 April 2009. Courtesy of *Alison McArthur*

Fig. 12.124. *Lycaena phlaeas* (Small Copper) basking, 1 April 2015. *Christodoulos Makris*

Fig. 12.125. *Lycaena phlaeas* (Small Copper) nectaring on *Helichrysum italicum*, July 2012. *Christodoulos Makris*

Fig. 12.126. *Lycaena phlaeas* (Small Copper) nectaring on *Eryngium creticum*, June 2016. *Christodoulos Makris*

Presence in the eastern Mediterranean

Widely distributed throughout the region, probably as nominotypical *L. phlaeas phlaeas*, although earlier works referred to the ssp. *L. phlaeas timeus*, (Cramer, 1777) TL: 'Smyrne' in the Middle East, synonymized in Hesselbarth *et al.* (1995: 459) with the nominotypical ssp.

Recorded as '*Polyommatus Phlaeas* L.' by Lederer (1855), '*Chrysophanus phlaeas* L.' and '... *form eleus Fabr.*' by Bucknill (1911), 'Rumicia phlaeas L. [race eleus Fab.; ab. (race) turcicus Gerh.; caeruleopunctata Stdgr.]' by Turner (1920) and as 'Chrysophanus phlaeas L.' by Rebel (1939).

Distribution in Cyprus

Altitudinal: found at all altitudes in Cyprus (Makris, 2003: 148).
Geographical: a common butterfly, rarely numerous, but occasionally seen in tens in favourable habitats. Present throughout the island although with fewer records from the Mesaoria Plain and eastern areas.

Phenology

Polyvoltine, with specimens observed in all months of the year.

Larval hostplants

Polygonaceae: *Rumex* spp., especially *R. cyprius* (pink sorrel) (Fig. 12.128), *R. cristatus* (Greek dock); probably *Polygonum equisetiforme* (horsetail knotweed) (Figs 12.137 and 12.138) and *Emex spinosa*.

Ecology and behaviour

Present in a variety of habitats, but perhaps most commonly found in the shade of pine forests in the hotter months. Along an approximately

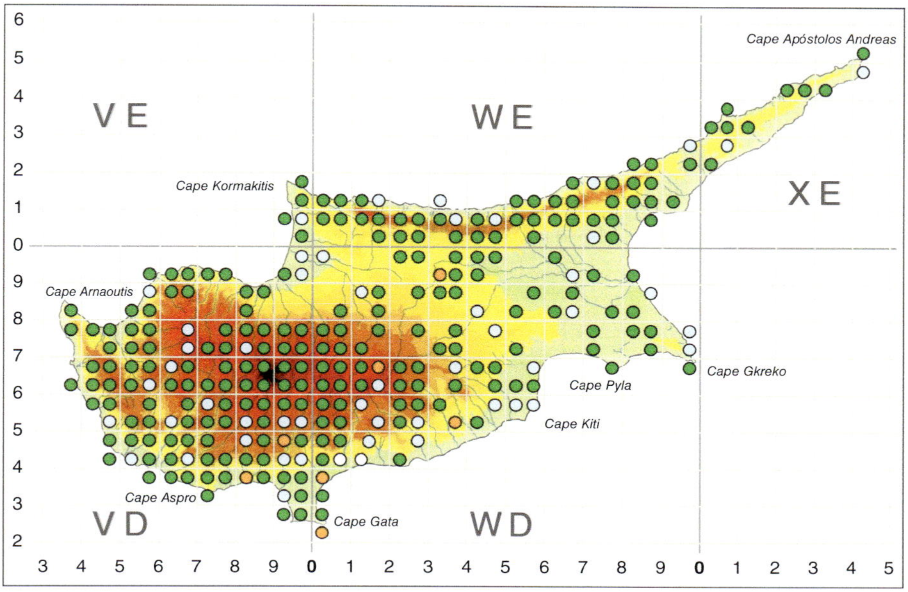

Fig. 12.127. *Lycaena phlaeas* distribution map showing 5 km² observations within 10 km² UTM squares. © *Eddie John*

Fig. 12.128. *Rumex cyprius*, 13 May 2011. *Eddie John*

two-hour long 'transect' (about 300–400 m asl) on 27 December 1997, after a period of three (warm) weeks when no adult *L. phlaeas* were seen, the first author located four discrete colonies where freshly emerged adults were on the wing. All were in an area in the Mosphiloti hills walked at least weekly over a 14-month period, providing an opportunity to follow the species' progress throughout a full year. Adults remained on the wing there from 27 December throughout the remaining winter and the following spring months, with worn and fresh specimens flying together on 14 February 1998. From the end of July 1998, in what had become a parched countryside, no further observations were made until fresh emergences began on 29 November. Although aggregated records show that *L. phlaeas* may be seen in Cyprus during the intervening months at unspecified elevations (see phenogram), the absence of observations over a *ca.* four-month period at a regularly walked transect suggested pupal diapause over this period, in this habitat, coinciding with absence of host-plants until the return of autumn rain.

Nectaring has been observed on a variety of flowers (see Figs 12.122, 12.125 and 12.126).

Description/identification aids

Wingspan: up to *ca.* 30 mm. Forewing **uppersides** are similar in both sexes, each showing a degree of seasonal dimorphism. In the spring brood, the forewing upperside ground colour is *bright orange* with a wide brown border and black markings postdiscal and in the cell; the hindwing ground colour is dark brown with a wide submarginal orange band, spotted with black at the margin. Some specimens, f. 'caeruleopunctata' Rühl, have blue postdiscal spotting on the hindwing. A short hindwing tail is enhanced in later broods, when both sexes have much of the forewing orange replaced by brown (referred to by earlier researchers as form or race 'eleus', as discussed above), especially in the ♂. **Undersides** in both sexes and in all broods are similar: forewing orange with black marks and a light brown border and apex; hindwing largely plain, with a light brown ground colour, very small black spots and faint orange submarginal band.

Similar species

May easily be separated from *Lycaena thersamon* (see below) by the *much-reduced and less distinct hindwing underside black spotting* in both sexes of *L. phlaeas* and the lack of spotting on the forewing upperside of *L. thersamon* (♂ only). See also *Deudorix livia* ♂.

Conservation status in Cyprus

Not threatened. Listed by Van Swaay *et al.* (2010a) and Numa *et al.* (2016) as 'LC'.

Lycaena thersamon (Esper, 1784) **Lycaenidae** (Lycaeninae)

Lesser Fiery Copper

TL: Sarepta, S Russia.

Fig. 12.129. *Lycaena thersamon* (Lesser Fiery Copper) first brood ♂ nectaring on *Hirschfeldia incana*, 3 May 2015. *Eddie John*

Fig. 12.130. *Lycaena thersamon* (Lesser Fiery Copper) ♂ nectaring on *Mentha longifolia* ssp. *cyprica*, 10 September 2016. *Christodoulos Makris*

Fig. 12.131. *Lycaena thersamon* (Lesser Fiery Copper) ♀ nectaring on *Glebionis coronaria*, 23 April 2010. Courtesy of *Alison McArthur*

Fig. 12.132. *Lycaena thersamon* (Lesser Fiery Copper) ♀ on *Polygonum equisetiforme*, 16 October 2012. *Christodoulos Makris*

Fig. 12.133. *Lycaena thersamon* (Lesser Fiery Copper) ♂ nectaring on *Polygonum equisetiforme*, 20 September 2010. *Christodoulos Makris*

Fig. 12.134. *Lycaena thersamon* (Lesser Fiery Copper) ♀ nectaring on *Polygonum equisetiforme*, 16 October 2012. *Christodoulos Makris*

Fig. 12.135. *Lycaena thersamon* (Lesser Fiery Copper) pairing, 27 August 2016. *Christodoulos Makris*

Presence in the eastern Mediterranean

Widely distributed throughout the Levant, probably as the nominotypical subspecies although names of other ssp., especially *L. thersamon omphale* Klug, 1834, have been used. Some, e.g. Turner (1920), regarded this as a form (f. 'omphale') of the nominotypical ssp., which they applied to second and subsequent broods. Olivier (1993: 202) regarded the Cyprus ssp. as *L. thersamon thersamon*, as did Hesselbarth *et al.* (1995: 485).

Early works refer to, for example, '*Chrysophanus thersimon. Esper*' (Bucknill, 1911), 'Loweia (Thersamonea) (Chrysophanus) thersamon Esp. [aest. g. omphale Klug.]' (Turner, 1920) and 'Chrysophanus thersamon Esp.' (Rebel, 1939).

Distribution in Cyprus

Altitudinal: unlike *L. phlaeas*, *L. thersamon* is rarely found above 800 m and is much more likely to be encountered in lowland areas. Nevertheless, *L. thersamon* has been observed at Almyrolivado (1600 m) in late July (second author, personal observation), a surprising elevation, since *Polygonum equisetiforme*, the hostplant in Cyprus, is not reported above 1100 m (Christofides, 2017: 264). Other hostplants, used in the wider region but not recorded for Cyprus, are found at elevations even lower than 800 m.
Geographical: a fairly common butterfly, widely distributed throughout Cyprus, but more commonly reported in the south-west.

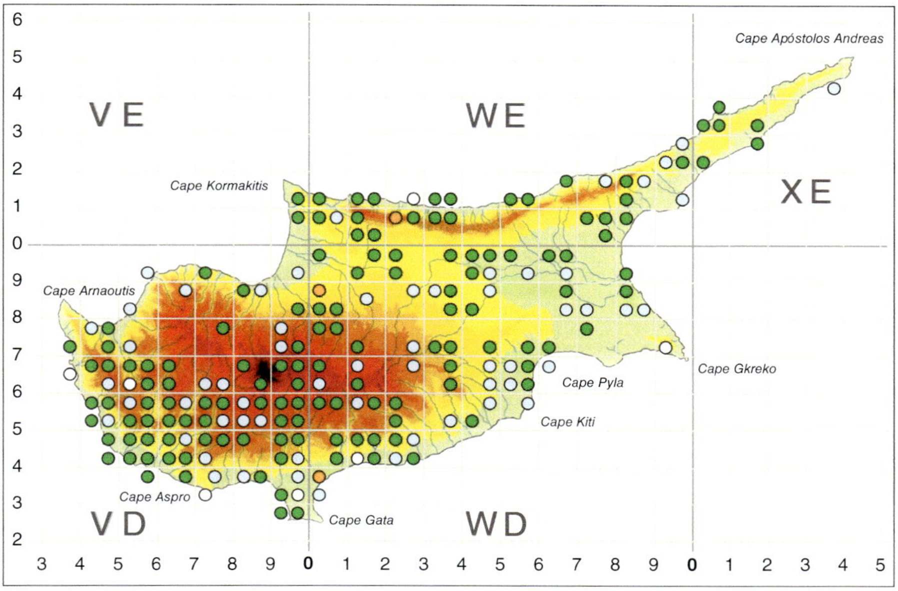

Fig. 12.136. *Lycaena thersamon* distribution map showing 5 km² observations within 10 km² UTM squares. © *Eddie John*

Fig. 12.137. *Polygonum equisetiforme*, 14 May 2011. *Eddie John*

Fig. 12.138. *Polygonum equisetiforme*, 29 April 2016. *Eddie John*

Phenology

Probably polyvoltine, with first emergences in March (occasionally earlier) and remaining on the wing in overlapping generations until November or later (CBRS data). Overwintering stage is uncertain, but probably in larval and pupal stages, subject to late season activity.

Larval hostplants

Polygonaceae: *Polygonum equisetiforme* (horsetail knotweed) (Figs 12.137 and 12.138), on which the egg is laid at the base of a flower. Reported on *Rumex cyprius* and *Emex spinosa* elsewhere in the Levant. In Cyprus, the species is very strongly associated with drifts of *P. equisetiforme.* (Annual *P. bellardii* [syn. *P. patulum*] has been given as a hostplant in the past (de Worms, 1967), but is considered to have been a misidentification for *P. equisetiforme.*)

Ecology and behaviour

Found principally in damp areas (fringes of wetlands, depressions in waste ground, ditches and regularly irrigated areas) that may support large, dense mats of the hostplant. Larsen and Nakamura (1983) noted the species' attraction to the flowers of the sea holly *Eryngium creticum*. In Cyprus, nectaring is frequently seen on flowers of the hostplant *P. equisetiforme*, but the species is also attracted to many other flowers, for example *Hirschfeldia incana, Mentha longifolia* ssp. *cyprica* and *Glebionis coronaria* (formerly *Chrysanthemum coronarium*).

Description/identification aids

Wingspan: up to 35 mm. ♂ and ♀ **uppersides** are orange, *forewing unmarked in the* ♂ except for a narrow black margin, well spotted with black in the ♀. Hindwings in both sexes are darker orange, with postdiscal and submarginal spotting, heavier in the ♀. Fore- and hindwing **underside** markings are similar in both sexes: ground colour orange with a submarginal band of darker orange, black spotting in the postdiscal and submarginal areas and in the cell. Both sexes of the first brood *lack tails*; later broods (previously referred to as f. 'omphale') may have fairly *long filamentous* tails, especially in the ♀ (see pairing in Fig. 12.135), but tail length is extremely variable, even within the same brood (see Hesselbarth *et al.*, 1995: 333).

Similar species

See *L. phlaeas* above.

Conservation status in Cyprus

Not threatened locally and considered as 'LC' by Van Swaay *et al.* (2010) and Numa *et al.* (2016).

Deudorix livia (Klug, 1834) **Lycaenidae** (Theclinae)

Pomegranate Playboy/Hairstreak

TL: Egypt: "Kineh; Assuan"; Saudi Arabia: "Arabia".

Fig. 12.139. *Deudorix livia* (Pomegranate Hairstreak) photographed in the Negev Desert, Israel, 5 May 2014. *Christodoulos Makris*

Fig. 12.140. *Deudorix livia* (Pomegranate Hairstreak) photographed in the Negev Desert, Israel, 5 May 2014. *Christodoulos Makris*

Fig. 12.141. The sole Cyprus specimen of *Deudorix livia* (Pomegranate Hairstreak), upperside of a ♀ caught by Gabriel Georghiou at Peristerona, 25 September 1994. *Christodoulos Makris*

Fig. 12.142. The sole Cyprus specimen of *Deudorix livia* (Pomegranate Hairstreak), underside of a ♀ caught by Gabriel Georghiou at Peristerona, 25 September 1994. *Christodoulos Makris*

Presence in the eastern Mediterranean

A migratory species of tropical Africa 'north of the rainforest belt and north of the Equator in the east' (Larsen, 1986), *D. livia* (also known as *Virachola livia*) was once considered resident in the Levant only in the

Nile Delta. However, in recent decades, migrants have regularly reached the Levant countries of Israel, Lebanon and Syria. Müller *et al.* (2005) reported that *D. livia* had 'established permanent colonies in suitable oases in the Rift Valley over several years and might be on its way to do so in the Mediterranean'.

Although no recent information is known, it is likely that the species reaches Syria more frequently than has been reported. Indeed, a published record from Syria (Talhouk, 1954; Mahran Zeity, Damascus, 2021, personal communication to first author) possibly refers to a specimen collected by F. Cremona (reported in Graves, 1910) and might account for Syria originally being listed as a country hosting this species, although numerous authors (e.g. Larsen and Larsen, 1980) include Syria within the species' range. That *D. livia* is making inroads into the Mediterranean basin is without question; Ksantini *et al.* (2011) reported on the arrival in 2006 of the species in Tunisia, where it was discovered on pomegranate. Since then, it has been discovered in that country on *Acacia farnesiana*, *A. arabica* and *Phoenix dactylifera* (date) (Abbes *et al.*, 2020).

Distribution in Cyprus

Altitudinal: found at 200 m (see below).
Geographical: *D. livia* has been observed just once in Cyprus, in an agricultural area at Peristerona in the Mesaoria Plain (Figs 12.141 and 12.142) (Makris, 2003: 146).

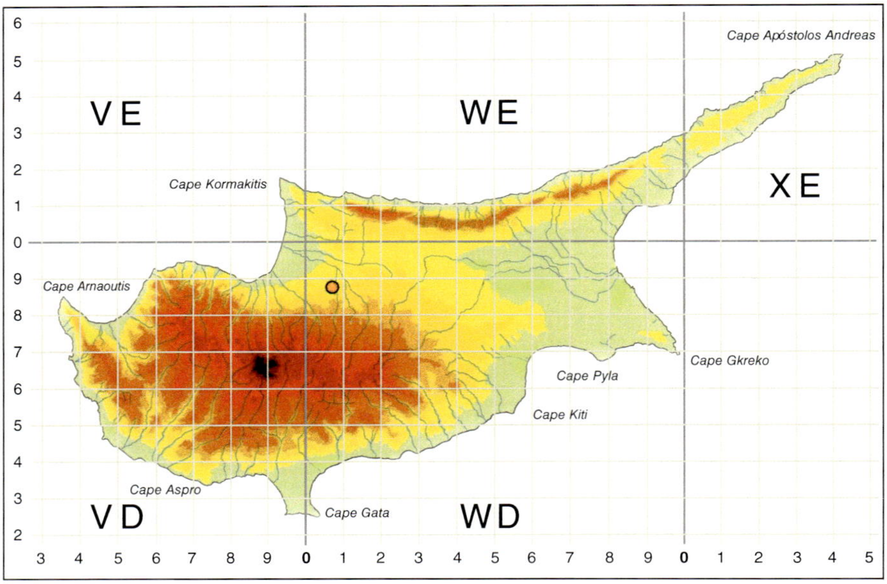

Fig. 12.143. *Deudorix livia* distribution map showing a single 5 km^2 observation within the 10 km^2 UTM square. © *Eddie John*

Fig. 12.144. *Punica granatum* (pomegranate), 12 October 2017. *Eddie John*

Phenology

Polyvoltine within usual range.

Larval hostplants

D. livia is highly polyphagous, using Fabaceae including *Acacia* spp., *Cassia* spp., *Prosopis farcta*; Mimosaceae: *Acacia* spp.; Punicaceae: *Punica granatum* (pomegranate) (Fig. 12.144), on which it has been reported as a pest (Talhouk, 1954), and numerous other hostplants, including *Phoenix dactylifera* (date). Similarly, Larsen (1976) stated, '*D. livia* is a pest on pomegranate and dates, feeding on a bewildering variety of other plants as well.'

Ecology and behaviour

A vigorous hilltopping, mud-puddling and moderately migratory species that in recent years has become common in Israel, but is infrequently reported from Lebanon. The single *D. livia* (♀) known for Cyprus (*in coll.* Gabriel Georghiou, Limassol) was captured after it was seen on 25 September 1994 nectaring on *Heliotropium hirsutissimum* in an area at Peristerona with numerous pomegranate trees (Makris, 2003: 146); this first observation for a country of the European Union was initially reported in Dennis *et al.* (2000). The possibility of this being an adventive, with early stages infesting imported plants or pomegranate fruit, seems likely. Considerably further west in the Mediterranean basin, four specimens of *D. livia* (three ♂♂ and one ♀) were collected in mid-September

2004 near Marathon in Greece, where final instar larvae were also found (Müller *et al.*, 2005). As with Cyprus, its presence might be explained by accidental introduction, even though the species is known to undertake occasional long-distance migrations (Larsen, 1984a: 88). Although breeding has been reported in coastal areas of Israel, *D. livia* is unlikely to survive the northern Mediterranean winter.

Description/identification aids

Wingspan: up to 35 mm. Sexes show *marked* sexual dimorphism, ♂ forewing **upperside** *a bright copper, lacking any spots*, with a narrow black margin and dark scaling at the apex; hindwing orange with darker basal scaling. ♀ forewing *unmarked brown with bluish basal scaling*, hindwing with *prominent anal lobe and large black anal angle ocellus*, separated by a less distinct ocellus. **Undersides** of both sexes are broadly similar: light brown ground colour, irregular banding in darker brown and with a *prominent anal lobe and large black anal angle ocellus ringed with light orange*. Both sexes have a pair of long tails.

Similar species

The absence of orange on the male underside sets it apart from either *Lycaena* spp. found in Cyprus; the female hindwing underside is much less heavily marked than *Lampides boeticus* and underside hindwings of both sexes lack the *broad, white post-discal band* of that species. Tails are also longer in *D. livia* and spotting in the hindwing anal angle area is larger. However, note that *D. livia* and *L. boeticus* are of similar size.

Conservation status in Cyprus

Rare migrant, not applicable. Considered as 'LC' at the Mediterranean level by Numa *et al.* (2016).

Favonius quercus (Linnaeus, 1758) Lycaenidae (Theclinae)

Purple Hairstreak

TL: Not stated, probably England.

Fig. 12.145. *Favonius quercus* (Purple Hairstreak) ♀, 7 May 2014. *Christodoulos Makris*

Fig. 12.146. *Favonius quercus* (Purple Hairstreak) ♀, 7 May 2014. *Christodoulos Makris*

Fig. 12.147. *Favonius quercus* (Purple Hairstreak) mud-puddling, 21 August 2013. *Christodoulos Makris*

Fig. 12.148. *Favonius quercus* (Purple Hairstreak) mud-puddling, 16 July 2017. Courtesy of *Elli Tzirkalli*

Presence in the eastern Mediterranean

Occurs in nearby mainland countries, possibly as nominotypical *Favonius quercus quercus*. There is uncertainty about the status of the Cyprus subspecies, although Olivier (1993: 2002) opted for *F. q. quercus*, as did

Hesselbarth *et al.* (1995: 503) for Turkey. Larsen (1974: 152), commenting on the single specimen available from Lebanon, somewhat reluctantly referred this to the nominotypical race, while noting that northern European specimens are markedly smaller. Turner (1920), however, stated (of Cyprus specimens), 'The few examples received are quite typical' (of the nominotypical ssp.). Subspecies *longicaudatus* (Riley, 1921) TL: "NW. Persia", appears an unlikely possibility but further investigation is required.

In the past, this species (not listed by Lederer, 1855) has variously been referred to under the generic names of *Zephyrus* (Bucknill, 1911), *Bithys* (Turner, 1920), *Thecla* (Rebel, 1939) and in recent years, *Quercusia* or *Neozephyrus quercus*. Wiemers *et al.* (2018, 2020) are among many authors favouring *Favonius quercus*.

Distribution in Cyprus

Altitudinal: generally found above 800 m in Cyprus, occasionally lower around specimen oaks.
Geographical: a species mainly restricted to the Troodos Mountains and foothills; there is just a single record (by Birtan Gokeri, 19 June 2018) from the north-western Pentadaktylos range, but none from the eastern half of the island. Although considered rare in Cyprus, the arboreal habits of *F. quercus*, with much time spent in the canopy, have probably led to the species being under-recorded – see below.

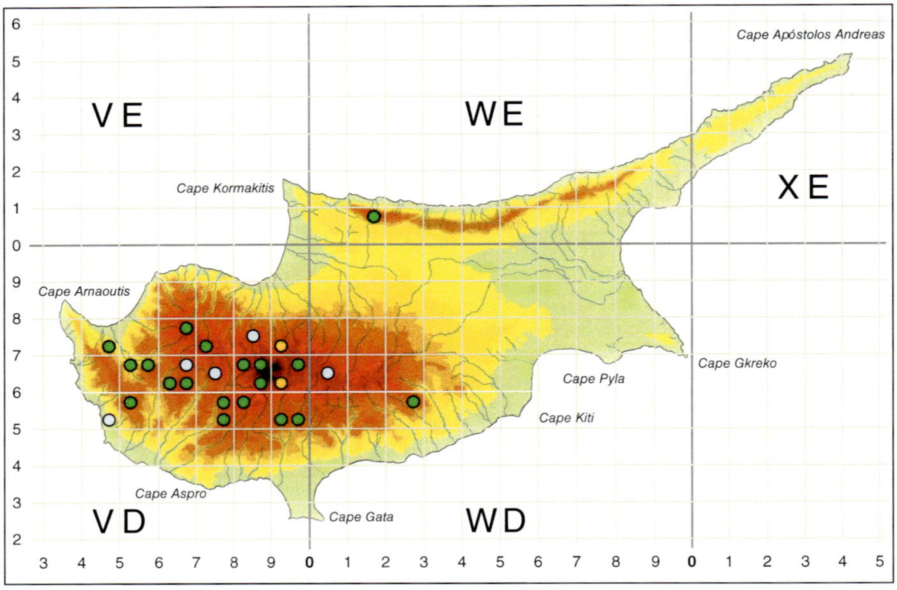

Fig. 12.149. *Favonius quercus* distribution map showing 5 km² observations within 10 km² UTM squares. © *Eddie John*

Phenology

Univoltine, with sporadic sightings from May to September (CBRS data). Overwinters in the egg stage.

Larval hostplants

Fagaceae: in Cyprus, *Quercus alnifolia* (golden oak, an endemic evergreen shrub or small tree) (Fig. 12.150) and the deciduous *Quercus infectoria* ssp. *veneris* (Fig. 12.151). The former is confined to igneous formations of the Troodos Mountains, indicating the use of *Q. infectoria* in the Pentadaktylos.

Fig. 12.150. *Quercus alnifolia*, 10 December 2011. Courtesy of *Yiannis Christofides*

Fig. 12.151. *Quercus infectoria* ssp. *veneris*, 7 May 2015. *Eddie John*

Ecology and behaviour

Normally a species of woodland, but also seen in open habitats, for example farmed countryside, in very close association with trees of *Q. infectoria* ssp. *veneris* (authors' personal observations). As in other countries, *F. quercus* in Cyprus may be seen high in the canopy, or circling a favoured tree. Although known to nectar on *Rubus sanctus*, *Mentha longifolia* and *Carduus* spp. (Makris, 2003: 138; Parker, 1983), it is a rare visitor to flowers, feeding instead on aphid secretions (honeydew) at treetop level, and is a particularly frequent mud-puddler.

Description/identification aids

Wingspan: generally 30–37 mm (Makris, 2003: 138), but up to 42 mm has been recorded (Graves, 1907). The *only* 'hairstreak' present in Cyprus. The vernacular name derives from the **upperside** iridescent purple-blue of the ♂; the ♀ is sparingly marked in purple-blue on the forewing, having a mainly dark brown upperside ground colour. **Undersides** of both sexes are similar, with a grey-brown ground colour, a white postdiscal line across fore- and hindwings (heavier in the latter), black forewing submarginal marks and two orange ocelli at the hindwing anal angle, where there is a short tail.

Similar species

None in Cyprus; the species should not be confused with any other.

Conservation status in Cyprus

Probably not threatened, but due to rarity in Cyprus and the species' often cryptic habit (time spent in the canopy) the extent of distribution is poorly defined. Listed as 'LC' by Van Swaay *et al.* (2010a) and, at the Mediterranean level, by Numa *et al.* (2016).

Lampides boeticus (Linnaeus, 1767) Lycaenidae (Polyommatinae)

Long-tailed Blue

TL: "Habitat in Barbaria" (Algeria).

Fig. 12.152. *Lampides boeticus* (Long-tailed Blue) ♂ nectaring on *Sinapis alba*, 22 February 2013. Courtesy of *Sami Tamson*

Fig. 12.153 *Lampides boeticus* (Long-tailed Blue) ♀ nectaring on *Polygonum equisetiforme*, 12 November 2017. *Christodoulos Makris*

Fig. 12.154. *Lampides boeticus* (Long-tailed Blue) nectaring on *Cistus creticus*, 25 June 2011. *Christodoulos Makris*

Fig. 12.155. Backlit *Lampides boeticus* at rest on *Medicago × varia*, 31 May 2013. *Eddie John*

Fig. 12.156. *Lampides boeticus* (Long-tailed Blue) pairing, 6 May 2021. *Christodoulos Makris*

Presence in the eastern Mediterranean

An ubiquitous species, distributed over much of the world and present in all countries of the Levant. It is the only member of its genus and no subspecies have been described.

L. boeticus was initially listed for Cyprus as '*Lycaena Boetica*' by Lederer (1855), then as *Polyommatus boeticus* by Bucknill (1911). Turner (1920) referred to the taxon under its current name, as did Rebel (1939).

Distribution in Cyprus

Altitudinal: from sea level to the highest elevations.
Geographical: a very common species, present throughout the island.

Phenology

Polyvoltine, with the main flight period between March and November, but occasional specimens may be seen in any week of the year (CBRS data). Overwintering stage is probably as a slowly developing larva or as a pupa.

Larval hostplants

The species is highly polyphagous, accepting many members of the pea family (Fabaceae) especially *Medicago sativa* (lucerne/alfalfa) (Fig. 12.158) and including *Alhagi maurorum*, *Astragalus* spp., *Vicia* spp., *Lotus* spp. (Fig. 12.159). *L. boeticus* may be an occasional pest on commercially grown pea and bean crops. Ovipositing has also been observed on the

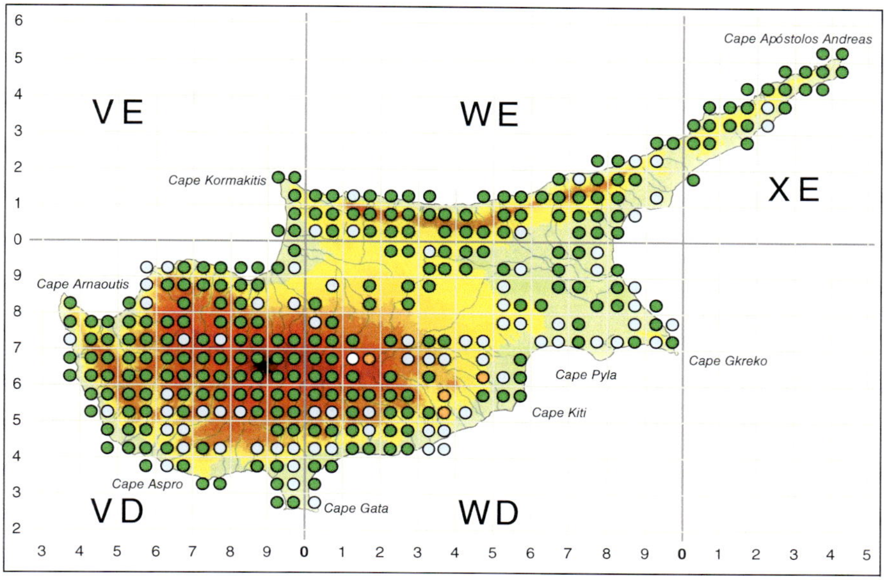

Fig. 12.157. *Lampides boeticus* distribution map showing 5 km² observations within 10 km² UTM squares. © *Eddie John*

Fig. 12.158. *Medicago sativa*, 23 June 2016. *Christodoulos Makris*

Fig. 12.159. *Lotus* sp., 10 May 2015. *Eddie John*

introduced *Spartium junceum* (Spanish broom) and on *Capparis spinosa* (first author, personal observations), the latter a hostplant for *L. boeticus* recorded in southern Spain by Barbudo *et al.* (1988), but not previously reported for Cyprus.

Ecology and behaviour

The species is highly migratory; there is no known evidence of *immigrant* activity relating to Cyprus, although this seems likely. However, a concerted movement of *L. boeticus* was evident at Evretou Dam in the Pafos region on 4 April 2013 when, over a period of two hours, *ca.* 150–200 of the species were seen heading roughly north along a 3 m wide track at a height of *ca.* 1–2.5 m (Nick Bowles, Buckinghamshire, 2013, personal communication to first author; Bowles, 2013). In the absence of similar reports from elsewhere in Cyprus, this appeared to indicate a local dispersal confined within the island and induced by harvesting of local lucerne fields. Alison McArthur (Peeblesshire, 2013, personal communication to first author), who visited the island shortly afterwards, confirmed that several lucerne fields near Pafos Airport (some 25 km directly south of Evretou), showed signs of recent harvesting.

Single individuals may appear almost anywhere, but *L. boeticus* is most likely to be encountered in irrigated pea- and bean-producing agricultural areas, or where alfalfa/lucerne is grown as a fodder crop, when numbers may become very large. The introduced *Melia azedarach* (Indian Bead Tree, Persian Lilac) has been seen to attract many individuals for the purpose of nectaring. Larvae have a facultative myrmecophilous relationship with many ant species (Fiedler, 2021: Appendix S1), but details are lacking for Cyprus.

Description/identification aids

Wingspan: up to 35 mm. Flight is *rapid and erratic*, setting it aside from other common 'blues' in Cyprus, with the exception of *Leptotes pirithous*. ♂ **upperside** is metallic blue, forewing unmarked other than a black margin broadening towards the apex; hindwing with a black margin and two black ocelli at the anal angle. ♀ uppersides are brown with blue basal scaling and variable white hindwing marks. **Undersides** of the sexes are similar, the hindwing having fine broken *lines* of white markings on a brown ground colour, an irregular *broad post-discal band of white* (*cf. L. pirithous*) and *two black ocelli, part encircled with silver within an area of deep orange* in the anal area. Both sexes have long tails.

Similar species

See *Deudorix livia* ♀, also *Leptotes pirithous*, which is generally smaller, but with markedly different undersides.

Conservation status in Cyprus

Not threatened. Listed by Van Swaay *et al.* (2010a) and Numa *et al.* (2016) as 'LC'.

Leptotes pirithous (Linnaeus, 1767) Lycaenidae (Polyommatinae)

Lang's Short-tailed Blue

TL: "Algiriae" (Algeria).

Fig. 12.160. *Leptotes pirithous* (Lang's Short-tailed Blue) ♂, 5 September 2011. Courtesy of *Colin Baker*

Fig. 12.161. *Leptotes pirithous* (Lang's Short-tailed Blue) ♀, 14 August 2015. *Christodoulos Makris*

Fig. 12.162. Attempted courting of ♀ (above) *Leptotes pirithous* (Lang's Short-tailed Blue), 5 September 2011. Courtesy of *Colin Baker*

Fig. 12.163. *Leptotes pirithous* (Lang's Short-tailed Blue) on *Ziziphus lotus*, 22 May 2013. *Eddie John*

Fig. 12.164. *Leptotes pirithous* (Lang's Short-tailed Blue), 8 September 2013. *Christodoulos Makris*

Fig. 12.165. *Leptotes pirithous* (Lang's Short-tailed Blue), 13 July 2013. *Christodoulos Makris*

Fig. 12.166. *Leptotes pirithous* (Lang's Short-tailed Blue) pairing, 11 September 2011. Courtesy of *Colin Baker*

Presence in the eastern Mediterranean

Wide-ranging and present throughout the Levant, this Afrotropical species originated in Madagascar from which country it spread throughout Africa into southern Europe (Fric *et al.*, 2019). Numbers, generally, in the Levant appear to have increased in recent decades, probably resulting from migration and possibly influenced by a warming climate. Whereas Larsen (1974: 164) stated, 'No recent collectors [in Lebanon], apart from myself, have recorded it', *L. pirithous* now seems common in the wider region, for example in Hatay, Turkey (Atahan *et al.*, 2018: 16). No subspecies are recognized.

Lederer (1855) referred to *L. pirithous* as '*Lycaena Telicanus* Herbst.', while Bucknill (1911) used '*Syntarucus telicanus. Lang. form Ægyptiacus. B. Baker.*' Turner (1920), expressed this as 'Syntarucus (Langia) telicanus Lang [race aegyptiacus B.-B.]', while Rebel (1939), using the confusing 'Tarucus telicanus Lang', considered the use of '*aegyptiacus*' to be without justification, a view that prevails. Wiemers *et al.* (2018, 2020) use the widely accepted *L. pirithous.*

Distribution in Cyprus

Altitudinal: found from sea level to above 1000 m or more, but most commonly encountered at lower elevations.
Geographical: nowadays a frequently observed butterfly, sometimes appearing in large numbers around its hostplants. CBRS data indicates a > 380% increase in distribution within 10 km² UTM squares since the preparation of an earlier distribution map by the first author (see Makris, 2003: 160). Furthermore, whereas Parker (1983) showed *L. pirithous* to be present in just five 10 km² squares, this number has now increased to 100. In part, this can be attributed to increased recording effort, but the many observations of breeding and the relative ease with which this species is observed confirm that this migratory/resident species is now well established in Cyprus and has extended its range throughout the island.

Phenology

Polyvoltine, with fresh specimens generally appearing in March and remaining on the wing in successive overlapping broods until November.

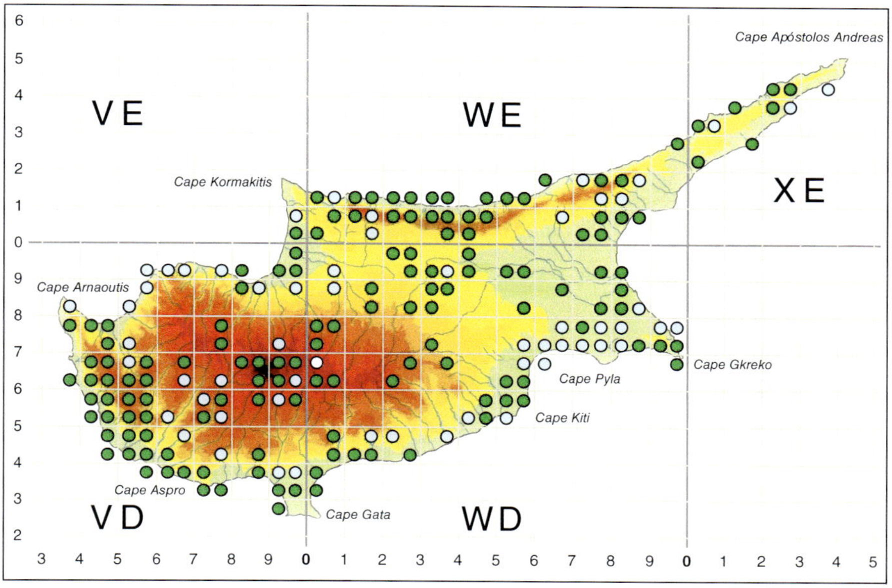

Fig. 12.167. *Leptotes pirithous* distribution map showing 5 km² observations within 10 km² UTM squares. © *Eddie John*

In coastal areas of Cyprus, *L. pirithous* has been observed on the wing throughout a mild winter (Nigel Cottle, Somerset, 2007; David and Rosalyn Sparrow, Cyprus, 2007, personal communications to first author), and may be encountered in any week of the year. Overwintering stage appears variable, according to circumstances; some may perish in less sheltered areas during winter.

Larval hostplants

The species is highly polyphagous, but in Cyprus *L. pirithous* is strongly attracted to an introduced South African garden ornamental, *Plumbago auriculata* (syn. *capensis*) (Plumbaginaceae) (Figs 12.168 and 12.169) around which the species has been seen in many hundreds and on which the species breeds. Among many other hostplants are members of the Fabaceae, especially *Alhagi maurorum* and *Medicago sativa* (lucerne/alfalfa) and, as elsewhere in the region, *Erophaca baetica*; also the ornamental *Medicago arborea, Genista fasselata* (prickly broom) (second author, personal observation) and occasionally *Prosopis farcta* (first author, personal observation). In 2006 (and since), the second author observed *L. pirithous* ovipositing on the locally common *Glinus lotoides* (Molluginaceae, formerly Aizoaceae) (Fig. 12.171), a hostplant used elsewhere in the Levant in preference to *Alhagi maurorum*. In May 2016, several *L. pirithous* were seen to be very active around a large, isolated flowering bush later identified by Rieks van Klinken (Brisbane, 2021, personal communication to first author) as the non-native *Parkinsonia aculeata* (Fabaceae), colloquially known as Jerusalem thorn among many other names, and presumably an escapee from the village nearby.

Fig. 12.168. *Leptotes pirithous* (Lang's Short-tailed Blue) ovipositing on *Plumbago auriculata*, a commonly used garden ornamental hostplant, 5 September 2011, Courtesy of *Colin Baker*

Fig. 12.169. *Leptotes pirithous* (Lang's Short-tailed Blue) ovipositing on *Plumbago auriculata*, 12 August 2015. Courtesy of *Anthony Simmons*.

Fig. 12.170. *Leptotes pirithous* (Lang's Short-tailed Blue) ovipositing on *Thymelaea hirsuta*, 10 November 2013. *Christodoulos Makris*

Fig. 12.171. *Leptotes pirithous* (Lang's Short-tailed Blue) ovipositing on *Glinus lotoides*, 17 August 2006. *Christodoulos Makris*

Ecology and behaviour

A fast-flying species, resembling *L. boeticus* on the wing and found in a range of habitats, but in recent decades is increasingly associated with urban locations (parks, gardens) and cultivated areas, often in company with *L. boeticus*. Unquestionably, the growing popularity of *Plumbago* as a garden ornamental has favourably influenced population numbers, particularly as these are often found in sheltered, large conurbations. No evidence of migration has been reported in Cyprus, but Larsen (1975) noted that in Lebanon the species 'cannot thrive without reinforcements'. In writing that 'The strong fluctuations and the phenology indicate that *S. pirithous* (L.) is not indigenous in Lebanon and that the maintenance of the population depends on immigration' (translated from the original German), Seizmair (2013) supported this view. In Cyprus, large population fluctuations have been experienced, with increases perhaps due to migration and population declines brought about by colder, wet winters or resulting from intensive parasitoid attack.

Nectaring on a variety of flowers has been reported, including those of *Dittrichia graveolens*, *Polygonum equisetiforme* and *Rubus sanctus*. Larvae are attended by ants of many species (Fiedler, 2021).

Description/identification aids

Wingspan: up to 25 mm. Like *L. boeticus*, flight is *rapid and erratic*, the butterfly often being seen in 'whirring' flight around its hostplant. **Uppersides** are not unlike *L. boeticus*, but are much darker. **Undersides** of both sexes are similar, with a light brown ground colour and a *mosaic* of patterning in white, resembling that of *L. boeticus*, but *lacking this species' hindwing white post-discal band*. Tails are shorter than in *L. boeticus*.

Similar species

L. boeticus – see previous species.

Conservation status in Cyprus

Not threatened. Listed by Van Swaay *et al.* (2010a) and Numa *et al.* (2016) as 'LC'.

Azanus jesous (Guérin-Méneville, 1849) Lycaenidae (Polyommatinae)

African Babul Blue

TL: Abyssinia (Ethiopia).

Fig. 12.172. *Azanus jesous* (African Babul Blue) photographed in Nicosia, July 2015. Courtesy of *Hasan Bağlar*

Fig. 12.173. *Azanus jesous* (African Babul Blue) photographed in Israel, 5 May 2014. *Christodoulos Makris*

Fig. 12.174. *Azanus jesous* (African Babul Blue) photographed in The Gambia, 23 July 2008. Courtesy of *David Penney*

Presence in the eastern Mediterranean

This migratory Afrotropical species is usually found sparingly in southern countries of the Levant (Larsen, 2015). In recent years, *A. jesous* has been photographed in the Turkish provinces of Hatay in 2013 (Atahan *et al.*, 2018: 147) and Adana in 2014 (AdaMerOs, 2021), presumably migrating northwards, undetected, as far as is known, through Syria.

Distribution in Cyprus

Altitudinal: known only from low elevations (< 200 m) in Cyprus.
Geographical: a species first found in Cyprus on 5 July 1940 by the entomologist Themistokles Shiakides (Makris, 2003: 180), who took four specimens in an area of Nicosia known to host a population of *L. galba* (which shares the same hostplant as *A. jesous*). Seventy-five years later, in July 2015, a singleton of *A. jesous* was photographed in the same 5 km^2 UTM square in Nicosia (Hasan Bağlar, Cyprus, 2015, personal communication to first author).

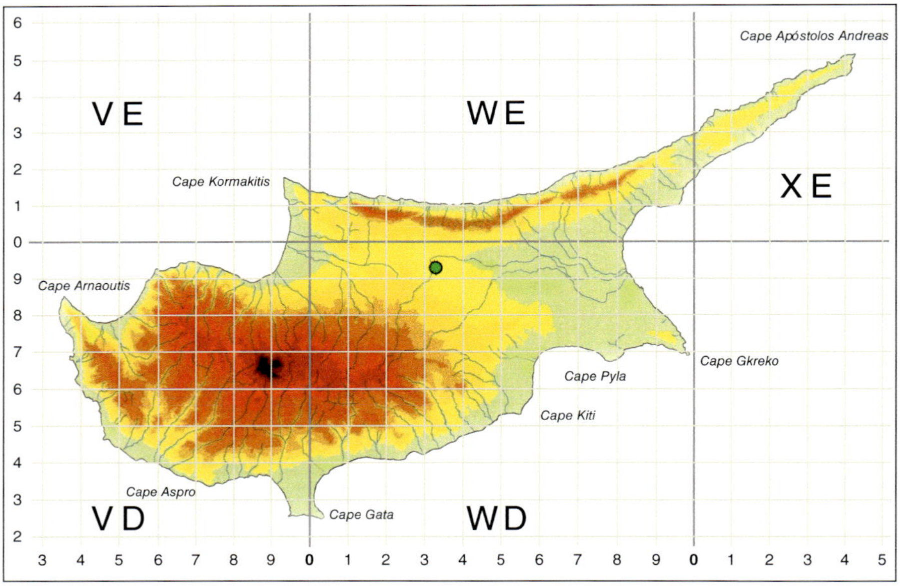

Fig. 12.175. *Azanus jesous* distribution map showing the location of the single 5 km^2 observation. © *Eddie John*

Phenology

Polyvoltine within its usual range.

Fig. 12.176. *Prosopis farcta*, 27 June 2006. *Eddie John*

Larval hostplants

Leguminosae/Fabaceae, including *Prosopis farcta* (Mediterranean mesquite) (Fig. 12.176) and *Acacia* spp.

Ecology and behaviour

It is not known if the 2015 singleton (see above) was a migratory specimen or whether a small colony has persisted undetected among the more common and very similar *L. galba* in the Nicosia area. If the opinion expressed by Larsen (1976) that *A. jesous* is not over-tolerant of cold winter temperatures is correct, survival from one season to the next seems unlikely. However, this view is challenged by Dubi Benyamini on the evidence of mid-winter observations of *A. jesous* in the Arabah and the Sinai Peninsula (Dubi Benyamini, Beit-Arye, 2015, personal communication to first author). Accidental introduction of early stages on ornamental *Acacia*, into the branches of which adults were seen to be attracted (Pittaway, 1985) and 'around which its whole life revolves' (Larsen 2005), is a more likely possibility for its appearance in Nicosia. Further detection of this species will require diligent observation due to the close similarity with *L. galba*; both species are *very closely* associated with another common hostplant, *Prosopis farcta*, and are known to fly together in other parts of the Levant. Larvae are reported to be attended by ants of the genus *Camponotus* (Fiedler, 2021: Appendix S1).

Description/identification aids

Wingspan: up to 25 mm. **Upperside** ground colour of the ♂ is mauve-blue, with a black marginal line, otherwise unmarked on the forewing, but with two black spots at the hindwing anal angle. ♀ brown, with blue basal scaling and similarly marked to the ♂. **Undersides** are similar in both sexes, with light brown ground colour and an *almost complete series of black marginal spots* (*cf. L. galba*) prominent over much of the hindwing, and smaller spots on the forewing. There are also basal, discal and postdiscal spots.

Similar species

See *Luthrodes galba*.

Conservation status in Cyprus

Not applicable – very rare migrant or adventive. Listed by Numa *et al.* (2016) as 'LC' at the Mediterranean level.

Tarucus balkanicus (Freyer, 1844)

Lycaenidae (Polyommatinae)

Little Tiger Blue

TL: Turkey.

Fig. 12.177. *Tarucus balkanicus* (Little Tiger Blue) ♂ nectaring on *Polygonum equisetiforme*, 26 September 2013. *Christodoulos Makris*

Fig. 12.178. *Tarucus balkanicus* (Little Tiger Blue) ♀, 28 May 2016. *Christodoulos Makris*

Fig. 12.179. *Tarucus balkanicus* (Little Tiger Blue) ♂ nectaring on *Polygonum equisetiforme*, 26 September 2013. *Christodoulos Makris*

Fig. 12.180. *Tarucus balkanicus* (Little Tiger Blue) ♂ nectaring on *Matricaria chamomilla*, 3 June 2017. *Christodoulos Makris*

Fig. 12.181. *Tarucus balkanicus* (Little Tiger Blue) ♀, 28 May 2016. *Christodoulos Makris*

Fig. 12.182. *Tarucus balkanicus* (Little Tiger Blue) pairing, ♀ left, ♂ right, 28 May 2016. *Christodoulos Makris*

Presence in the eastern Mediterranean

An eremic species, widely distributed throughout the Levant as the nominotypical subspecies. Referred to as '*Lycaena Balcanica* Freyer' by Lederer (1855), but thereafter (e.g. Bucknill, 1911; Turner, 1920; Rebel, 1939) as *Tarucus balkanicus*.

Distribution in Cyprus

Altitudinal: found from sea level to 400 m (rarely above).
Geographical: a common but localized butterfly with a disjunct distribution principally restricted to the Mesaoria Plain, in which the hostplant is plentiful, but with isolated pockets elsewhere.

Phenology

Polyvoltine, with first specimens appearing in late March/early April. Overlapping generations keep the species on the wing until November, rarely December. Overwinters as a pupa.

Larval hostplants

Rhamnaceae (buckthorn family): *Ziziphus lotus* (jujube) (Figs 12.184 and 12.185), a Tertiary invasive of former tropical savannah. Wiltshire (1948) gave *Paliurus spina-christi* as a hostplant, but although used elsewhere in the Levant, records of *P. spina-christi* in Cyprus relate to *Z. lotus* (Christofides, 2017: 282).

Ecology and behaviour

So strongly associated is this species with the larval hostplant *Z. lotus* that during the autumn build-up of numbers, it is common to find the butterfly

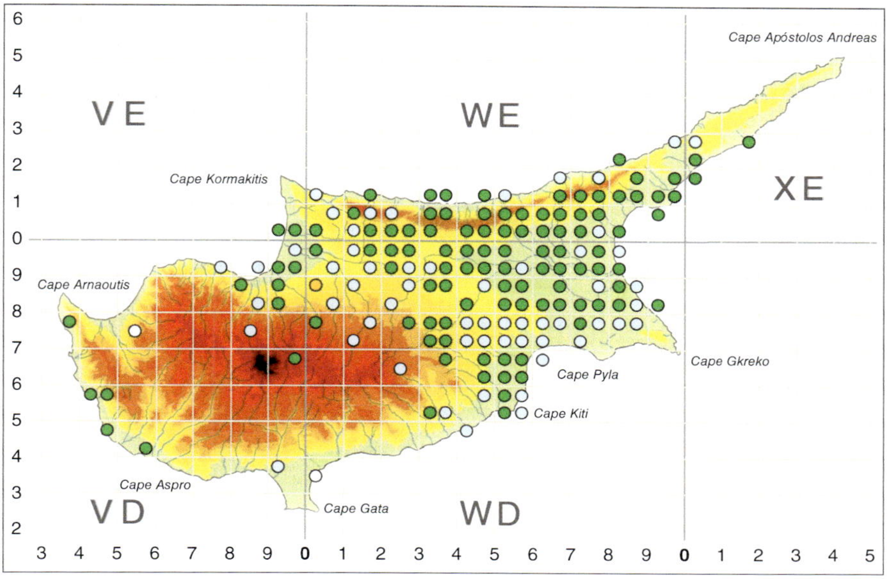

Fig. 12.183. *Tarucus balkanicus* distribution map showing 5 km² observations within 10 km² UTM squares. © *Eddie John*

Fig. 12.184. *Ziziphus lotus* in typical habitat on the Mesaoria Plain, 11 May 2011. *Eddie John*

Fig. 12.185. *Ziziphus lotus*, 11 June 2012. *Eddie John*

in very close attendance. *Z. lotus* is also used by the species as one of its nectar sources, as are *Polygonum equisetiforme*, *Phagnalon rupestre*, *Cardopatium corymbosum* and *Helichrysum stoechas* ssp. *barrelieri*. The absence of the hostplant from intermediate and higher levels confines the butterfly to well-defined areas of occupancy, mainly in the Mesaoria Plain and remnants of former habitat remaining after urbanization, for example at the coastal strip of the Pafos headland.

At Mia Milia on 24 May 2013, over 100 *T. balkanicus* were observed puddling and group nectaring, with no *Z. lotus* present nearby. In common with many other lycaenids, *T. balkanicus* often rests head down, occasionally displaying a 'rubbing' movement of the wings, which is possibly intended to deceive predators into mistaking the moving tails for antennae, especially as the large hindwing anal spot mimics an eye. Larvae are attended by several species of ant (Wiltshire, 1948; Lafranchis, 2019; Fiedler, 2021: Appendix S1).

Description/identification aids

Wingspan: up to 23 mm. ♂ **upperside** is mauve, with indistinct black markings at the end of the forewing cell and in the postdiscal area. ♀ upperside is brown, with similar black markings. The **underside** fore- and hindwings have distinctive black marginal and postdiscal lines, with *silver and black spotting in the submarginal band of the hindwing.* Basal and discal areas are spotted black, *creating an impression of stripes.* There is a moderately long tail on the hindwing. A specimen with very short tails, photographed in November 1994, is illustrated in Makris (2003: 165), but this appears to have been an anomalous rarity as Turner (1920), reported, 'The long series received show scarcely any aberration.'

Similar species

None in Cyprus.

Conservation status in Cyprus

Not threatened. Categorized as 'LC' in Europe (Van Swaay *et al.*, 2010a) and in the Mediterranean (Numa *et al.*, 2016).

Zizeeria karsandra (Moore, 1865) Lycaenidae (Polyommatinae)

Dark Grass Blue

TL: "Plains of N.W. India".

Fig. 12.186. *Zizeeria karsandra* (Dark Grass Blue) ♂ on the hostplant *Glinus lotoides*, 10 September 2016. *Christodoulos Makris*

Fig. 12.187. *Zizeeria karsandra* (Dark Grass Blue) ♂, 24 October 2017. *Eddie John*

Fig. 12.188. *Zizeeria karsandra* (Dark Grass Blue) ♀ nectaring on *Spergularia* sp., 23 April 2014. *Eddie John*

Fig. 12.189. *Zizeeria karsandra* (Dark Grass Blue) on *Polygonum equisetiforme*, 3 September 2016. Christodoulos Makris

Fig. 12.190. *Zizeeria karsandra* (Dark Grass Blue) nectaring on *Trifolium fragiferum*, 3 July 2011. *Christodoulos Makris*

Fig. 12.191. *Zizeeria karsandra* (Dark Grass Blue) nectaring on *Mentha longifolia* ssp. *cyprica*, 10 September 2016. *Christodoulos Makris*

Fig. 12.192. *Zizeeria karsandra* (Dark Grass Blue) pairing, 3 July 2011. *Christodoulos Makris*

Presence in the eastern Mediterranean

This non-migratory species is present in all countries of the Levant as nominotypical *Z. karsandra karsandra* (Larsen, 2015). However,

Kudrna (2002: 300), Kudrna *et al.* (2011: 483, 2015: 322) and Kudrna (2019: 255) treated the Cyprus data for this taxon (submitted as *Z. karsandra* by the first author) as *Zizeeria knysna knysna* (Trimen, 1862). In Kudrna (2019: 255), the presence of *Z. karsandra* in Crete (Rowlings and Cuvelier, 2018) was referred to as 'ssp. *karsandra*', with a further comment that '[*Z. karsandra*] is provisionally regarded as conspecific with *Z. knysna*'. Larsen (2015) noted the need for molecular examination, adding that 'there is clear genital difference' between the two species. Vazrick Nazari (Ottawa, 2012, personal communication to first author) advised: 'Several *Z. karsandra* and *Z. knysna* specimens have been barcoded, and they form three well defined clusters: 1) *Z. knysna* (Spain, Tunisia, Morocco, Tanzania, South Africa, Madagascar), 2) *Z. karsandra* (Egypt, Cyprus, Malta, Pakistan, Bangladesh), and 3) *Z. karsandra* (Australia). ... There is a single *karsandra* specimen from Cyprus, and it matches perfectly with other specimens in its group' (see BOLDSYSTEMS, 2014). *Z. karsandra* and *Z. knysna* are listed as separate, valid European species by most authors, e.g. Wiemers *et al.* (2018, 2020).

The species was originally listed for Cyprus as '*Lycaena lysimon* Hb' (= *Zizeeria knysna*) by Lederer (1855) and Bucknill (1911), and later referred to as '*Zizeeria lysimon*' by Turner (1920). Surprisingly, given the extent to which *Z. karsandra* is found nowadays in Cyprus, Bucknill (1911) noted 'we have not met with it'. Similarly, Wilson (quoted in Turner, 1920) stated (of '*Zizeeria lysimon*'), 'Recorded, but I have never seen it.' The species does not appear to have been formally documented between the publication of Lederer (1855) and that of Turner (1920), but appears in Rebel (1939) as 'Zizera lysimon (hb.) karsandra Moore'.

Colloquial names are often inconsistent: this species has variously been called 'Small Mauve Blue' (Larsen 1974: 196; Parker, 1983), 'Oriental Grass Blue' (Larsen, 1984a: 70), 'Asian Grass Blue' (Larsen, 1974: 126; Katbeh-Bader *et al.*, 2004) or 'African Grass Blue' (John, 2000; Makris, 2003: 174). However, the last-mentioned species' name is now regarded by most authors to refer to the closely related *Zizeeria knysna*, e.g. Larsen (1996: 238), a species not present in the Levant.

Distribution in Cyprus

Altitudinal: principally a species of lower elevations, less frequently found above 400 m.
Geographical: a *very* common butterfly distributed across all lowland areas of the island.

Phenology

Polyvoltine, with records from all months (CBRS data), but most commonly seen in the second half of the year, as numbers build in successive generations.

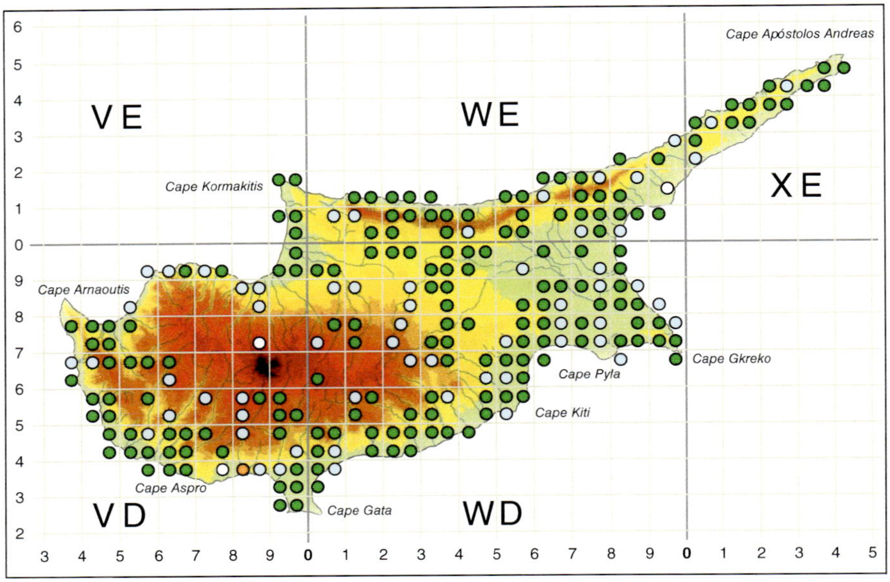

Fig. 12.193. *Zizeeria karsandra* distribution map showing 5 km^2 observations within 10 km^2 UTM squares. © *Eddie John*

Larval hostplants

A polyphagous species. In Cyprus, principally Polygonaceae: *Polygonum equisetiforme* (Figs 12.137 and 12.138), but also Aizoaceae: *Glinus lotoides* (Fig. 12.194); Amaranthaceae: *Amaranthus blitum*; Fabaceae: *Alhagi maurorum*, *Medicago* spp., *Trifolium fragiferum*; Zygophyllaceae: *Tribulus terrestris*.

Ecology and behaviour

While individuals of this tiny unobtrusive butterfly may easily be overlooked, it is rarely encountered as a solitary species, so the lack of numbers a century or more ago (as discussed above) is difficult to explain. Possibly, increased irrigation during the intervening years has created conditions that have enabled significant spread of its main hostplant, thereby influencing population growth of the butterfly. Certainly, *Z. karsandra* is usually found in close association with *P. equisetiforme* in roadside ditches, damp depressions in waste ground or irrigated areas, and the species demonstrates a remarkable ability to thrive in well-populated areas, including gardens and roadsides (Fig. 12.195). Consequently, *Z. karsandra* is to be found in many 5 km^2 UTM squares and in some locations it flies in company with *Luthrodes galba*.

Fig. 12.194. *Glinus lotoides*, 13 July 2021. *Christodoulos Makris*

Fig. 12.195. Roadside growth of *Polygonum equisetiforme* on which ovipositing by *Z. karsandra* was observed, 12 October 2017. *Eddie John*

In recent years, *G. lotoides* has been shown to be another very important hostplant for *Z. karsandra*; in summer and autumn when water levels recede at dams, extensive areas of exposed ground may become carpeted with this hostplant, encouraging the build-up of large numbers of the butterfly (second author, personal observation). Human activity (there has been extensive dam building in the past 100 years) appears to have greatly increased population numbers since the days of the early researchers.

Three dwarf specimens, even smaller in size than *Freyeria trochylus*, were encountered on 23 April 2014 at a coastal site in Karpasia (first author, unpublished data). Overlapping generations frequently produce large aggregations around the hostplants in autumn. Larvae are attended by ants of various genera (Makris, 2003: 77; Galanos, 2020).

Description/identification aids

Wingspan: up to 25 mm. One of the *smallest butterflies* found in Cyprus. ♂ **uppersides** are mauve, with a *wide black band over the submarginal and marginal areas* of fore- and hindwings, otherwise unmarked. ♀ uppersides are grey-brown, with blue scaling in the basal and discal areas. **Undersides** of both sexes are light grey-brown, both wings with a brown marginal line and brown submarginal marks. *A distinctive series of black spots*, ringed off-white in the discal (hindwing) and post-discal (fore- and hindwing) areas, separates this species from other small blues.

Similar species

L. galba is of similar size and coloration, but has distinctive silver/black spots at the hindwing anal angle – see entry under this species.

Conservation status in Cyprus

Not threatened (see Chapter 14). Not assessed in Europe by Van Swaay *et al.* (2010a), but described as 'LC' in the Mediterranean list (Numa *et al.*, 2016).

Celastrina argiolus (Linnaeus, 1758) Lycaenidae (Polyommatinae)

Holly Blue

TL: England.

Fig. 12.196. *Celastrina argiolus* (Holly Blue) ♀, 26 April 2014. Courtesy of *Adrian Riley*

Fig. 12.197. *Celastrina argiolus* (Holly Blue) nectaring on *Cistus creticus*, 17 June 2012. *Christodoulos Makris*

Fig. 12.198. *Celastrina argiolus* (Holly Blue) nectaring on *Rubus sanctus*, 29 June 2006. *Eddie John*

Fig. 12.199. *Celastrina argiolus* (Holly Blue) nectaring on *Rubus sanctus*, 4 July 2012. Courtesy of *Sami Tamson*

Fig. 12.200. *Celastrina argiolus* (Holly Blue) on *Asphodelus ramosus*, 12 May 2016. *Christodoulos Makris*

Fig. 12.201. *Celastrina argiolus* (Holly Blue) mud-puddling, 17 June 2012. *Christodoulos Makris*

Presence in the eastern Mediterranean

Present in all nearby mainland countries, probably as nominotypical *C. argiolus argiolus*; the validity of *Celastrina argiolus* var. 'paraleuca' (Röber, 1897) listed for Lebanon by Larsen (1974: 197; Benyamini, 2002a: 196) is unconfirmed and was synonymized with the nominotypical subspecies by Hesselbarth *et al.* (1995: 557) and had previously been so regarded by Olivier (1993: 202).

This was another species overlooked by Zach and hence not listed for Cyprus in Lederer (1855). Bucknill (1911) referred to the species as '*Cyaniris argiolus L. The Azure Blue*', Turner (1920) as 'Celastrina argiolus L. [ab. hypoleuca, ♀, Koll.]', while Rebel (1939) reverted to 'Cyaniris argiolus L.'

Distribution in Cyprus

Altitudinal: occasionally present close to sea level, but usually found at intermediate and higher elevations, sparingly above 1800 m (Makris, 2003: 287).

Geographical: a fairly common butterfly, but seemingly largely confined to the Troodos Massif, which accords with Wilson's statement, 'confined to the Troödos range of mountains, where it is common during June and July', as reported in Turner (1920). The CBRS distribution map for this species has changed little in recent years, other than showing a slight eastwards extension of range (possibly an artefact attributable to increased recording effort).

Phenology

Probably trivoltine, with the possibility of a fourth brood in some years. March (even February) to September, possibly later. Overwinters as a pupa.

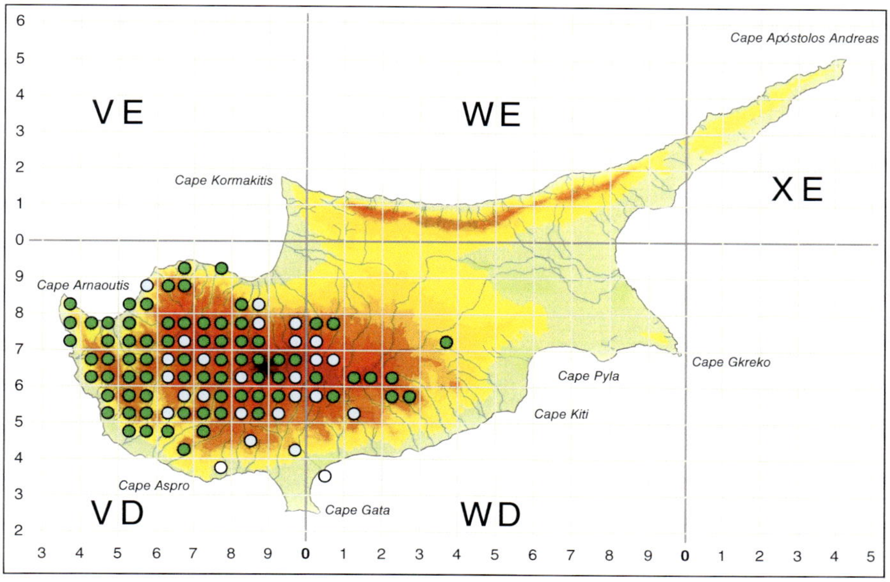

Fig. 12.202. *Celastrina argiolus* distribution map showing 5 km² observations within 10 km² UTM squares. © *Eddie John*

Larval hostplants

Larvae are polyphagous and among the plants used are Araliaceae: *Hedera helix*; species of Fabaceae including *Genista fasselata* (Fig. 12.203); elsewhere, Rhamnaceae: *Rhamnus* spp.; Rosaceae: *Rubus* spp.

Ecology and behaviour

Usually found singly in well-vegetated bushy areas, often in shaded valleys and glades in forested areas, especially near water and most frequently encountered in the mountains of the Troodos range, where it may be locally very common. Flight over the ground is generally much higher than other blues found in Cyprus and the species is a known wanderer. *C. argiolus* is particularly attracted to the flowers of *Rubus sanctus*, on a single flower of which a group of six was photographed by Andy Grant in July 2015 (personal communication to first author). Larvae are attended by ant species of numerous genera (Kitching and Luke, 1985; Makris, 2003: 168).

Description/identification aids

Wingspan: up to *ca.* 35 mm. ♂ **upperside** is blue and largely unmarked. Forewing margins appear chequered where the veins cross into the white wing fringes. ♀ upperside is blue, with variable degrees of brown banding, in later broods confining the blue to the basal and discal areas. **Undersides** of sexes are similar, *the silvery-blue/white fore- and hindwing undersides*, largely devoid of markings other than *scattered small black spots* (that

Fig. 12.203. *Genista fasselata*, 12 October 2017. *Eddie John*

appear to be less distinct than western and northern populations), are characteristic of the species.

Similar species

Undersides are unlike those of any other species found in Cyprus.

Conservation status in Cyprus

Not threatened. Listed by Van Swaay *et al.* (2010a) and Numa *et al.* (2016) as 'LC'.

Pseudophilotes vicrama (Moore, 1865) Lycaenidae (Polyommatinae)

Eastern Baton Blue

TL: "Cheeni in Middle Kunavur, and the Alps above the Chinese village of Shipkee in Tibet."

Fig. 12.204. *Pseudophilotes vicrama* (Eastern Baton Blue) ♂, 13 April 2012. *Christodoulos Makris*

Fig. 12.205. *Pseudophilotes vicrama* (Eastern Baton Blue) ♂ nectaring on *Helichrysum stoechas*, 26 March 2016. *Christodoulos Makris*

Fig. 12.206. *Pseudophilotes vicrama* (Eastern Baton Blue) ♀, 12 March 2013. *Christodoulos Makris*

Fig. 12.207. *Pseudophilotes vicrama* (Eastern Baton Blue) ♀ nectaring on *Phagnalon rupestre*, 24 April 2014. *Eddie John*

Fig. 12.208. *Pseudophilotes vicrama* (Eastern Baton Blue) ♀, 12 March 2013. *Christodoulos Makris*

Fig. 12.209. *Pseudophilotes vicrama* (Eastern Baton Blue) ♂ nectaring on *Frankenia hirsuta*, 23 April 2014. *Eddie John*

Fig. 12.210. *Pseudophilotes vicrama* (Eastern Baton Blue) ♂, 13 April 2012. *Christodoulos Makris*

Fig. 12.211. *Pseudophilotes vicrama* (Eastern Baton Blue) ♂, 12 March 2013. *Christodoulos Makris*

Presence in the eastern Mediterranean

Present in all countries of the Levant, where the subspecific status is in need of confirmation. As with many other species, early taxonomic names have been inconsistent: Turner (1920) referred to 'Scolitantides baton Brgstr', Rebel (1939) to 'Lycaena baton (Brgstr.) schiffermilleri Hem.', whereas Wiltshire (1948) listed the species as '*Philotes astabene*, Hemming (= *clara*, Stgr. praeocc.)', adding, 'I took both forms (i.e. *schiffermulleri* and *astabene*) together, in iii [March, 1947] and iv [April, 1947] ... the male genitalia of the two forms were identical'. However, it is clear from genitalic work done by Hemming (1929) that *baton* and *vicrama* are not conspecific. Olivier (1993: 99) gave *P. vicrama clara* (Staudinger, 1901) for Cyprus, E. Turkey and the Levant, adding the cautionary remark '(if really distinct subspecifically!)'. Larsen and Nakamura (1983) referred to *Pseudophilotes vicrama astabene*, adding 'Subsp. *astabene* flies in the Levant.' Kudrna *et al.* (2011: 38) and Kudrna (2019: 43) treated *Pseudophilotes* as a distinct subgenus and a junior subjective synonym of *Scolitantides*, whereas Wiemers *et al.* (2018, 2020) retained the genus name of *Pseudophilotes*.

Not reported for Cyprus by Lederer (1855) or Bucknill (1911), although both Gibbs and Wilson (in Turner, 1920) reported observations in 1916 and 1918 respectively.

Distribution in Cyprus

Altitudinal: found from coastal locations to 400 m, very rarely higher.

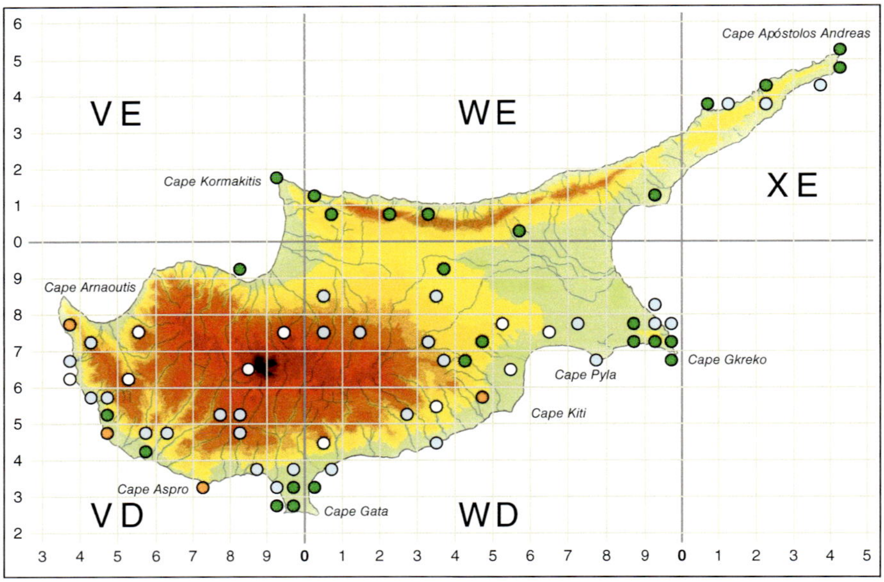

Fig. 12.212. *Pseudophilotes vicrama* distribution map showing 5 km² observations within 10 km² UTM squares. © *Eddie John*

Geographical: a local species found in small, widely scattered colonies across Cyprus. Generally an uncommon butterfly, but occasional reports from areas in which this species had not previously been reported suggest that it is an under-recorded species in Cyprus, perhaps due to the species' superficial similarity to *Polyommatus icarus*.

Phenology

Bivoltine to partially trivoltine, with first specimens appearing on the wing in March, possibly earlier. Until 2011, there were no records later than the end of June (CBRS data), but a 1 August sighting (and photographing) of a fresh specimen (Hasan Bağlar, Cyprus, 2011, personal communication to first author) points to a partial third brood in favourable years. Pamperis (2009: 224) stated for Greece, 'March to October in two generations', Hesselbarth *et al.* (1995: 561) referred to three generations in favourable locations, while Larsen and Nakamura (1983) wrote (of Jordan) 'in the neighbouring countries there is a brood also in September'. Overwinters as a pupa.

Larval hostplants

In Cyprus, principally associated with Lamiaceae: *Thymbra capitata* (Mediterranean thyme) (Fig. 12.213) and possibly *Calamintha* and *Mentha* spp., as in Greece (Lafranchis, 2019). Reported elsewhere on Convolvulaceae: *Cuscuta palaestina*; Rosaceae: *Sarcopoterium spinosum*; Caryophyllaceae: *Paronychia argentea*.

Fig. 12.213. *Thymbra capitata*, 14 May 2013. *Eddie John*

Ecology and behaviour

A sedentary, fairly rare and local species, found in contrasting habitats from exposed coastal areas to hillside locations, in association with the hostplant. Most often encountered singly, but large groups of adults may also be seen. Parker (1983) failed to record the species during his three-season stay on the island in the 1970s, perhaps suggesting that population numbers have increased in recent decades. Larvae are attended by ants (Fiedler, 2021), and adults are attracted to the flowers of *Phagnalon rupestre*, *Helichrysum stoechas* ssp. *barrelieri* and *Frankenia hirsuta*.

Description/identification aids

Wingspan: up to 27 mm. ♂ **upperside** is silvery-blue, bordered by a line of black, with a small black mark at the end of the forewing cell and small black submarginal spots on the hindwing. ♀ uppersides are similarly marked, colour is brownish with much basal and discal blue scaling. Both sexes have contrasting white fringes, *strongly chequered where the wing veins cross the cilia*. ♂ and ♀ **underside** fore- and hindwings have a *silvery-grey ground colour* and are similarly marked in both sexes: forewing with *contrasting prominent black spotting* especially in the submarginal and discal areas; hindwing with *strong orange submarginal marks*, with inner black lunules and black marginal spotting and a series of black discal and postdiscal spots. Unlike other blues in Cyprus, *wing fringe chequering* is pronounced.

Similar species

A slight similarity to the *larger Polyommatus icarus* ♂, but black spotting is *larger* in *P. vicrama*, which also *lacks* any forewing underside orange.

Conservation status in Cyprus

Not seriously threatened, although there is pressure on colonies near areas of urban development, placing these in need of protection. Coastal, second-brood populations may be at some increased risk from habitat desiccation due to climate change, but the hostplant is drought tolerant. A species of European conservation concern where it is considered 'NT' (Van Swaay *et al.*, 2010a), but listed as 'LC' in the Mediterranean by Numa *et al.* (2016).

Glaucopsyche paphos Chapman, 1920 **Lycaenidae (Polyommatinae)**

Paphos Blue

TL: Cyprus.

Fig. 12.214. *Glaucopsyche paphos* (Paphos Blue) ♂, 29 March 2014. Courtesy of *Elli Tzirkalli*

Fig. 12.215. *Glaucopsyche paphos* (Paphos Blue) ♂, 26 April 2014. Courtesy of *Gerrit Ros and Jeannette Dewaerheijt*

Fig. 12.216. *Glaucopsyche paphos* (Paphos Blue) ♀, 26 March 2015. *Christodoulos Makris*

Fig. 12.217. *Glaucopsyche paphos* (Paphos Blue) ♂ nectaring on *Teucrium micropodioides*, 4 May 2016. *Eddie John*

Fig. 12.218. *Glaucopsyche paphos* (Paphos Blue) ♂ nectaring on *Crepis foetida*, 15 April 2016. *Christodoulos Makris*

Fig. 12.219. *Glaucopsyche paphos* (Paphos Blue) ♀ nectaring on *Cistus creticus*, 26 May 2019. *Christodoulos Makris*

Fig. 12.220. *Glaucopsyche paphos* (Paphos Blue) ♀, 26 March 2015. *Christodoulos Makris*

Fig. 12.221. *Glaucopsyche paphos* (Paphos Blue) group nectaring on *Asphodelus ramosus*, 27 April 2012. Courtesy of *Alison McArthur*

Fig. 12.222. *Glaucopsyche paphos* pairing, 19 March 2007. *Christodoulos Makris*

Fig. 12.223. *Glaucopsyche paphos* ovipositing on *Genista fasselata*, 19 March 2014. *Christodoulos Makris*

Presence in the eastern Mediterranean

An endemic species to Cyprus. Turner (1920), perhaps stimulated by a communication from P. P. Graves, wrote, 'Not being satisfied with the received determination of the Cyprian series as *G. melanops* Bdv., I submitted them to Dr. Chapman … [who] considered that he was justified in announcing it a new species …' (*Glaucopsyche paphos* – see Chapman, 1920). Despite the apparent acceptance of the name *Glaucopsyche paphos* at the time, the species was referred to by Rebel (1939) as *Lycaena paphos*, perhaps based on Chapman's previously held view (in 1910) that *Glaucopsyche melanops* (Boisduval, 1828) should be placed in the genus *Lycaena*.

Distribution in Cyprus

Altitudinal: present at all altitudes. Wiltshire (1948) wrote, 'Is not, as Rebel says, a mountain species, but a maquis species.' Rebel's opinion would have been influenced by the high-altitude location (Troodos) of Turner's holotype, collected on 10 July. In fact, both Rebel and Wiltshire were correct (see 'Phenology', below).
Geographical: a very common endemic species found over much of the island, but with few records from the Mesaoria Plain and especially the south-east.

Phenology

Univoltine or possibly partially bivoltine, the first specimens usually appearing in coastal areas in February (rarely, late January). CBRS data records the species on the wing until mid-July; these (sparse) later records

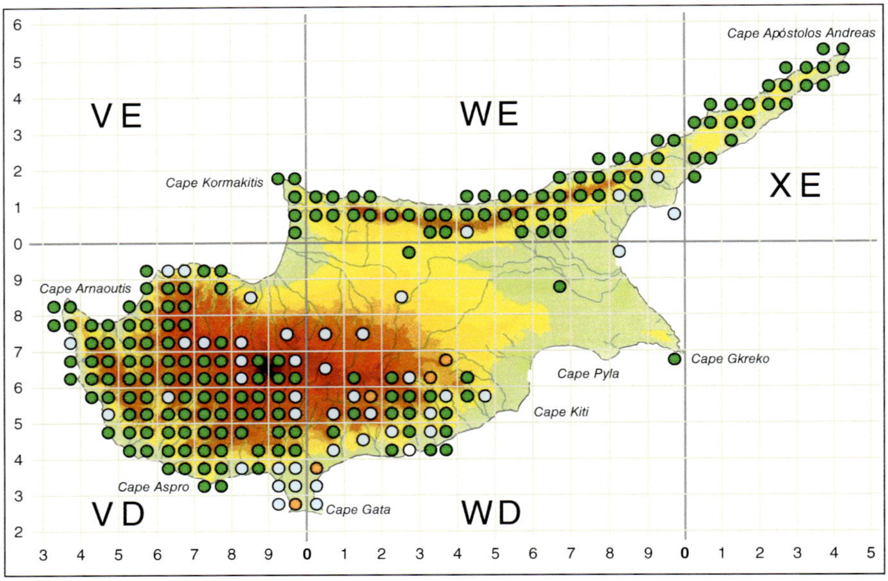

Fig. 12.224. *Glaucopsyche paphos* distribution map showing 5 km^2 observations within 10 km^2 UTM squares. © *Eddie John*

were from higher, cooler elevations (1800 m+) where emergence is delayed. However, during a visit to Cyprus in June 2012, the first author's observation of numerous *fresh* individuals of *G. paphos* at two northern coastal sites introduced doubt about the species being uniquely univoltine. At one coastal location (*ca.* 8 km south-east of Cape Kormakitis, 35°22.071′N; 32°59.871′E, 5 m asl) on 12 June 2012, two pairs of *G. paphos* were seen *in cop.* some *three months* later than the expected emergence time (February/March) at this coastal habitat. Further coastal observations followed on 18 June, *ca.* 80 km to the east, where individuals were seen at 30 m asl, approximately 300 m from the shoreline. Moreover, at Oreites Forest (*ca.* 400 m) freshly emerged individuals (see Fig. 12.219) were seen on the wing by the second author on 26 May 2019, an observation repeated at times and in many localities elsewhere on the island, when fresh specimens were least expected. These observations, in terms of timing and locations, appear inconsistent with univoltinism and point to the likelihood of a partial second brood (John, 2013; John and Skule, 2016: 306; Benyamini *et al.*, 2018).

The first of the northern coastal locations was revisited by the first author on 23 May 2013, when six individuals of *G. paphos* were seen in flight; this date also suggests partial bivoltinism. However, pupal diapause needs to be considered a possibility, as also discussed in Benyamini *et al.* (2018), in which paper rearing results demonstrated that pupae subjected to *dry* conditions have the ability to enter a second year of diapause, as reported for *G. melanops* in captivity (Tolman and Lewington, 1997: 92).

However, extended dry conditions are unlikely to prevail naturally in Cyprus, especially over the much wetter periods (autumn–spring) of two successive years. Moreover, *Genista fasselata* is a robust perennial shrub that grows to 1–2 m or higher and is well capable of producing foliage in dry conditions. Indeed, far from being arid, the coastal habitats described above are well vegetated, and it is difficult to envisage a weather-related reason for extended pupal diapause in such locations. Further investigations are required.

Larval hostplants

Fabaceae: *Genista fasselata* var. *fasselata* (broom) (Fig. 12.225); *G. fasselata* var. *crudelis* (Troodos broom – a dwarf endemic variety) (Fig. 12.226). *Calycotome villosa* (hairy thorny broom) requires confirmation as a hostplant.

Ecology and behaviour

Found in close association with the various hostplant brooms in maquis, garigue and rocky hillside areas. Group nectaring and mud-puddling is frequently seen.

Prior to 2017, *G. paphos* had not been confirmed from southern coastal areas east of Alethriko. Surprisingly, an observation by Ray Atkinson (Greater Manchester, 2017, personal communication to first author, supported by a photograph) of a ♂ *G. paphos* at Cape Gkreko on 10 March 2017, an area from which *Genista* was thought to be absent, suggested the possible use of other members of the Fabaceae family as hostplants. A further isolated observation in May of the same year at Kouklia (WD68 NE), *ca.* 15 km south of known locations (Daniel and Hilary Haines, London, 2017, personal communication to first author) led to further thoughts about the use of an alternative hostplant, as *Genista* had been unrecorded from both areas (Christofides, 2017: 172). However, since the publication of Benyamini *et al.* (2018), in which the Cape Gkreko observation was discussed, the presence of *Genista fasselata* near Protaras, some 4 km

Fig. 12.225. *Genista fasselata* var. *fasselata. Christodoulos Makris*

Fig. 12.226. *Genista fasselata* var. *crudelis. Yiannis Christofides*

from Cape Gkreko, has been confirmed (Kyriakos Kefalas, Cyprus, 2021, personal communication to second author). This would therefore appear to end further speculation about an alternative hostplant, at least at this location.

Larvae are attended by the ant species *Tapinoma simrothi,* and it is likely that species of ant from other genera are also attracted to the larvae (Benyamini *et al.*, 2018). This paper also reported on larval attack by a species of braconid wasp and by *Anisobas seyrigi*, a pupal parasitoid belonging to the family Ichneumonidae.

Description/identification aids

Wingspan: up to 30 mm. ♂ **uppersides** are unmarked turquoise blue with a black border; ♀ very dark brown. **Undersides** are similar in both sexes: grey-brown ground colour (darker in the ♀), the forewing underside with a characteristic series of *five or six distinctive large spots ringed with white.* Hindwing underside has a *complete series of postdiscal black spots ringed with white.* No other species is so marked in Cyprus.

Similar species

The superficially similar *Glaucopsyche alexis* (Poda, 1761) and *Iolana* (*Glaucopsyche*) *iolas* (Ochsenheimer, 1816) are found in nearby countries, but these species are *not known* for Cyprus.

Conservation status in Cyprus

Not threatened, despite its endemic status. Van Swaay *et al.* (2010a) and Numa *et al.* (2016) correctly listed *G. paphos* as 'LC'.

Luthrodes galba (Lederer, 1855) — Lycaenidae (Polyommatinae)

Small Desert Blue

TL: Beirut, Lebanon.

Fig. 12.227. *Luthrodes galba* (Small Desert Blue) ♂, 25 June 2017. *Christodoulos Makris*

Fig. 12.228. *Luthrodes galba* (Small Desert Blue) mud-puddling, 9 September 2018. *Christodoulos Makris*

Fig. 12.229. *Luthrodes galba* (Small Desert Blue), 15 September 2014. *Christodoulos Makris*

Fig. 12.230. *Luthrodes galba* (Small Desert Blue), 28 May 2016. *Christodoulos Makris*

Fig. 12.231. *Luthrodes galba* (Small Desert Blue) pairing, 16 June 2006. *Eddie John*

Fig. 12.232. *Luthrodes galba* (Small Desert Blue) pairing on the hostplant, 9 June 2012. *Christodoulos Makris*

Presence in the eastern Mediterranean

Present in all countries of the Levant as nominotypical *L. galba galba*. This species was the subject of genitalic examination by which means Bethune-Baker (1913), using the original combination *Chilades galba*, separated this species from *Chilades phiala*. Elsewhere, Larsen (1983: 415) suspected the involvement of a subspecies, and this was soon to be described (comb. n.) by Nekrutenko (1984), as *Lachides galba phiala* (Groum-Grshimaïlo), along with *Lachides galba galba* (Lederer), comb. n. In referring to the species as 'Chilades phiala Gr.-Gr, = [Zizera galba Led.]', Turner (1920) quoted Bucknill as stating in 1916, 'Recorded in the Annals of the Cyprus Natural History Society (1912–1913) as having been taken for the first time in 1912 or 1913.' Nevertheless, given the (current) extent of distribution in Cyprus, it appears inexplicable that the species was not recorded by Zach (Lederer, 1855) or by Bucknill (1911), especially as Wilson (in Turner, 1920) is quoted in 1918 as stating: 'Fairly common in the plains.'

Luthrodes galba is listed in some publications as *Lachides galba*, but *Lachides* Nekrutenko, is treated by Kudrna *et al.* (2011: 36) and Kudrna (2019: 40) as, 'a junior subjective synonym of *Chilades* Moore, [1881] (E. Balletto personal communication)'. In a revision of *Polyommatus* blue butterflies using molecular data, Talavera *et al.* (2012) reassigned *galba* and *trochylus* (see entry for this species), both previously listed as belonging to *Chilades*, to separate genera – *galba* to the genus *Luthrodes*, *trochylus* to *Freyeria*. Wiemers *et al.* (2018, 2020) adopted these combinations.

Distribution in Cyprus

Altitudinal: strictly a species of lowland areas to about 400 m.
Geographical: a very local butterfly, mainly found near the salt lakes and on the Mesaoria Plain. Occasional observations from disjunct locations (e.g. John and Mérit, 2008), may reflect the unintentional introduction of *Prosopis farcta* (a weed of cultivation) into such areas by human activity and the subsequent arrival of *L. galba*.

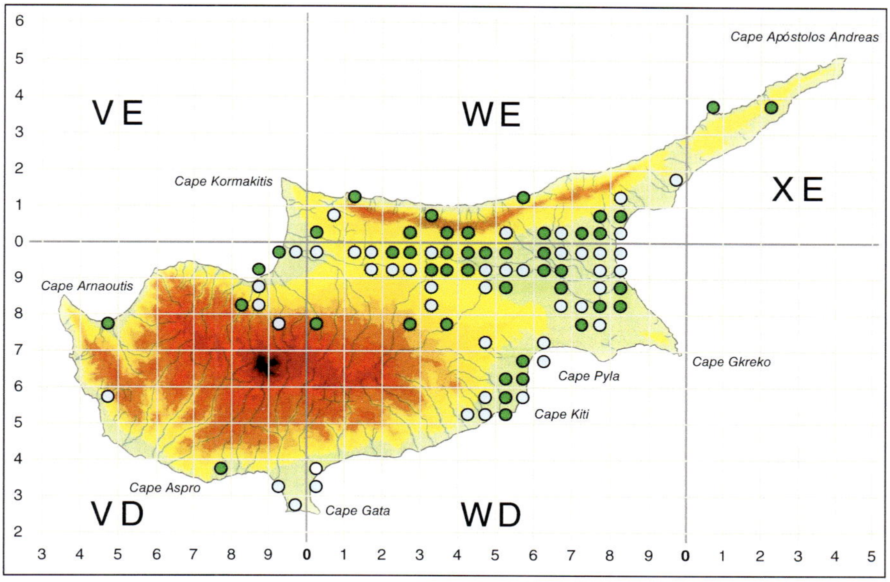

Fig. 12.233. *Luthrodes galba* distribution map showing 5 km² observations within 10 km² UTM squares. © *Eddie John*

Phenology

Trivoltine to polyvoltine. One of the later spring species to emerge, usually late April or early May (although sightings have been reported from late February and early March – CBRS data) and remaining on the wing in successive generations until October or early November. Overwinters as a pupa, or possibly a slowly developing larva.

Larval hostplants

Fabaceae: in Cyprus, *Prosopis farcta* (Mediterranean mesquite) (Fig. 12.234); elsewhere, *Alhagi maurorum* and Mimosaceae: *Acacia* spp. In his 1983 paper, Parker inadvertently referred to *Prosopis spicigera*, but this is a related small tree, whereas *P. farcta* is a small shrub, growing to 40 cm (Christofides, 2017: 194) and referred to as *P. stephaniana* (syn.) by Tshikolovets (2011: 189).

Ecology and behaviour

Numbers increase into the autumn when *L. galba* may become temporarily abundant in locations where its hostplant is present in quantity, but even small isolated groups of *P. farcta*, as occasionally found growing on roadside verges, may play host to *L. galba*. This low-flying species is *very strongly* associated with the hostplant, from which areas it is not known to stray any distance, even nectaring on the hostplant's small, yellow flowers. The species is known to mud-puddle. *L. galba* overwinters as a pupa in colder areas, but the late spring emergence in Cyprus suggests

Fig. 12.234. *Prosopis farcta*, 27 June 2006. *Eddie John*

slow larval development during winter followed by early spring pupation. Larvae are attended by ants at the third and fourth instars (Makris, 2003: 182; Galanos, 2020); Wiltshire (1957: 27) specified the small ant species *Monomorium gracillimum* (Hymenoptera: Myrmicinae), a species known for Cyprus (Edwards *et al.*, 2016: 420).

Description/identification aids

Wingspan: up to 23 mm. One of the smallest of Cyprus's butterflies. ♂ **uppersides** are dark blue with a broad black submarginal band and a black spot at the anal angle of the hindwing; ♀ uppers have a brown ground colour. **Undersides** of the sexes have a similar grey-brown ground colour, the hindwing with three black basal spots (that on the abdominal fold smaller), a black discal spot near the inner margin and, at the anal angle, *two larger black spots, highlighted with silver*. Across fore- and hindwings is a series of brown discal, postdiscal and submarginal marks outlined in white.

Similar species

Z. karsandra (but which species *lacks* the distinctive silver/black spots at the hindwing anal angle), and especially the very rare migrant *Azanus jesous* (which has an *almost complete series of hindwing black marginal spots*).

Conservation status in Cyprus

Not threatened. While urbanization and farming practices have destroyed much of the established habitat in areas of the Mesaoria Plain, it is also considered that farming activity is beneficial in assisting spread of the hostplant *P. farcta*. Not assessed by Van Swaay *et al.* (2010a), whereas Numa *et al.* (2016) listed *L. galba* as 'LC' in the Mediterranean.

Freyeria trochylus (Freyer, 1844) Lycaenidae (Polyommatinae)

Grass Jewel

TL: Turkey: "europäischen Türkey".

Fig. 12.235. *Freyeria trochylus* (Grass Jewel) ♂, 9 October 2006. *Christodoulos Makris*

Fig. 12.236. *Freyeria trochylus* (Grass Jewel) ♀, 27 September 2015. *Christodoulos Makris*

Fig. 12.237. *Freyeria trochylus* ♂ (Grass Jewel), 9 August 2012. Courtesy of *Clive Burrows*

Fig. 12.238. *Freyeria trochylus* (Grass Jewel) ♀ nectaring on *Verbena officinalis*, 28 July 2019. Christodoulos Makris

Fig. 12.239. *Freyeria trochylus* (Grass Jewel) ♀ on *Heliotropium hirsutissimum*, 10 September 2015. Christodoulos Makris

Fig. 12.240. *Freyeria trochylus* (Grass Jewel) pairing, ♀ on left, 1 May 2017. *Christodoulos Makris*

Presence in the eastern Mediterranean

Found in all countries of the Levant as nominotypical *F. trochylus trochylus*. In a revision of *Polyommatus* blue butterflies using molecular data, Talavera *et al.* (2012) reassigned *trochylus* and *galba* (see entry for this species), both previously listed as belonging to *Chilades*, to different genera. *C. trochylus* was placed in the genus *Freyeria*, stating that 'molecular data demonstrate that *Freyeria* is not closely related to *Chilades* and represents a distinct clade that cannot possibly be subsumed within *Chilades* ...'. However, de Freina (2014) elected to reverse this change: '*Freyeria* syn. rev. is again reconfirmed as a junior subjective synonym of *Chilades*.' Here, we follow Talavera *et al.* (2012) and Wiemers *et al.* (2018, 2020).

Recorded as '*Lycaena Trochylus* Freyer' by Lederer (1855), this was listed by Bucknill (1911) as '*Chilades trochilus. Freyer. The Tiny Blue*', as 'Chilades trochylus Frr.' by Turner (1920) and 'Chilades trochilus Frr.' by Rebel (1939).

Distribution in Cyprus

Altitudinal: sea level (where the species can be locally numerous) to biotopes high in the Troodos massif.
Geographical: a fairly common butterfly in Cyprus, and widely distributed across all parts of the island.

Phenology

Polyvoltine, March to November. In favourable conditions, first specimens may appear in February, the earliest known on 10 February 2013 (Yiannis Christofides, Cyprus, 2013, personal communication to first author). Overwinters as a pupa.

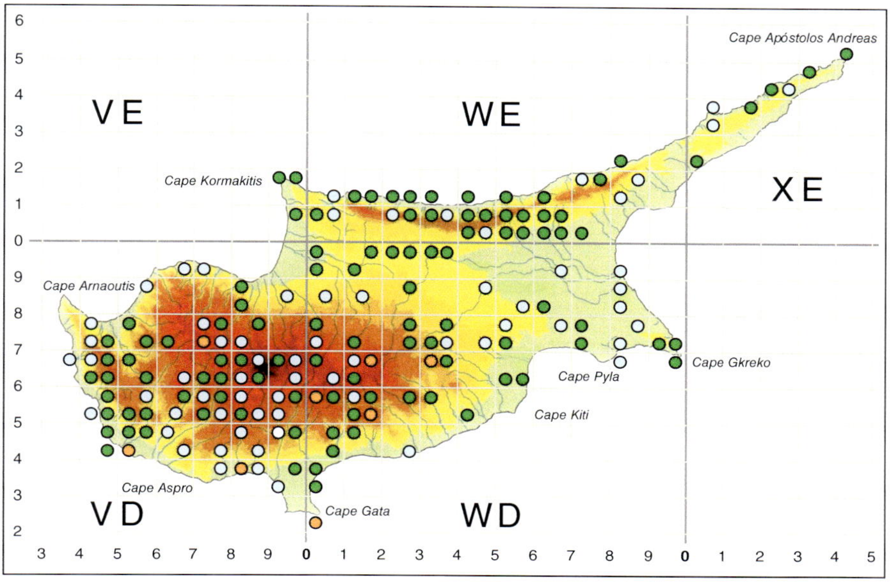

Fig. 12.241. *Freyeria trochylus* distribution map showing 5 km² observations within 10 km² UTM squares. © *Eddie John*

Larval hostplants

In Cyprus, *F. trochylus* is known to use *Andrachne telephioides* (Figs 12.242 and 12.243) of the spurge family (Euphorbiaceae), but other plants might act as hosts. Tolman and Lewington (1997: 101) stated, 'Captive larvae feed readily on flowers and leaves of *Heliotropium hirsutissimum* [Boraginaceae]', but there is no evidence of this (a very common plant in Cyprus) being used naturally on the island. *H. hirsutissimum* and *H. kotschyi* are listed by Tshikolovets (2011: 189) and *H. europaeum* (also present in Cyprus) and *H. digynum* are used elsewhere. Similarly, Parker (1983) referred to 'heliotrope' as the larval foodplant, which is also listed as a hostplant in Qatar (Pittaway, 1980) and Saudi Arabia (Pittaway, 1985).

Ecology and behaviour

A sedentary species, found in a variety of open, sometimes very exposed locations at coastal habitats and in many other types of habitat, usually in close association with the hostplant. *F. trochylus* appears to be very adaptable to a variety of habitats – from coast to high in the Troodos Mountains, where it has been seen as high as 1499 m on 13 June 2001 and 1900 m on 7 August 2005 (authors' personal observations). Colony sizes in favourable years may number several tens; at a hot, dry stony riverbed in Soloi in June 2012, for example, a population numbering *ca.* 50 was seen, but the species is usually encountered in smaller numbers. *F. trochylus* is possibly under-recorded in Cyprus, due to its drab colours and tiny size;

Fig. 12.242. *Andrachne telephioides*, 19 May 2007. *Eddie John*

Fig. 12.243. *Andrachne telephioides*, 14 May 2012. Courtesy of *Yiannis Christofides*

indeed, Larsen and Nakamura (1983) noted the ease with which localized colonies may be overlooked. Nectaring has been frequently observed on *Thymbra capitata* and *Helichrysum stoechas* ssp. *barrelieri* (authors' personal observations) and on *Heliotropium hirsutissimum* (Makris, 2003: 188). Parker (1983) referred to attendance by ants of the genus *Pheidole*, a species of which (the invasive *P. indica*), has recently been confirmed for Cyprus (Salata *et al.*, 2019).

Description/identification aids

Wingspan: up to 20 mm. A tiny butterfly – the smallest butterfly species in Europe. Hindwing **upper**- and **undersides** of both sexes have a similar ground colour of brown, with a *characteristic submarginal band of orange*, within which are *three or four large black ocelli highlighted with blue scaling*. Note that *C. trochylus* is the *only* member of the very small blues to fly at higher altitudes in Cyprus.

Similar species

None in Cyprus.

Conservation status in Cyprus

Not threatened. Listed by Van Swaay *et al.* (2010a) and Numa *et al.* (2016) as 'LC'.

Aricia agestis ([Denis & Schiffermüller], 1775) Lycaenidae (Polyommatinae)

Brown Argus

TL: Vienna, Austria.

Fig. 12.244. *Aricia agestis* (Brown Argus) nectaring on *Heliotropium dolosum*, 26 August 2018. *Christodoulos Makris*

Fig. 12.245. *Aricia agestis* (Brown Argus), 20 June 2019. *Christodoulos Makris*

Fig. 12.246. *Aricia agestis* (Brown Argus) nectaring on *Cistus creticus*, 21 May 2016. *Christodoulos Makris*

Fig. 12.247. *Aricia agestis* (Brown Argus) nectaring on *Helichrysum italicum*, 3 August 2015. *Christodoulos Makris*

Fig. 12.248. *Aricia agestis* (Brown Argus) pairing, 9 May 2015. *Christodoulos Makris*

Presence in the eastern Mediterranean

Widely distributed throughout the Levant, probably as the nominotypical subspecies. With reference to material from Lebanon, Larsen and Nakamura (1983) discussed the historical use of subspecies *calida* (used as a form or 'ab.' name in Cyprus, by Turner, 1920), but stated the need for a full revision of the species.

The species has been known by many names over the years, including '*Lycaena Alexis* Hufn. (*agestis* S. V.)' by Lederer (1855), '*Cupido astrarche. Bergstr. The Brown Argus.*', (Bucknill, 1911), '*Aricia medon* Hufn (*astrarche* Brgstr.)' (Turner, 1920) and 'Lycaena astrarche Brgstr.', by Rebel (1939). The genus is referred to as *Plebejus* (*Plebeius*) by some authors, whereas others regard *Aricia* as a subgenus of *Polyommatus*. Here, we follow Talavera *et al.* (2012) and Wiemers *et al.* (2018, 2020).

Distribution in Cyprus

Altitudinal: from sea level to the highest elevations. It is at these higher biotopes that *A. agestis* appears to have a greater presence in Cyprus.
Geographical: a relatively uncommon butterfly, especially in coastal locations, with no records east of the Lefkonoiko area and five only from the entire northern region (CBRS data), from which it is probably under-recorded. Where present, it is rarely numerous. Although the distribution map in Parker (1983) shows a presence in far fewer (10 km^2) UTM squares, his text suggested the species to have been encountered more frequently than in recent years.

Phenology

Probably trivoltine, April to October, with early specimens possible in late March. Overwinters as a pupa.

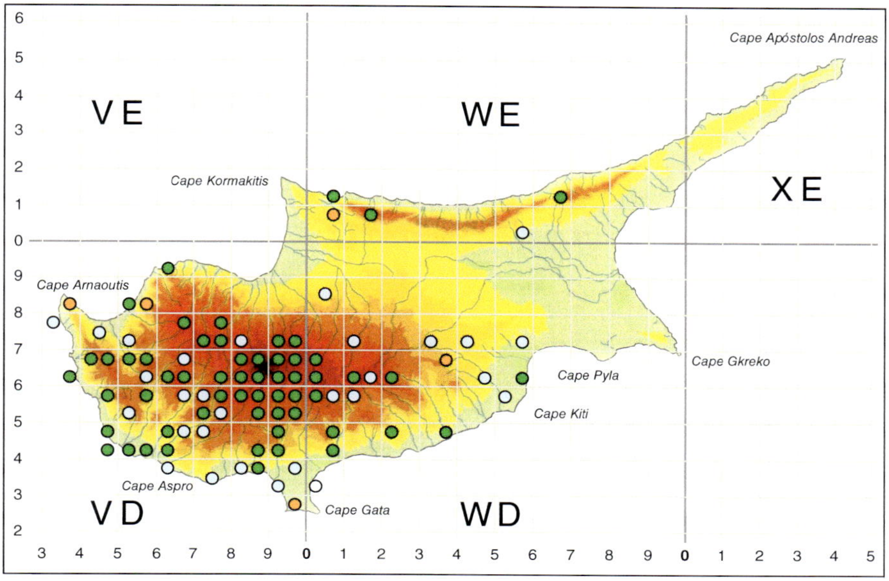

Fig. 12.249. *Aricia agestis* distribution map showing 5 km² observations within 10 km² UTM squares. © *Eddie John*

Larval hostplants

Little is known of the hostplants used in Cyprus, but elsewhere the species uses Geraniaceae: *Geranium* spp. (Fig. 12.250), *Erodium* spp. and Cistaceae: *Helianthemum* spp.

Ecology and behaviour

Found in a variety of habitats: maquis, garigue and rocky areas with high levels of vegetation, occasionally in damp areas near water. The species has been observed nectaring on a range of flowers, including *Cistus creticus*, *Helichrysum italicum* and *Rubus sanctus*. Larvae are attended by ants.

Description/identification aids

Wingspan: up to 30 mm. ♂ and ♀ **uppersides** have a similar brown ground colour, with well-developed submarginal orange lunules, ending in a black spot near the margin. **Undersides** of both sexes are light brown to grey-brown, with strong lunules as on the uppersides, and well-defined fore- and hindwing black postdiscal (forewing only) and discal spots encircled with white – see 'Similar species'.

Similar species

A. agestis is probably under-recorded in Cyprus, due to its similarity with ♀ *Polyommatus icarus* and with which it shares similar habitat. On the hindwing underside, an arc of postdiscal spots is common to both species, but in *A. agestis* the space between the two adjacent discal spots near the costa is usually closer and nearly touching (*cf.* encircled areas in Figs 12.251 and 12.252), with the lower spot *usually displaced slightly inwards* from the costa, *outside* the arc of spots and forming the so-called 'colon'. *A. agestis* lacks the forewing underside basal cell spots, usually typical of *P. icarus* (but see entry under that species). A further characteristic of *A. agestis*, but *not* of *P. icarus*, is that the black wing vein markings on both fore- and hindwings may occasionally create a *slightly chequered* effect as they cross into the cilia, but not strikingly so, as in *Pseudophilotes vicrama*.

Conservation status in Cyprus

Not considered threatened. Listed by Van Swaay *et al.* (2010a) and Numa *et al.* (2016) as 'LC'.

Fig. 12.250. *Geranium tuberosum*, 11 March 2011. Courtesy of *Yiannis Christofides*

Fig. 12.251. *Aricia agestis* (Brown Argus), with the 'colon' circled in white, 28 June 2010. *Christodoulos Makris*

Fig. 12.252. *Polyommatus icarus* (Common Blue) ♀, 15 October 2016. Note that the spot on the right in the circle is not displaced, and the arrowed basal spot is absent in *A. agestis*. *Christodoulos Makris*

Polyommatus icarus (Rottemburg, 1775) Lycaenidae (Polyommatinae)

Common Blue

TL: Saxony, Germany.

Fig. 12.253. *Polyommatus icarus* (Common Blue) ♂, 12 November 2017. *Christodoulos Makris*

Fig. 12.254. *Polyommatus icarus* (Common Blue) first brood ♀, 11 April 2013. Courtesy of *David and Rosalyn Sparrow*

Fig. 12.255. *Polyommatus icarus* (Common Blue) ♀ nectaring on *Picris rhagadioloides*, 13 May 2017. *Christodoulos Makris*

Fig. 12.256. *Polyommatus icarus* (Common Blue) ♂ nectaring on *Dittrichia viscosa*, 12 November 2017. *Christodoulos Makris*

Fig. 12.257. *Polyommatus icarus* (Common Blue) ♀ nectaring on *Polygonum equisetiforme*, 30 September 2006. *Christodoulos Makris*

Fig. 12.258. *Polyommatus icarus* (Common Blue) pairing, ♀ left, ♂ right, 3 July 2011. *Christodoulos Makris*

Presence in the eastern Mediterranean

This species is commonly found throughout the region, probably as nominotypical *P. icarus icarus* in northern parts of the Levant. Larsen (1974: 179, 180) listed *P. icarus zelleri* Verity, 1919 and *P. icarus juno* Hemming, 1933 for Lebanon, with each taxon generally found to be occupying different elevational ecological niches, a separation strongly defended in Larsen (1980). However, Larsen and Nakamura (1983) stated, 'The name *zelleri* is conventionally applied to the Mediterranean populations', with no mention made of *juno*. *P. i. zelleri* was later synonymized with nominotypical *P. i. icarus* by Hesselbarth *et al.* (1995: 676), while *P. i. juno* is viewed as a valid ssp. of *P. icarus* by, for example, Wiemers *et al.* (2010) and Tshikolovets (2011: 254). Zorkot (2016: 288) treated *P. juno* Hemming, 1933 as a valid endemic species 'confined to Lebanon and Palestine'.

P. icarus was initially listed for Cyprus as '*Lycaena Icarus* (*alexis* S. V.)' by Lederer (1855), then as '*Lycaena icarus. Rott. The Common Blue*' by Bucknill (1911), 'Polyommatus icarus Rott. [ab. minor Ckrll.]' by Turner (1920) and later as 'Lycaena icarus Rott.' by Rebel (1939).

Distribution in Cyprus

Altitudinal: recorded at all elevations.
Geographical: a very common butterfly, found throughout the island.

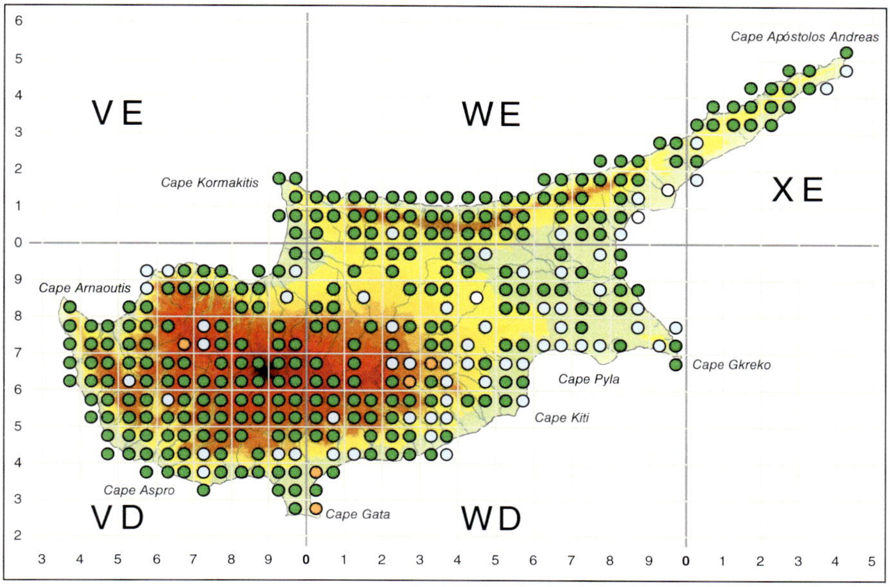

Fig. 12.259. *Polyommatus icarus* distribution map showing 5 km² observations within 10 km² UTM squares. © *Eddie John*

Phenology

Polyvoltine, mainly March to November, but with occasional sightings in all months except January (CBRS data). Overwinters as a pupa.

Larval hostplants

Fabaceae: *P. icarus* is highly polyphagous – in Cyprus, using *Lotus* spp., *Medicago sativa* (lucerne/alfalfa), *Ononis spinosa* (spiny rest-harrow) (Fig 12.107), *Trifolium* spp. (Fig. 12.260), *Alhagi maurorum* (Fig. 12.261), probably *Genista* spp. and many other legumes. Repeated ovipositing on *Capparis spinosa* (Capparaceae) was noted on 22 May 2013 by the first author (unpublished data).

Ecology and behaviour

Found in diverse habitats, but likely to be most numerous in coastal areas, waste ground and field margins, where adults nectar on a wide variety of flowers. On 19 May 2013, *ca.* 50 *P. icarus* (mainly ♂♂) were observed nectaring on *Centaurea hyalolepis*. The species is especially common in agricultural areas where *M. sativa* (lucerne/alfalfa) is cultivated as a forage crop. Larvae are attended by ants of various species (Makris, 2003: 198; Fiedler, 2021: Appendix S1).

Description/identification aids

Wingspan: up to 35 mm. ♂ **uppersides** are a uniform bright blue with a narrow black marginal line and white cilia, otherwise unmarked. ♀♀ of the first brood may have some (or much) blue suffusion on the uppersides but generally the ground colour is brown with well-developed submarginal orange lunules, ending in a black spot near the margin. ♂ **underside** ground colour is grey with blue basal scaling, the ♀ grey-brown. Both sexes have well defined fore- and hindwing submarginal orange lunules, bordered

Fig. 12.260. *Trifolium fragiferum*, 12 June 2012. Courtesy of *Yiannis Christofides*

Fig. 12.261. *Alhagi maurorum*, 27 July 2021. *Christodoulos Makris*

with black, and black postdiscal and discal spots encircled with white, larger in the ♀. The basal cell spots on the forewing underside, usually characteristic of *P. icarus*, are *not always present* in southern European specimens (the late Torben Larsen, 2015, personal communication to first author; John, 2000).

Similar species

A. agestis (♀ only) – see previous species. See also *P. vicrama.*

Conservation status in Cyprus

Not threatened. Listed by Van Swaay *et al.* (2010a) and Numa *et al.* (2016) as 'LC'.

Family: NYMPHALIDAE

Libythea celtis (Laicharting, 1782)	Nettle-tree Butterfly	Libytheinae
Danaus chrysippus (Linnaeus, 1758)	Plain Tiger/African Queen	Danainae
Charaxes jasius (Linnaeus, 1767)	Two-tailed Pasha	Charaxinae
Limenitis reducta Staudinger, 1901	Southern White Admiral	Limenitidinae
Hypolimnas misippus (Linnaeus, 1764)	False Plain Tiger	Nymphalinae
Nymphalis polychloros (Linnaeus, 1758)	Large Tortoiseshell	
Vanessa atalanta (Linnaeus, 1758)	Red Admiral	
Vanessa cardui (Linnaeus, 1758)	Painted Lady	
Argynnis pandora ([D & S], 1775)	Cardinal	Heliconiinae
Issoria lathonia (Linnaeus, 1758)	Queen of Spain Fritillary	
Hipparchia syriaca cypriaca (Staudinger, 1879)	Eastern Rock Grayling	Satyrinae
Hipparchia cypriensis (Holik, 1949)	Cyprus Grayling	
Chazara briseis larnacana (Oberthür, 1909)	The Hermit	
Chazara persephone (Hübner, [1805])	Great Steppe Grayling	
Pseudochazara anthelea (Hübner, [1824])	White-banded Grayling	
Maniola cypricola (Graves, 1928)	Cyprus Meadow Brown	
Hyponephele lupina cypriaca Riley, 1921	Oriental Meadow Brown	
Pararge aegeria (Linnaeus, 1758)	Speckled Wood	
Lasiommata megera (Linnaeus, 1767)	Wall Brown	
Lasiommata maera (Linnaeus, 1758)	Large Wall Brown	
Kirinia roxelana (Cramer, 1777)	Lattice Brown	
Ypthima asterope (Klug, 1832)	African Ringlet	

Libythea celtis (Laicharting, 1782) Nymphalidae (Libytheinae)

Nettle-tree Butterfly

TL: South Tyrol, Italy.

Fig. 12.262. Reared *Libythea celtis* (Nettle-tree Butterfly). *Eddie John*

Fig. 12.263. *Libythea celtis* (Nettle-tree Butterfly), 22 June 2019. *Christodoulos Makris*

Fig. 12.264. *Libythea celtis* (Nettle-tree Butterfly) nectaring on *Cistus creticus*, 6 May 2018. Courtesy of *Nigel Peace*

Fig. 12.265. *Libythea celtis* (Nettle-tree Butterfly) on *Sorbus graeca*, 6 June 2020. *Christodoulos Makris*

Fig. 12.266. *Libythea celtis* (Nettle-tree Butterfly) nectaring on *Rubus sanctus*, 27 July 2019. *Christodoulos Makris*

Fig. 12.267. *Libythea celtis* (Nettle-tree Butterfly) mud-puddling, 1 June 2013. *Eddie John*

Fig. 12.268. *Libythea celtis* (Nettle-tree Butterfly) well camouflaged among litter into which it had retreated upon the onset of rain in the Troodos Mountains, 22 June 2019. *Christodoulos Makris*

Presence in the eastern Mediterranean

L. celtis is the only member of the subfamily Libytheinae ('snout' butterflies) known in Europe, and within the Levant is found only in Turkey and Cyprus, where it is present as nominotypical *L. celtis celtis*. Cyprus represents the southernmost extent of distribution in the eastern Mediterranean.

Surprisingly, this species was encountered by Zach in 1853, on behalf of Lederer (1855) ('surprisingly', because several common species, such as *C. croceus*, were apparently missed). It was listed by Lederer (1855) as '*Libythea Celtis* Fab.', and later as '*Libythea celtis Fuessl. The Snout*', by Bucknill (1911). Turner's (1920) brief entry for 'Libythea celtis Laich.' noted that Bucknill had found it 'though not abundantly, on the southern mountains', and included Wilson's comment, 'Confined to the Troödos range of mountains. Rare. I have only taken it four times in ten years, during July and August.' *L. celtis* appears never to have been numerous in Cyprus. Parker (1983) noted the absence of any records later than those of Wilson, and failed to find the species during his stay in Cyprus. Although known for Cyprus since its listing by Lederer (1855), such is the rarity of *L. celtis* in Cyprus that no records of the species' presence can be traced from the early 1900s to June 1995, when the species was reported from the region of Trooditissa Monastery (Fuchs, 1996). While listed by Georghiou (1977: 206), no other details were given, suggesting reference to the earlier aforementioned observations.

Distribution in Cyprus

Altitudinal: almost all confirmed sightings in Cyprus have been from 1000 m to 1500 m, probably reflecting the main areas of distribution of its hostplant.

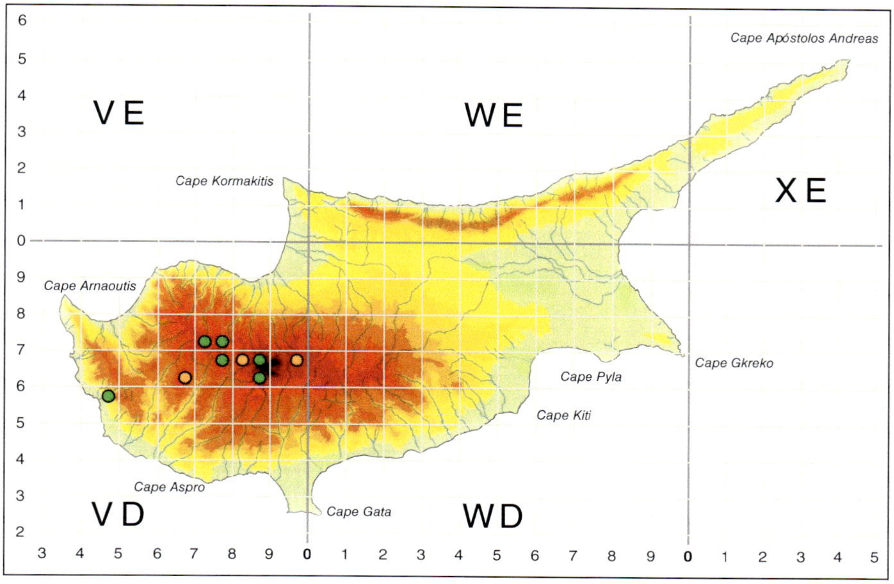

Fig. 12.269. *Libythea celtis* distribution map showing 5 km² observations within 10 km² UTM squares. © *Eddie John*

Geographical: a rare butterfly, known from only nine 5 km² UTM squares in Cyprus, and mainly confined to the Troodos mountains. In 2013, the finding by the first author of a small stand of *Celtis australis* (at *ca.* 600 m) in the far west of the island helped to explain the presence there of a single *L. celtis* in April 2011, reported by CBSG member Eddie Harding. *C. tournefortii* is present on the Pentadaktylos mountain range, but *L. celtis* has yet to be recorded there.

Phenology

Univoltine, April to September, rarely later. Post-hibernation adults appear in April, the new brood taking to the wing in June. Adults generally enter early hibernation, but have been seen on the wing as late as 26 October at 884 m in the Troodos Mountains (Matt Smith, Cyprus, 2020, personal communication to first author). Kawahara (2006) stated that more than a single generation might be possible in warmer parts of the species' distribution, but there is no evidence of this in Cyprus. Wiltshire (1957: 30) reported *L. celtis* as bivoltine in Iraq, adding 'flying in April and August' (see 'Ecology and behaviour', below), whereas Tshikolovets *et al.* (2014: 237), for Iran and Iraq, stated 'Univoltine (records for second generation need clarification)'. Without providing further details, Döring and Hoffmann (2004) referred to 'up to three generations annually, one in the winter and one to two in the summer' for the Mediterranean region; a partial second generation was stated by Polak *et al.* (2009) for Slovenia.

Larval hostplants

Cannabaceae: *Celtis tournefortii* (Oriental hackberry) was confirmed in 1996 as the hostplant in Cyprus, and on which the second author has found larvae and pupae (John and Makris, 2001; Makris, 2003: 204). However, a further population, with the potential for being the largest colony of *L. celtis* in Cyprus, has been found at 1100 m in the Troodos Mountains (unpublished data, first author), using as hostplant the introduced *Celtis australis* (nettle tree or hackberry) (Fig. 12.270), a tree native to the Mediterranean basin and Western Asia (Magni and Caudullo, 2016). *Celtis tournefortii* is an indigenous rare plant (Red Data Book category 'Endangered' (EN)), with small stands at Panagia, Agios Nikolaos-Omodos, Madari, Lefkara, Kionia peak and Kornos peak on the Pentadaktylos range.

L. celtis is not monophagous on *C. australis*, as suggested by Kudrna *et al.* (2011: 277), or seemingly, on *Celtis* spp. exclusively, as Toper Kaygin *et al.* (2006) reported the use of *Alnus glutinosa* (Betulaceae) (common alder/European alder) in Turkey, despite the presence of *C. australis* just 50 m away.

Ecology and behaviour

Breeding in Cyprus was unconfirmed until the second author's observation of ovipositing on *Celtis tournefortii* at 1500 m in the Troodos Mountains in May 1996, with the later discovery there in June 1997 of larvae and pupae (John and Makris, 2001). The presence of breeding populations above 1000 m is at variance with the experience in Catalonia, where *L. celtis*

Fig. 12.270. *Celtis australis* (nettle tree), Troodos Mountains, 20 May 2013. *Eddie John*

Fig. 12.271. *Libythea celtis* larva on *Celtis australis*, 20 May 2007. *Eddie John*

was reported by Stefanescu (2006) to occupy lowland and mid-elevation mountain areas, with breeding observed at 700 m. In Valencia, Montagud-Alario and García-Alamá (2010: 88) reported similar activity, noting also the species' propensity to disperse to higher elevations in summer, a characteristic emphasized by Rabl and Rabl (2015). Whereas in Greece, *L. celtis* was reported from sea level to 2300 m (Pamperis, 2009: 145), the species has never been encountered in Cyprus at elevations below 600 m. It is not known if the singleton reported from VD45 NE (as discussed under 'Distribution in Cyprus') belonged to a small colony, or whether individuals from that area disperse to higher elevations in summer.

Considered by some as a migratory species ('dispersal' might be a more appropriate term – see also Schaider, 1976; Eitschberger *et al.*, 1991), the Cyprus populations do not appear even to qualify as significant vagrants, remaining at higher elevations and dispersing within a small area – mainly, it would appear, in search of water or moist areas. In discussing migration, Feltwell (2008) reported on a mass emergence of *L. celtis* in southern France following the extensive defoliation of local *C. australis*, and raised the question of the benefits of migration to areas with less competition for food resources. However, because of small population size of *L. celtis* in Cyprus, there is no evidence of defoliation of host trees and certainly no pressure on the species to disperse to other areas in search of suitable host trees.

L. celtis is generally encountered in light woodland in the Troodos Mountains, usually close to water and occasionally in areas with an anthropogenic connection and the associated planting of *Celtis*. The species has been observed nectaring on *Mentha longifolia*, *Rubus sanctus* and *Cistus creticus*; it is also attracted to tree sap and is an avid puddler, taking nutrients from mud, rainwater pools, and springs crossing mountain tracks (John and Makris, 2001; authors' personal observations). In one such observation, a single *L. celtis* was seen puddling in company with five *Hipparchia cypriensis* and *Limenitis reducta*. In Catalonia, Stefanescu (2006) reported puddling by worn, post-hibernation female *L. celtis*, perhaps as a means of enhancing fecundity, as proposed for female *Chazara briseis* and *Maniola cypricola* in Cyprus (John and Dennis, 2019).

Mating and egg-laying take place post-hibernation, the eggs being deposited in April and May on buds that are close to opening; males and females may be found around host trees. Kawahara (2006) provided further information on the biology of this species. In Cyprus, most observations have been of newly emerged individuals in May and June, after which period *L. celtis* might enter a light summer diapause, resuming activity in late summer/early autumn prior to hibernation (see 'Phenology' above). As with other hibernating nymphalids, *L. celtis* is very well camouflaged when at rest (see Fig 12.268).

Hesselbarth *et al.* (1995: 762) reported on attack at the pre-pupal stage of *L. celtis* by the gregarious endoparasitoid wasp *Pteromalus puparum* (Hymenoptera: Chalcidoidea, Pteromalidae), but it is not known if this wasp affects populations in Cyprus.

Description/identification aids

Wingspan: up to 40 mm. A highly distinctive butterfly, belonging to the 'snout' or 'beak' family, so named because of the *long labial palpi*, which gives the head its characteristic shape. Forewings of both sexes are *markedly dentate*, creating a profile unlike that of any other species in Cyprus. ♂ and ♀ **uppersides** have a brown ground colour, with orange markings, extensive on the forewing, reduced to the postdiscal area on the hindwing. In both sexes, there are two small white marks near the forewing costa. The forewing **underside** largely mirrors the upperside, while the hindwing underside is plain grey-brown.

Similar species

None in Cyprus.

Conservation status in Cyprus

Listed as 'LC' in Europe and at the Mediterranean level (Van Swaay *et al.*, 2010a; Numa *et al.*, 2016), but considered vulnerable in Cyprus, where it is in need of protection and should not be collected in *any* of its stages.

Danaus chrysippus (Linnaeus, 1758) Nymphalidae (Danainae)

Plain Tiger/African Monarch/African Queen

TL: Canton, China.

Fig. 12.272. *Danaus chrysippus* (Plain Tiger) ♂, 28 November 2016. *Eddie John*

Fig. 12.273. *Danaus chrysippus* (Plain Tiger) ♀, 16 November 2014. *Christodoulos Makris*

Fig. 12.274. *Danaus chrysippus* (Plain Tiger) ♂, 10 May 2018. Courtesy of *Ray Atkinson*

Fig. 12.275. *Danaus chrysippus* (Plain Tiger) ♂, 30 November 2014. *Christodoulos Makris*

Fig. 12.276. Mating pair of *Danaus chrysippus* (Plain Tiger), ♂ above, 16 November 2014. *Christodoulos Makris*

Fig. 12.277. *Danaus chrysippus* (Plain Tiger) f. 'alcippus', 7 December 2018. Courtesy of *Matt Smith*

Fig. 12.278. Late-season ovipositing by *Danaus chrysippus* (Plain Tiger), 28 November 2016. *Eddie John*

Presence in the eastern Mediterranean

A regular migrant to Cyprus, nominotypical *D. chrysippus chrysippus* is known from all countries of the Levant. Adults are polymorphic and, very rarely, f. 'dorippus', f. 'alcippoides' and f. 'alcippus' (a form having a white hindwing) may appear (Larsen, 1986; John and Vane-Wright, 2019). The taxonomic complexities of *D. chrysippus* are beyond the scope of this book and the reader is referred to, for example, Smith (2014: 169, 182), Braby *et al.* (2015), John and Vane-Wright (2019), Vane-Wright and John (2019) and Vane-Wright (2020).

D. chrysippus was listed by Lederer (1855) as '*Danais Chrysippus* L.' along with the (rare) accompanying note: 'From middle May through the entire summer' (translated from the original German). Bucknill (1911), who referred to the species as '*Danais chrysippus L. The Black-veined Brown*', failed to see the species but added 'It ... is said to have occurred abundantly during summer near Larnaca in 1853', a remark evidently referring to Zach's observations in that year. Turner (1920) recorded this as 'Danaida chrysippus L.', noting Wilson's puzzlement in 1918 over the sporadic nature of the species' appearance: 'The records of this species are curious. It was recorded in 1853 and in 1888, after which date it appears to have vanished from the Island until 1911, when a few made their appearance at Kyrenia near the sea.' (However, John & Vane-Wright (2019) noted that the BMNH has specimens of *D. chrysippus* collected in Cyprus by Dorothea Bate in 1901.) Rebel (1939), using the species' name 'Danais chrysippus L.', noted 'The Museum received three very large ♂ (70 mm Exp.) [one having] a deep nut brown colour ... with the label "Limassol 26 December '26 (Mavr.)" [i.e. 1926, and referring to captures by George Mavromoustakis]. 'They are of the form *aegypiius* [*sic*] Schreb.' (translated from the original German; see Vane-Wright and John, 2019).

Distribution in Cyprus

Altitudinal: mostly seen at coastal locations, rarely higher in Cyprus.
Geographical: records are known from many coastal locations across the island, most within 5–10 km of the coast, with far fewer inland observations.

Phenology

Polyvoltine within the species' normal range. Migrants may be seen from March to November, occasionally well into December, but it is likely that winter conditions in Cyprus prevent *D. chrysippus* establishing permanent colonies. Local breeding has frequently been observed, but it is anticipated that areas in which the hostplant is found are re-colonized annually (John *et al.*, 2019a). Further work is required to determine the likelihood of overwintering success.

Larval hostplants

Larvae are oligophagous, accepting as many as 55 species of hostplant, but with the majority belonging to the Apocynaceae (Smith, 2021); in Cyprus:

Cynanchum acutum (Figs 12.280 and 12.281) and probably *Gomphocarpus fruticosus* (Makris, 2003: 208; John *et al.*, 2015b, 2019a). An earlier suggestion of ovipositing on a possible species of *Convolvulus* in Turkey was later corrected to *C. acutum* (Schurian *et al.*, 1993; Schurian, 1995; Klaus Schurian, Kelkheim, 2017, personal communication to first author). Bucknill's (1911) reference to the use of fennel appears erroneous.

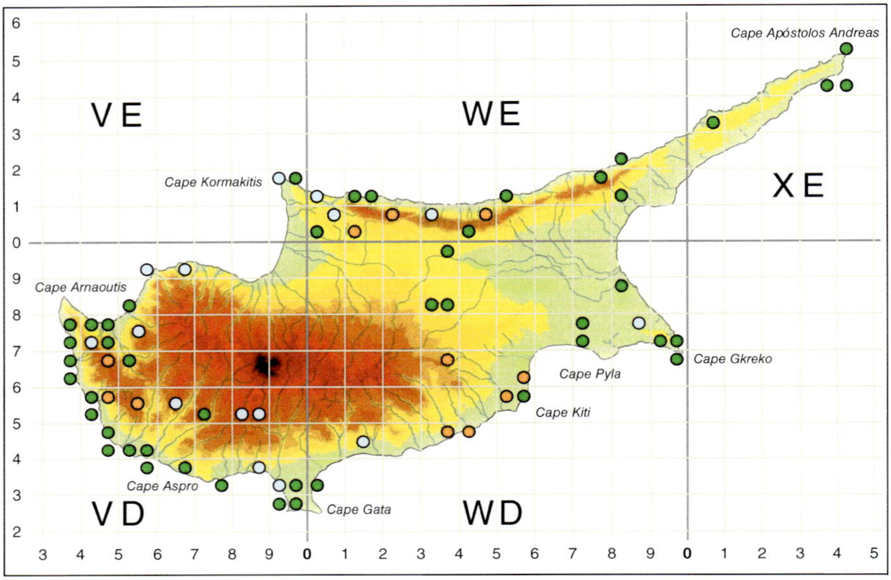

Fig. 12.279. *Danaus chrysippus* distribution map showing 5 km² observations within 10 km² UTM squares. © *Eddie John*

Fig. 12.280. *Cynanchum acutum* twining up *Phragmites australis*, 11 May 2016. *Eddie John*

Fig. 12.281. *Cynanchum acutum* in flower, 5 April 2015. *Christodoulos Makris*

Ecology and behaviour

A migrant to Cyprus that heads northwards during spring into Turkey, and probably has its origins in the Nile Delta or tropical Arabia (the late Torben Larsen, personal communication to first author), or the Jordan Valley, where it is a permanent resident (Larsen, 1976; John *et al.*, 2019a). A return migration of progeny from a later brood occurs in autumn, usually in larger numbers.

In the spring/early summer of 2014, first specimens were first reported at a coastal site on 19 and 24 March, with further sporadic observations in most weeks until 6 June. A further sighting on 19 September in the same area was followed by the observation of two more individuals on 7 November, leading to speculation of transient local breeding. This was confirmed on 9 November when a group of fewer than ten adults, along with small numbers of eggs and larvae, were found at an extensive patch of the hostplant, *Cynanchum acutum*. Ovipositing was observed at this site as late as 5 December (2014), but adults were no longer present by 15 December (Mike Hardman, Cyprus, 2014, personal communication to first author). At the same location on 28 November 2016, ovipositing (and mating) was again observed late in the season, by which time natural dieback meant that hostplant leaves of (barely) acceptable quality to the female were present only at ground level (Fig. 12.278). On 30 August 2017, in the fourth of a five-year study at the locality by members of the CBSG, *ca.* 40 individuals were seen by Matt Smith (John *et al.*, 2019a). This was the largest known aggregation in Cyprus since 1998, when the second author encountered similar numbers at sites on the Akrotiri Peninsula, when breeding was also confirmed for the first time in Cyprus (Makris, 2003: 208). Matt Smith (Cyprus, 2021, personal communication to first author) reported very small numbers present at the study site in late October 2021, but noted also that, as has happened in most years, attempts at eradication of *C. acutum* had left the biotope largely devoid of hostplant.

D. chrysippus migrant numbers arriving in Cyprus are usually small. Indeed, unlike the related *D. plexippus* (Monarch) in North America, the species is rarely known to migrate in large groups. *D. chrysippus* is most likely to be seen at the coast or in nearby fields and waste ground, nectaring on *Echium angustifolium* and *Limbarda* (*Inula*) *crithmoides* (Asteraceae) (Makris, 2003: 208), and is especially attracted to *Dittrichia viscosa* in the autumn, but also to animal droppings and even rotting carcases. *D. chrysippus* (and other members of the subfamily) are noted for sourcing pyrrolizidine alkaloids pharmacophagously from living, withered and dead parts of alkaloid-producing plants (e.g. Boppré, 1986; Smith, 2014: 399; Lawson *et al.*, 2021) that provide chemical defence and from which chemicals sex pheromones are also produced. Although not necessarily pertaining to *D. chrysippus*, other danaines have been observed apparently scratching at, and imbibing alkaloids from, wounded larvae of the family (Tea *et al.*, 2021). In common with some other species of butterfly, *D. chrysippus* displays thanatosis (feigning death) during the later stages of an interaction with a predator, an action that Larsen (1991) also associated with unpalatability in aposematic butterflies. In an extensive discourse on thanatosis, Humphreys and Ruxton (2018) preferentially described the anti-predator behaviour of thanatosis, as 'tonic immobility' (see also Shreeve *et al.*, 2006; Dennis, 2020: 302, 305 Fig. D41).

Uncharacteristically for a migrant species, the flight of almost all specimens of *D. chrysippus*, observed on numerous occasions by the authors since the 1998 migration, appeared to be graceful and leisurely, completely untypical of migratory behaviour. Southerly autumn migrations, with migrants in typically purposeful flight, *have* been noted on the Greek islands of Paros and Spetses, near Korinthos (Corinth) and also along the north-east coast of the Pelopónnese, and in Israel (John Coutsis, Athens, 2007, and Dubi Benyamini, Beit-Arye, 2014, personal communications to first author). However, to date, such activity has yet to be clearly documented in Cyprus.

Description/identification aids

Wingspan: up to 80 mm. ♂ and ♀ **upperside** ground colour is normally (see forms above) a *distinctive orange with black margins dotted with small white spots* on all wings. The forewing apex is *black with large white markings*. Hindwings are orange, with three black marks at the end of the cell, and (in the ♂) a broadly oval androconial patch below. **Undersides** are also similar in both sexes, with markings mirroring the uppersides except for the black wing margins, where the white spots are *more pronounced*, the forewing apex that is orange-brown, and the ♂ androconial patch that is white-centred.

Similar species

D. plexippus bears a strong resemblance, but this has not been recorded as a migrant to Cyprus (see under 'Adventive species' in Chapter 13). See also female *Hypolimnas misippus*.

Conservation status in Cyprus

Not assessed by Van Swaay *et al.* (2010a); listed as 'LC' by Numa *et al.* (2016). Temporary coastal colonies in Cyprus are threatened by flooding or attempts at eradication of the invasive hostplant by ploughing.

Charaxes jasius (Linnaeus, 1767) Nymphalidae (Charaxinae)

Two-tailed Pasha

TL: "Barbaria" (Algeria).

Fig. 12.282. *Charaxes jasius* (Two-tailed Pasha), 3 August 2010. Courtesy of *Michael Field*

Fig. 12.283. *Charaxes jasius* (Two-tailed Pasha), 9 August 2006. *Christodoulos Makris*

Fig. 12.284. *Charaxes jasius* (Two-tailed Pasha), 3 August 2010. Courtesy of *Michael Field*

Fig. 12.285. *Charaxes jasius* (Two-tailed Pasha), 21 August 2021. *Christodoulos Makris*

Presence in the eastern Mediterranean

Present in nearby Levant countries as nominotypical *C. jasius jasius*. The species is rare in the Turkish provinces immediately to the north and north-east of Cyprus (Hesselbarth *et al.*, 1995: Map 298; Atahan *et al.*, 2018), whereas in Mediterranean Syria, both *Arbutus andrachne* and *C. jasius* are locally common (Mudar Salimeh, Latakia, 2021, personal communication to first author). Larsen (1974: 111) concluded that the species was probably extinct in Lebanon, but in 2004, *C. jasius* was rediscovered in southern Lebanon (Bury, 2014), close to the border with Israel, in which country it is also known (Benyamini, 2002a: 112). In view of the wide-ranging habits of this species, it appears possible that individuals seen in southern Lebanon had strayed from Israel.

Overlooked by Zach (Lederer, 1855), but recorded by Bucknill (1911) using the scientific name with which we are familiar today, and the intriguing colloquial epithet '*The Attic Engraved*'. Turner (1920) noted that both Bucknill and Wilson had resorted to 'sugaring' in order to catch specimens.

Distribution in Cyprus

Altitudinal: sea level to above 1500 m.
Geographical: *C. jasius* is generally uncommon, but more numerous (sometimes significantly so) in localities where the hostplant is plentiful, especially in the Pentadaktylos Mountain range. Other than the Mesaoria Plain, the species may appear almost anywhere in Cyprus.

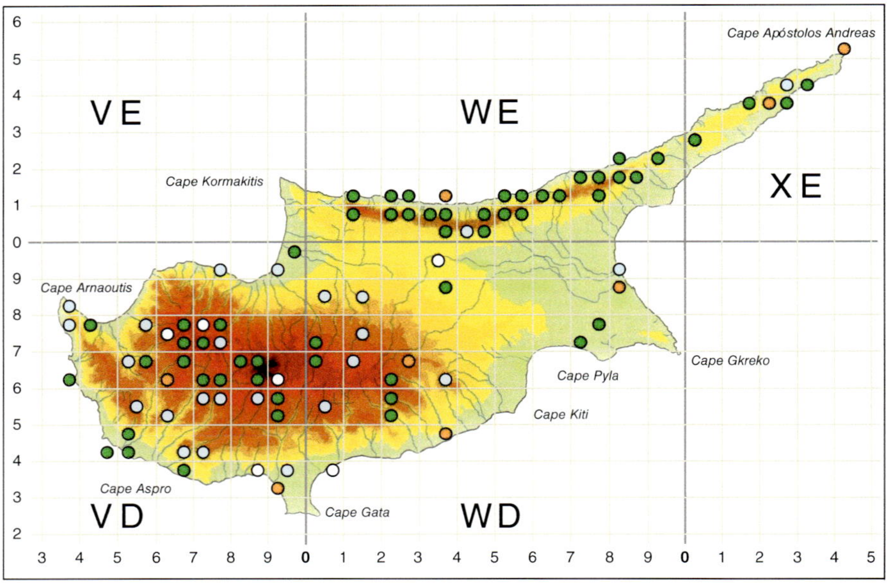

Fig. 12.286. *Charaxes jasius* distribution map showing 5 km² observations within 10 km² UTM squares. © *Eddie John*

Phenology

Bivoltine, May to November, occasionally earlier and later. Overwinters as a young larva.

Larval hostplants

Elsewhere over much of the Mediterranean, the hostplant is *Arbutus unedo* (strawberry tree) (Ericaceae), a rare plant in Cyprus, known from the Akamas (Makris, 2003: 216) and into which area it was possibly introduced (Christofides, 2017: 158). In Cyprus, the usual hostplant is the related *A. andrachne* (eastern strawberry tree) (Fig. 12.287) which grows abundantly in some areas, as in the Kantara region; *A. unedo* is categorized as critically endangered (CR) (for Cyprus) in the Red Data Book.

In low-lying areas, *C. jasius* has also been observed using Rutaceae: *Citrus* spp. (Makris 2003: 216; Mike Lloyd, Cyprus, 2014, personal communication to first author) and Solanaceae: *Nicotiana glauca*, although larvae have seldom reached pupal stage on the latter (Makris 2003: 216; Roger White, Cyprus, 2011, personal communication to authors).

Ecology and behaviour

C. jasius is easily recognizable, having a distinctive profile and fast, vigorous flight that enables it to range widely, and quickly cover distances from mountainous areas to citrus orchards in coastal regions. The ♂ is territorial and the species is a renowned hilltopper.

Although mainly attracted to mud and ripe or fermenting fruit, the species has been observed at *Ceratonia siliqua* (carob) flowers, possibly seeking fermenting nectar (the late Torben Larsen, personal communication to first author). Indeed, Bucknill (1911) had earlier noted its

Fig. 12.287. *Arbutus andrachne* (strawberry tree), 10 March 2016. *Christodoulos Makris*

Fig. 12.288. Early instar larva of *Charaxes jasius*, 21 November 2015. This was one of many found on *Citrus limon* (lemon) in a garden near the coast in northern Cyprus. Courtesy of *Mike Lloyd*

attraction to 'putrid carobe', an observation confirmed by Daniel Haines (London, 2010, personal communication to first author) who noted that after feeding on carob, the behaviour of the butterflies created an impression of intoxication, as they 'flop down from the tree, rather like falling leaves' where they become easy prey for lizards. A predilection for alcohol of a somewhat different nature (beer) has also been reported, while Larsen (1974: 111) informed that the males 'are very fond of rotting shrimp or crab'. *C. jasius* is also attracted to other noxious substances, for example carrion and animal droppings, but less malodorous attractants also have appeal; for example, on 1 June 2013, the first author's lunchtime cheese sandwich! Certainly, the species' chemosensory system is well developed.

Description/identification aids

Wingspan: up to 85 mm. A large butterfly with a vigorous flight and unmistakable profile. Sexes are very similarly marked: forewing **upperside** is a dark brown with *a wide orange submarginal/marginal band*; hindwing upperside is similarly marked, but with *silvery-blue post-discal spotting and two pairs of characteristic tails*. **Undersides** *mirror the orange band of the upperside*, next to which is a brown post-discal band with complex patterning and an irregular white discal band. The remaining discal, and the basal areas, have *a brown ground colour with a mosaic of white markings*.

Similar species

None in the region.

Conservation status in Cyprus

Not threatened. Van Swaay *et al.* (2010a) and Numa *et al.* (2016) listed *C. jasius* as 'LC'.

Limenitis reducta Staudinger, 1901 Nymphalidae (Limenitidinae)

Southern White Admiral

TL: Hankynda (Armenia).

Fig. 12.289. *Limenitis reducta* (Southern White Admiral) ♂, 14 July 2019. *Christodoulos Makris*

Fig. 12.290. *Limenitis reducta* (Southern White Admiral) nectaring on *Echinops spinosissimus*, 30 July 2011. *Christodoulos Makris*

Fig. 12.291. *Limenitis reducta* (Southern White Admiral) nectaring on *Echinops spinosissimus*, 30 July 2011. *Christodoulos Makris*

Fig. 12.292. *Limenitis reducta* (Southern White Admiral), 6 May 2008. *Christodoulos Makris*

Fig. 12.293. *Limenitis reducta* (Southern White Admiral), 26 June 2006. *Eddie John*

Presence in the eastern Mediterranean

Present in most countries of the Levant, where Olivier (1993: 201) and Hesselbarth *et al.* (1995: 979) recognized the ssp. *Limenitis reducta herculeana* Stichel, 1908 for northern parts of the region (Turkey and Cyprus). Elsewhere in the Levant, nominotypical *L. r. reducta* and ssp. *L. r. schiffermülleri* Higgins, 1933 have been reported (Larsen and Nakamura, 1983).

Lederer (1855) referred to the Cyprus species as '*Limenitis Camilla* S. V.', Turner (1920) and Rebel (1939) as 'Limenitis rivularis Scop. (*camilla* Schiff.)'. It is unknown whether *Limenitis anonyma* (Lewis, 1872), as discussed in Larsen and Nakamura (1983), and which name followed *rivularis*, has ever applied to Cyprus populations.

Distribution in Cyprus

Altitudinal: intermediate to higher elevations (1800 m).
Geographical: an uncommon, local butterfly, widely distributed throughout much of the Troodos Massif, although never numerous. In accord with Parker (1983), CBRS data suggests a much-reduced presence in the Pentadaktylos Mountains, with records from only three 5 km^2 UTM squares. A report of *L. reducta* in a Protaras hotel garden (near Cape Gkreko) in 2003 requires confirmation; the location of the sighting is a considerable distance from the species' normal range and habitat (altitudinal and geographical), explaining its exclusion from the distribution map.

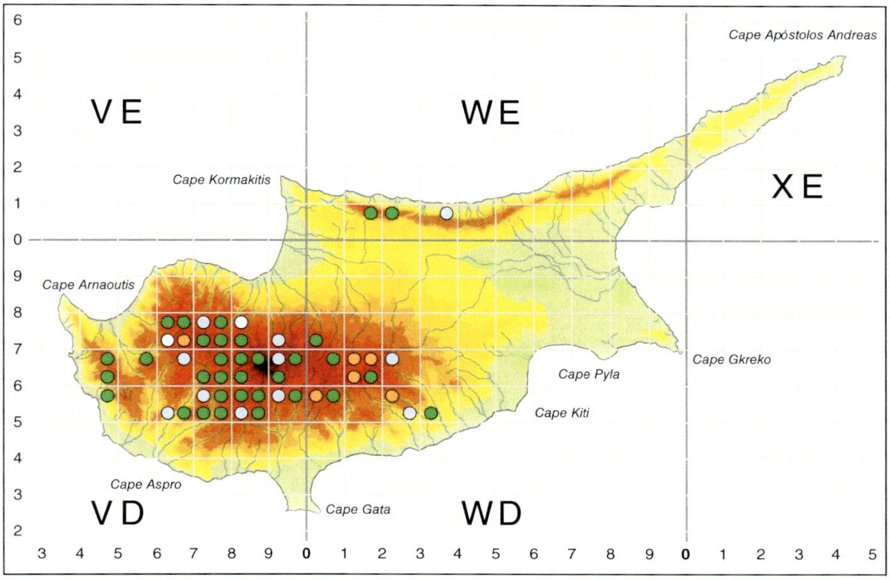

Fig. 12.294. *Limenitis reducta* distribution map showing 5 km^2 observations within 10 km^2 UTM squares. © *Eddie John*

Phenology

Thought to be trivoltine in Cyprus. April to October, rarely later. Overwinters as a young larva, protected within a hibernaculum.

Larval hostplants

Caprifoliaceae: *Lonicera etrusca* (Etruscan honeysuckle) (Figs 12.295 and 12.297).

Ecology and behaviour

Perhaps the most elegant in flight of all Cyprus's butterflies, *L. reducta* inhabits sunny forest glades and woodland or valley areas where the hostplant is found, and especially where the biotope supports both the hostplant and *Rubus sanctus*, a favoured source of nectar. *L. reducta* is a known puddler and is a frequent visitor to mud at springs and streams (Makris, 2003: 222). It is a territorial species, often returning repeatedly to a selected sunny perch. Although well distributed across the Troodos Massif and foothills from the west to central areas of Cyprus, colony sizes appear to be small, but this may be somewhat illusory if the species shares the same behaviour as that of the related *Limenitis camilla* (Linnaeus, 1764), in seeking honeydew in the tree canopy. Migratory behaviour, as documented by Bálint *et al.* (2016), has not been reported in Cyprus.

Fig. 12.295. *Lonicera etrusca*, 22 June 2012. Courtesy of *Yiannis Christofides*

Fig. 12.296. Young larva of *L. reducta*, prior to construction of hibernaculum. *Eddie John*

Fig. 12.297. *Lonicera etrusca* climbing on *Quercus alnifolia*, 25 June 2011. *Christodoulos Makris*

Description/identification aids

Wingspan: up to *ca.* 60 mm. The characteristic flight pattern of 'flap, flap, glide' (Parker, 1983) makes this species easily recognizable on the wing. **Uppersides** are similar in both sexes, *bluish-black, with a series of large white submarginal markings* forming a pattern across fore- and hindwings. Another white mark appears in the forewing cell. **Undersides** are similar in both sexes, with a *brown and ochre ground colour* and with *white markings* similar to the upperside; the basal area of the hindwing underside is *white with brown veining.*

Similar species

None in the Levant.

Conservation status in Cyprus

Not threatened. Listed by Van Swaay *et al.* (2010a) and Numa *et al.* (2016) as 'LC'.

Hypolimnas misippus (Linnaeus, 1764) Nymphalidae (Nymphalinae)

False Plain Tiger/ Diadem/ Danaid Eggfly

TL: [Indonesia, Java] (see Corbet, 1949).

Fig. 12.298. *Hypolimnas misippus* (False Plain Tiger) ♂ (photographed in The Gambia). Courtesy of *David Penney*

Fig. 12.299. *Hypolimnas misippus* (False Plain Tiger) ♀ (photographed in The Gambia). Courtesy of *David Penney*

Fig. 12.300. *Hypolimnas misippus* (False Plain Tiger) ♂ (photographed in The Gambia). Courtesy of *David Penney*

Fig. 12.301. *Hypolimnas misippus* (False Plain Tiger) ♀ (photographed in The Gambia). Note the absence of a black forewing apex. Courtesy of *David Penney*. (Photos reproduced with permission from *Field Guide to Wildlife of The Gambia 2nd Edition* (2012).

Presence in the eastern Mediterranean

H. misippus is a Palaeotropical migrant of sub-Saharan Africa, Arabia and most of the Oriental Region, and one of the most widely distributed butterflies in the world (Larsen, 2005: 335). The species was considered by Larsen (1974: 114) to have reached its northernmost distribution in Lebanon, where it was known in the early 1900s, probably having arrived from the Nile Delta in Egypt. A single male later became known from Limassol, Cyprus, in August 1936, when a specimen, now in the Steinhardt Museum of Natural History, Tel Aviv, was taken by Prof. Bytinski-Salz (Benyamini, 2021a), and a female singleton was recorded at Kızkalesı, İçel (Mersin) Province, Turkey, in April 1993 (Hesselbarth *et al.*, 1995: 1015). The possibility of accidental transfer by boat at the harbour of Kızkalesı, the location of the Turkish capture, is believed likely (Onat Başbay, Ankara, Turkey, 2021, personal communication to first author).

In these more northern areas of the Levant, the most recently known record is from Kfar Aabîda, in April 2003 in the north of Lebanon (Mérit and Mérit, 2007). Such is the species' local rarity that this record, along with those from Cyprus and Turkey, represents all known evidence of this species in the region over the past 100 years or so.

Although occasionally common in Egypt and known to have maintained a presence over several years in the early 1900s near Beirut in Lebanon (Graves, 1911), the species is not resident in Egypt (Gilbert and Zalat, 2007: 105), contributing to the sporadic nature of the Levant appearances. There are recent records from Israel; these and others within the Levant are comprehensively summarized in Benyamini (2021a).

Distribution in Cyprus

Altitudinal: data deficient.
Geographical: data deficient in Cyprus; known only from Limassol on the south coast.

Phenology

Polyvoltine. Breeding is unconfirmed from the northern regions of the Levant.

Larval hostplants

African clades of *Hypolimnas* are reported to be confined to Urticaceae, while Euphorbiaceae, Malvaceae, Moraceae, Portulacaceae and Acanthaceae, in addition to Urticaceae, are reported as hostplants elsewhere (Ackery *et al.*, 1995: 337; Larsen, 1996: 340; Sahoo *et al.*, 2018). Potentially, Cyprus could therefore offer opportunities for transient breeding. *H. misippus* is strongly polyphagous.

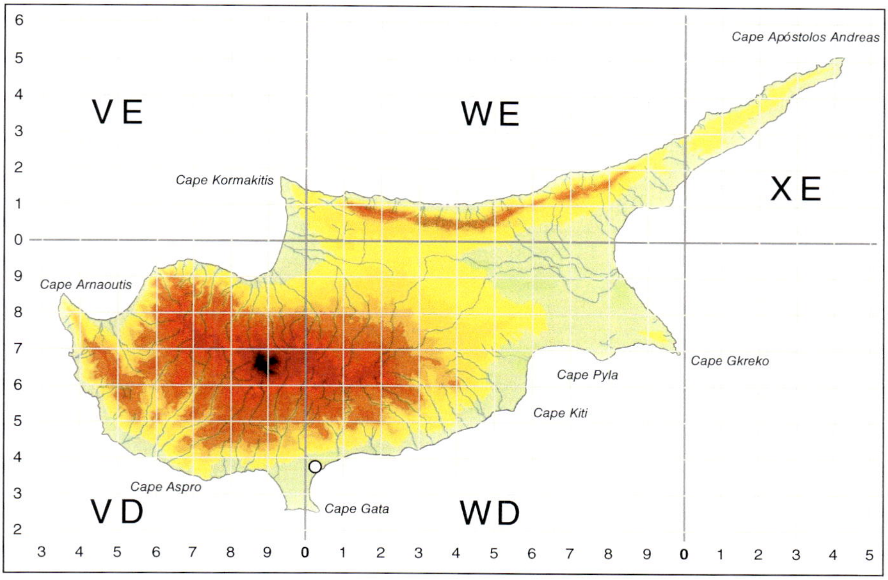

Fig. 12.302. *Hypolimnas misippus* distribution map showing the location of the only known appearance of *H. misippus* in Cyprus. © *Eddie John*

Ecology and behaviour

The most striking characteristic of this species is the remarkable degree of sexual dimorphism displayed (see Figs 12.298 and 12.299), by far the most extensive among any species of butterfly recorded in Cyprus. Furthermore, the female is a near-perfect mimic of *Danaus chrysippus* and, in the Australian region, of *D. petilia* (Stoll, 1790; Tennent *et al.*, 2013). However, there appears to be no universal agreement as to the precise mimetic status of *H. misippus*. The female's mimicry has been described by some as Müllerian (e.g. Tennent *et al.*, 2013), a form of mimicry where both model and mimic are unpalatable, inedible or harmful. However, among other reasons, Gordon (1987) argued that *H. misippus* is not known to be distasteful, and favoured 'the interpretation that this species is a Batesian mimic', whereby an edible species obtains protection by resembling the appearance of an inedible species. All morphs (colour forms) of the model (*D. chrysippus*) are mimicked, meaning that, for example, *D. chrysippus* ff. 'alcippus' and 'alcippoides' have their markings mimicked by different female forms of *H. misippus* (Owen, 1971: 136; Smith and Gordon, 1987). Furthermore, while the flight of the male is fast and strong, the female is able to mimic the leisurely, languid flight of the model, but can revert to swift and erratic flight as necessary (Larsen, 1996: 340; Tennent *et al.*, 2013). Males are non-mimetic.

Although a 'definite long-range migrant' (Larsen, 2005: 335), the very rare appearances of the species in the northern areas of the Levant suggest that its presence there is 'probably more accidental than truly migratory' (Larsen 1975). While the rare presence in Cyprus and on the southern coast of Turkey might be due to adventive means, arrival might alternatively have resulted from a migration along coastal countries of the eastern Mediterranean, or a direct trans-Mediterranean migration from the Nile Delta. Long sea crossings are easily within the species' capability: *H. misippus* has been recorded 600 miles out at sea in the South Atlantic Ocean (Larsen, 1996: 340) and is known to have been resident on the Cape Verde Islands in the North Atlantic Ocean, for many years (Tennent *et al.*, 2013).

It is questionable whether establishment is possible in the northern regions of the Levant, as the species appears unable to withstand the current Mediterranean winter.

Description/identification aids

Wingspan: male: 68–80 mm; female: 7–80 mm (Tshikolovets, 2011: 460). There is marked sexual dimorphism: ♂ forewing **uppersides** are very dark blue, almost deep velvet black, with subapical and discal elliptical white patches, while the hindwing has a single, larger discal white patch. ♀ uppersides strongly resemble those of *D. chrysippus* but lack the three irregular black marks at the end of the cell. ♂ forewing **undersides** have a light brown ground colour, with white forewing patches matching the upperside, while the hindwing patch extends across the wing from costa to inner margin. ♀ undersides again strongly mimic those of *D. chrysippus* but the most significant difference is found in the forewing apex, which is *light grey/brown in colour, contrasting with the black apex of the model species.*

Similar species

Males: none; females: *D. chrysippus*, as described above.

Conservation status in Cyprus

Not applicable – infrequent vagrant/migrant. Not assessed by Numa *et al.* (2016). 'DD' in Cyprus.

Nymphalis polychloros (Linnaeus, 1758) Nymphalidae (Nymphalinae)

Large Tortoiseshell

TL: Sweden.

Figs 12.303 and 12.304. *Nymphalis polychloros*. Specimen reared from a larva found on *Pyrus communis* (pear) at Gerakies, May 1931 (images reproduced from Makris, 2003: 225). *Christodoulos Makris*

Fig. 12.305. *Nymphalis polychloros* (photographed in Crete), 2 June 2012. *Christodoulos Makris*

Fig. 12.306. *Nymphalis polychloros* mud-puddling (photographed in Crete), 2 June 2012. *Christodoulos Makris*

Presence in the eastern Mediterranean

Sparingly represented in the Levant as nominotypical *N. polychloros*, where it is more common in northern parts of the region (Turkey); Atahan *et al.* (2018: 48), for example, stated that *N. polychloros* is present from 300 m to 1700 m in Hatay to the north-east of Cyprus. Larsen (1975) considered the possibility that the species' rare presence in Lebanon could be linked with migrants from populations in the Taurus Mountains of southern Turkey; Benyamini (1993) reported on a single migrating female in northern Israel. Larsen (1974: 115) hypothesized that, given the adequacy of hostplants in Lebanon, the species' minimal presence in the region might be due to marginal ecological conditions and it appears likely that this also applies in Cyprus and along the Lebanese border with Israel, where *N. polychloros* is very rare (Benyamini, 2002a: 232). Nevertheless, Beale and Sprenger (2006: 4) described *N. polychloros* as 'particularly frequent in the pine woods above Aammiq village' in the western Beqaa district of Lebanon.

In view of Zach's failure to record a number of the more common species, it is perhaps surprising that '*Vanessa Polychloros* L' was included among Lederer's list of 1855. The species was referred to as 'Eugonia polychloros L. [race fervida Stndfs.]', by Turner (1920), but reverted to 'Vanessa polychloros L' in Rebel (1939). Turner (1920) added, 'Probably the specimens seen were immigrants from the Taurus, where the race *fervida* is well known to occur.'

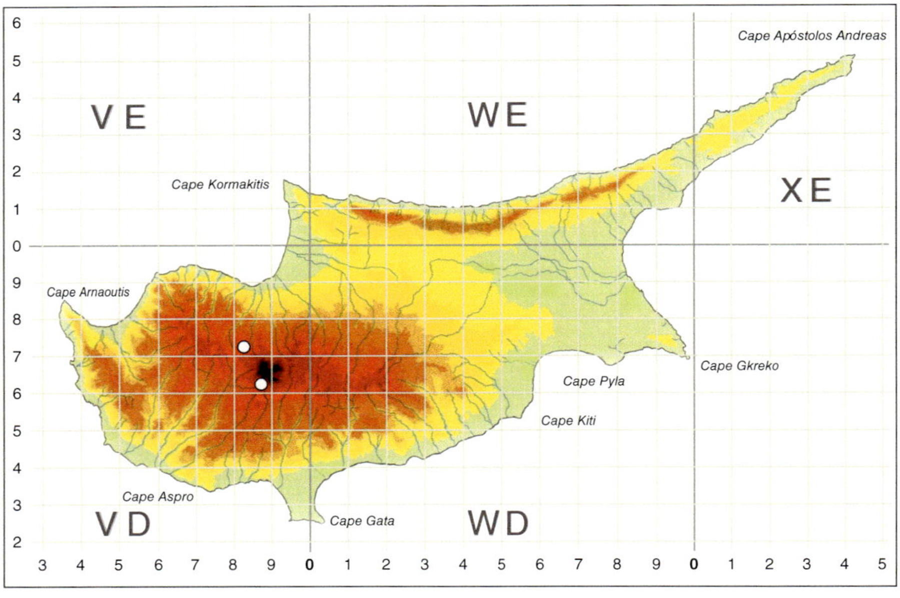

Fig. 12.307. *Nymphalis polychloros* distribution map showing 5 km² observations within 10 km² UTM squares. © *Eddie John*

Distribution in Cyprus

Altitudinal: historically known from areas above 1000 m.
Geographical: a species known only in Cyprus from the Troodos Massif, and first listed by Lederer (1855) – not 1875 as erroneously listed in earlier publications (e.g. Bucknill, 1911) and perpetuated by Turner (1920) and Rebel (1939) – and now likely to be extinct. Later records date from 1918 (Wilson, in Turner, 1920), 1929 and 1931 (Shiakides' collection, Agricultural University of Athens), but *N. polychloros* has not been recorded since then (see Makris, 2003: 224 for details).

Phenology

Univoltine. Little is known about the species' phenology in Cyprus. Elsewhere, overwintered adults emerge early in the year, with the new generation remaining on the wing until late summer, before entering hibernation.

Larval hostplants

In Cyprus, *N. polychloros* is known to have used cultivated *Pyrus communis* (common pear) (Fig. 12.308) (Rosaceae). Elsewhere in the Levant, the species uses Rosaceae: *Crataegus* spp., *Prunus* spp., *Pyrus* spp., Ulmaceae: *Celtis australis*, *Ulmus* spp.

Ecology and behaviour

A known migrant species, making it possible that the handful of records from Cyprus up to 1931 were due to occasional migrations, as speculated by Turner (1920), although the collection of larvae from *Pyrus communis* in May (1931) demonstrated at least transient breeding. There have been no known observations of the species since.

Description/identification aids

Wingspan: up to 60 mm. ♂ and ♀ **uppersides** have a similar orange ground colour with shades of brown overlaid with yellow and black markings especially near the forewing costa. Forewing submarginal

Fig. 12.308. *Pyrus communis*, 12 August 2009. Courtesy of *Georgios Hadjikyriakou*

black bands are repeated on the hindwing, where they are enhanced with blue markings. **Underside** markings have patterns of dark and light brown, with muted submarginal bands mirroring those on the upperside, the overall appearance greatly assisting camouflage during a prolonged hibernation.

Similar species

N. polychloros is similar to *Aglais urticae* (Linnaeus, 1758), the Small Tortoiseshell, a species not actively recorded in Cyprus, although CBSG members Gary Hulbert and Mark Easterbrook reported the finding of forewing fragments at a remote area of Achna Dam on 25 April 2013. No explanation for the presence of *A. urticae* was discovered, but it might have been an unlikely adventive.

Conservation status in Cyprus

Considered probably extinct. Listed by Van Swaay *et al.* (2010a) and Numa *et al.* (2016) as 'LC' in Europe and in the Mediterranean, respectively.

Vanessa atalanta (Linnaeus, 1758) **Nymphalidae** (Nymphalinae)

Red Admiral

TL: Sweden.

Fig. 12.309. *Vanessa atalanta* (Red Admiral) nectaring on *Cistus parviflorus*, 21 March 2013. *Christodoulos Makris*

Fig. 12.310. *Vanessa atalanta* (Red Admiral) nectaring on *Limonium sinuatum*, 7 May 2011. Courtesy of *David and Rosalyn Sparrow*

Fig. 12.311. *Vanessa atalanta* (Red Admiral) imbibing tree sap, 22 May 2010. Courtesy of *Sami Tamson*

Fig. 12.312. *Vanessa atalanta* (Red Admiral) nectaring on *Berberis cretica*, 1 June 2019. *Christodoulos Makris*

Fig. 12.313. *Vanessa atalanta* (Red Admiral) puddling at mud, 21 August 2013. *Christodoulos Makris*

Presence in the eastern Mediterranean

Present throughout the Levant as nominotypical *V. atalanta atalanta*. Originally listed for Cyprus as *Vanessa atalanta* L. by Lederer (1855), this was later referred to as *Pyrameis atalanta* L. by Bucknill (1911), Turner (1920) and Rebel (1939) before reverting to the genus as currently named.

Distribution in Cyprus

Altitudinal: all elevations in Cyprus.
Geographical: widely distributed throughout the island, although rarely numerous.

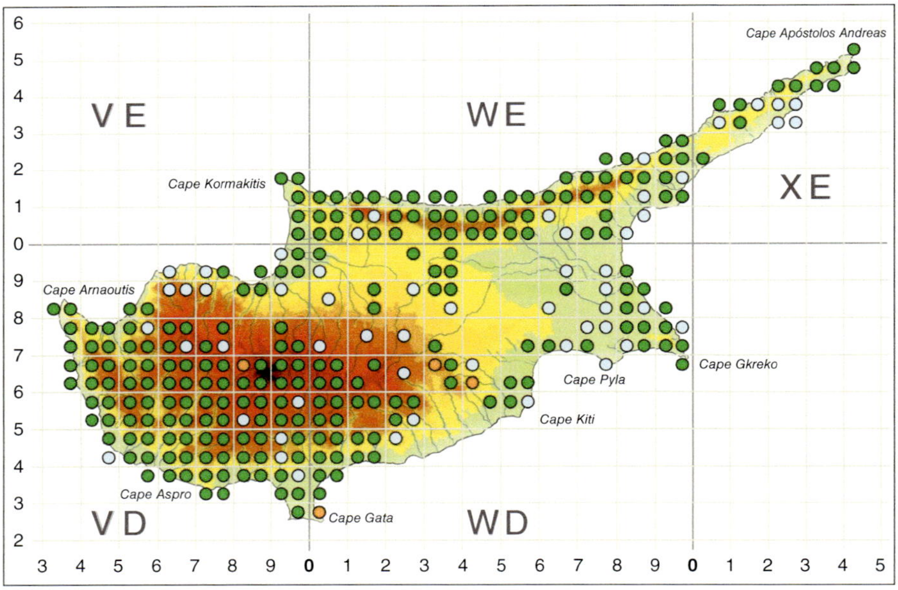

Fig. 12.314. *Vanessa atalanta* distribution map showing 5 km² observations within 10 km² UTM squares. © *Eddie John*

Phenology

Polyvoltine in parts of the Levant, but probably trivoltine in Cyprus. Recorded in all weeks during autumn, winter and spring, but becoming very scarce in summer (CBRS data).

Larval hostplants

Urticaceae: principally *Urtica urens* (small nettle), also *U. pilulifera* (Roman nettle) (Fig. 12.315); *Parietaria judaica* (pellitory-of-the-wall) is used in late spring and early autumn when nettle is not available (Makris, 2003: 230).

Fig. 12.315. *Urtica pilulifera*. Courtesy of *Yiannis Christofides*

Ecology and behaviour

A return southerly autumn migration from northern Europe results in a (small) build-up of *V. atalanta* numbers from late September/October. In similar fashion to that reported in Catalonia and Mediterranean southern Spain (Stefanescu, 2001; Cuadrado, 2017), the species breeds throughout late autumn, winter and spring in Cyprus, seasons when hostplant growth is at its most prolific, stimulated by seasonal rains. During these periods, the second author reported at least three broods in lowland areas between November and May (Makris, 2003: 289).

With evidence of similar migratory patterns extending from the far west of the Mediterranean basin to the near-Asian east, it is probable that seasonal south–north/north–south movements are likely throughout much of the Mediterranean. *V. atalanta* migrates northwards from Cyprus in spring, sometimes accumulating in noticeable numbers in the north of the island where nectaring occurs prior to migration. Rarely, very large numbers may be seen, as in April 2009 when 'thousands' were observed nectaring on *Arbutus andrachne*, and feeding from fermenting fruit and tree sap (Daniel Haines, London, 2009, personal communication to first author). Unlike migrations of the far more prolific *Vanessa cardui*, reports involving such movement of *V. atalanta* in Cyprus are very rare, as the latter is present in fewer numbers, but has been documented (John, 2001a).

As with *V. cardui*, small numbers of non-migratory *V. atalanta* may be seen throughout the year but are unlikely to be encountered in low-lying areas in summer, thus appearing to mirror the experience in Greece, where 'butterflies leave the low-lying areas and move to the mountains where they tend to stay inactive in trees' (Lafranchis, 2019). Similarly, Comay *et al.* (2020), when considering the species phenology in Israel, noted *V. atalanta* as migrating and aestivating.

V. atalanta is a frequent garden visitor in search of nectar and, although common, butterflies are usually seen singly, other than during the spring build-up. Nectaring is seen on a variety of flowers and the species is also known to puddle at mud (authors' personal observations). Hilltopping activity has been reported in Crete (Hardy, 1994).

Description/identification aids

Wingspan: up to 60 mm. A large, beautifully marked butterfly: ♂ and ♀ **uppersides** have a *near-black ground colour* on both fore- and hindwing. Sexes are similar: an *orange-red band spans the forewing from anal angle to costa*, with contrasting white markings on an otherwise black apical area; on the hindwing is a wide submarginal orange-red band with black spotting becoming mauve at the anal angle. Forewing **underside** largely mirrors the upperside, but the hindwing underside has an obscure dark patterning.

Similar species

None in the Levant.

Conservation status in Cyprus

Not threatened. Listed by Van Swaay *et al.* (2010a) and Numa *et al.* (2016) as 'LC'.

Vanessa cardui (Linnaeus, 1758) **Nymphalidae** (Nymphalinae)

Painted Lady

TL: "Europae, Africae".

Fig. 12.316. *Vanessa cardui* (Painted Lady) nectaring on *Dittrichia viscosa*, 24 October 2021. *Christodoulos Makris*

Fig. 12.317. *Vanessa cardui* (Painted Lady) nectaring on *Onopordum cyprium*, 19 May 2019. *Christodoulos Makris*

Fig. 12.318. *Vanessa cardui* (Painted Lady) nectaring on *Onopordum cyprium*, 19 May 2019. *Christodoulos Makris*

Presence in the eastern Mediterranean

Ubiquitous: one of the most widely distributed of all butterflies and found in all countries of the Levant. In similar fashion to *V. atalanta*, this was originally listed for Cyprus as *Vanessa cardui* L. by Lederer (1855), later to be referred to as *Pyrameis cardui* L. by Bucknill (1911), Turner (1920) and Rebel (1939) before reverting to the genus as currently named.

Distribution in Cyprus

Altitudinal: found from sea level to the highest elevations.
Geographical: a very common butterfly recorded from *ca*. 90% of 5 km^2 UTM squares in Cyprus (CBRS data) and will be seen in all.

Phenology

Polyvoltine, and one of a number of species recorded in every week of the year in Cyprus. Most commonly seen in spring and autumn.

Larval hostplants

Larvae are polyphagous, in Cyprus favouring members of the Malvaceae, especially *Malva multiflora* (small tree mallow) (Fig. 12.320). *V. cardui* also uses members of the thistle family, Asteraceae: *Echinops spinosissimus*, *Carduus* spp., *Onopordum cyprium*.

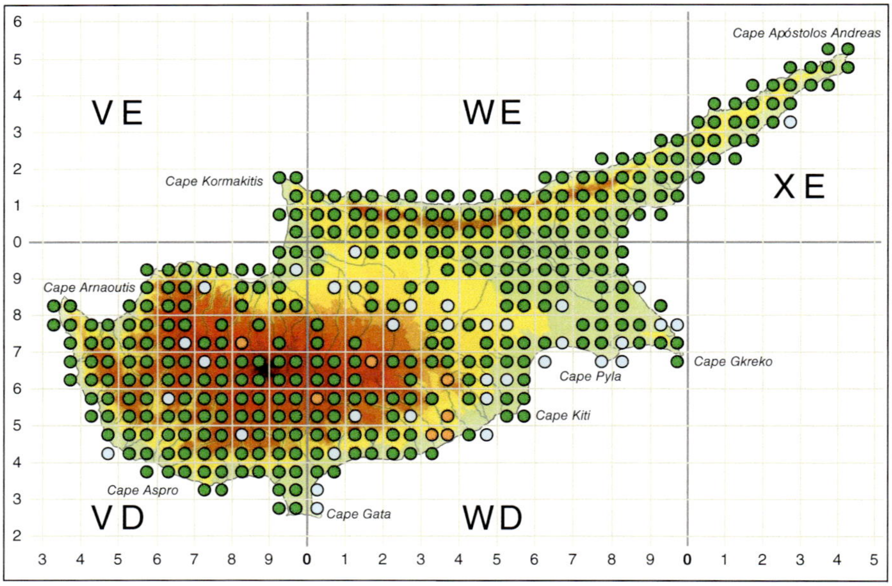

Fig. 12.319. *Vanessa cardui* distribution map showing 5 km^2 observations within 10 km^2 UTM squares. © *Eddie John*

Fig. 12.320. *Malva multiflora*, 18 May 2007. *Eddie John*

Fig. 12.321. *Vanessa cardui* (Painted Lady) larva on *Malva multiflora*, 21 December 2014. *Christodoulos Makris*

Ecology and behaviour

Found in all types of habitat. *V. cardui* is very frequently seen at coastal areas in spring and early summer, where arriving migrants are particularly attracted to the nectar of *Centaurea aegialophila* and *Limonium sinuatum*. Autumn/early winter aggregations also occur, with large numbers observed on *Lantana* (David Whaley, Cyprus, 2006, personal communication to first author). Periodically, migrations of huge numbers of *V. cardui* pass through the island, with an estimated 40+ million butterflies migrating across the island in 2001 (John, 2001a; see also Chapter 6). Very large migrations of *V. cardui* also took place in 2014 and especially in 2019 (John *et al.*, 2015a, 2019b). Sightings reduce as summer approaches, when *V. cardui* becomes scarce during the hottest months, before becoming common again with a return southerly migration of offspring in autumn. No published reports are known detailing an autumn (return) migration in Cyprus, because numbers of returning migrants are very much lower, but this phenomenon is well documented in the wider region (Stefanescu *et al.*, 2016; Talavera *et al.*, 2018). However, on 28 October 2015, a constant light stream numbering tens of *V. cardui* was noted at Randi Forest, heading south-east in the direction of Pissouri near the south coast (David Whaley, Cyprus, 2015, personal communication to first author).

Occasionally, reports are received suggesting the possibility of a 'reverse migration' to the south-east in spring, but it is difficult to assess if this activity involves immigrant or local populations of *V. cardui*. As an example, at around midday on 8 April 2019, during an exceptionally large spring migration numbering billions of *V. cardui* in the wider region, a cold south-westerly wind at Cape Drepano appeared to encourage hundreds of migrants to interrupt their passage to 'refuel' on coastal flowers. Later, some were seen to fly low towards the sea, but on finding conditions unfavourable (it is reasoned), reversed their north-westerly course to fly away to the south-east (Alison McArthur, Peeblesshire, 2019, personal communication to first author). The late Torben Larsen succinctly encapsulated his experiences with *V. cardui* thus: 'This is a butterfly whose migratory habits are as easy to observe as they are difficult to interpret' (Larsen, 1975). The same author wrote, 'In *Vanessa cardui*, on which more data are available than for other species, there is no clear correlation between the direction of flights and season in the Mediterranean, nor is it clear that any one direction is a priori better than another' (Larsen, 1982a). However, in Cyprus, there is no room for doubt that migratory routes cross the island on a northerly or north-westerly bearing carrying migrants in the direction of Turkey. Departure points most frequently reported are from Cape Drepano, Cape Arnaoutis and Cape Kormakitis. Migrants travel through the island at a height of 1–3 m, but on leaving the coast have been observed to gain height to 20–30 m, heading on a 315-degree course towards Turkey (McArthur and John, 2019; see also Flint in John, 2001a). There is no true seasonal diapause; in mild, low-lying areas, larval development continues throughout the winter (Fig. 12.321). While Parker (1983) referred to the likelihood of migration accounting for influxes of *V. cardui*, it is very surprising that neither he, nor any of the earlier researchers, referred to any form of migration having been witnessed for this species (or *V. atalanta*) in Cyprus.

Nectaring on a wide variety of flowers has been reported, including use of numerous members of the Asteraceae family, especially *Dittrichia viscosa*, *Onopordum cyprium* and *Centaurea aegialophila*, as well as members of other families, for example *Pterocephalus multiflorus*, *Heliotropium hirsutissimum*, *Limonium sinuatum*, *Plocama calabrica*, *Ceratonia siliqua* and cultivated *Pelargonium crispum*.

Description/identification aids

Wingspan: up to 65 mm. A large, handsome butterfly; sexes are similar, having an **upperside** ground colour of *delicate orange-brown* and a postdiscal area of *black with white spotting extending into the forewing apex*, and black discal markings. The hindwing has a postdiscal band of black marks, a smaller line of submarginal black spots, becoming blue at the anal angle, and a third row of black spots along the margin. Forewing **underside** ground colour is similar, markings mirroring those on the upperside, but with the apex, buff. Hindwing underside is intricately marked, and with *four distinct submarginal ocelli of variable size and coloration*, including dark blue scaling. Occasionally, long-lived individuals remain capable of vigorous flight even after losing much of the wing area and scaling.

Similar species

None in Cyprus.

Conservation status in Cyprus

Not threatened. Listed by Van Swaay *et al.* (2010a) and Numa *et al.* (2016) as 'LC'.

Argynnis pandora ([Denis & Schiffermüller], 1775) Nymphalidae (Heliconiinae)

Cardinal

TL: Surroundings of Vienna.

Fig. 12.322. *Argynnis pandora* (Cardinal) ♂, 4 August 2018. *Christodoulos Makris*

Fig. 12.323. *Argynnis pandora* (Cardinal) ♂ nectaring on *Cirsium vulgare*, 2 October 2017. *Christodoulos Makris*

Fig. 12.324. *Argynnis pandora* (Cardinal) ♀, 23 September 2018. *Christodoulos Makris*

Fig. 12.325. *Argynnis pandora* (Cardinal) ♂, 4 August 2018. *Christodoulos Makris*

Presence in the eastern Mediterranean

Present in countries of the Levant (south to the Lebanon/Israel border region) as nominotypical *A. pandora pandora*, but is considered scarce in the Turkish province of Hatay (Atahan *et al.*, 2018: 57). In contrast, *A. pandora* is common above 1000 m in the Latakia mountain range of Syria (Mudar Salimeh, Latakia, 2021, personal communication to first author) and on Cedar Mountain, Lebanon (second author, personal observation).

Unrecorded for Cyprus by Lederer (1855), this was stated by Bucknill (1911) to be 'confined in Cyprus to Troödos where it is tolerably abundant in July and August'. Numerous nomenclatural changes over the years have seen the species referred to as, for example, 'Dryas pandora Schiff.' by Turner (1920), 'Argynnis pandora Schiff.' by Rebel (1939) and as 'Pandoriana pandora pandora' (using the subgeneric name), by Larsen (1974: 120).

Distribution in Cyprus

Altitudinal: usually found above 1000 m in Cyprus, but with one reported observation considerably lower at Monagri (395 m) on 9 May 2020 (Elli Tzirkalli, Cyprus, 2020, personal communication to first author).
Geographical: nowadays a rather rare and local butterfly confined mainly to the higher elevations of the Troodos mountains. *A. pandora* was unconfirmed from northern mountain ranges until 2006, when the species

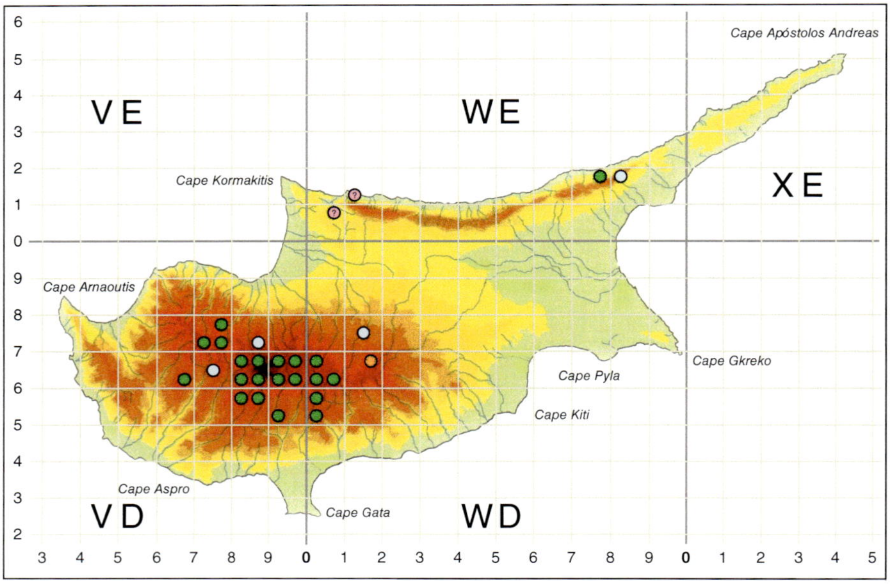

Fig. 12.326. *Argynnis pandora* distribution map showing 5 km² observations within 10 km² UTM squares. © *Eddie John*

was seen at *ca.* 500 m near Kantara (Daniel and Hilary Haines, London, 2006, personal communication to first author), and again in an adjacent 5 km^2 UTM square on 25 May 2013 at 566 m, by the first author. On 12 June 2012, an unidentified fritillary was seen at a distance of *ca.* 30 m in the north-west of the island (WE00 NE) that in appearance and size resembled *A. pandora* (first author, personal observation). This was followed on 22 April 2016 by an unconfirmed coastal observation of a possible *A. pandora*, *ca.* 5 km to the north-east in WE10 SE. The last-mentioned sighting is unusually early (for Cyprus), as is the coastal location; however, the species is known from sea level to 2300 m in Greece (Pamperis, 2009: 399). These observations suggest that *A. pandora* might be more widely distributed in the northern regions of the island than is indicated by the distribution map.

Phenology

Generally on the wing from May to October, but earlier and later records have been reported on 27 April 2010 (Alison McArthur, Peeblesshire, 2010, personal communication to first author) and on 2 December 2006 (Yiannis Christofides, Cyprus, 2006, personal communication to first author). *A. pandora* is considered by some authors to be univoltine (Makris, 2003: 234; Lafranchis, 2004: 201; Pamperis, 2009: 399), with aestivation or a prolonged emergence accounting for apparent longevity of lifespan (Hesselbarth *et al.*, 1995: 1073). For populations in central Spain, García-Barros (2000) also elected for univoltinism and discussed the interesting possibility of female *A. pandora* undergoing delayed ovarian maturation during a period of aestivation – as with some species of *Hipparchia* that mate prior to aestivation, but delay ovipositing until the re-growth of hostplants in the autumn (García-Barros, 1988; John and Parker, 2002); see ‘Ecology and behaviour’, below. As with *Hipparchia* spp., *A. pandora* has been observed in all weeks of the Cyprus summer. Bivoltinism appears to be a possibility, as ♂ and ♀ *A. pandora* having a fresh appearance were seen on 29 September 2000 and 2 October 2017 by the authors (Fig. 12.323); furthermore, Higgins and Riley (1970: 89), Tennent (1996: 49) and Tshikolovets (2011: 475) are among authors to refer to two annual broods. Overwinters in the larval stage.

Larval hostplants

Violaceae: *Viola* spp. Ovipositing has not been documented in Cyprus, but *Viola sieheana* (Troodos violet) (Fig. 12.327), *V. odorata* (sweet violet) (Fig. 12.328) and *V. alba* (white violet) are known for the island. Whilst almost all sources of reference give *Viola* spp., Higgins and Riley (1970: 90) added ‘rarely rue’ (*Ruta* sp.), while Montagud-Alario and García-Alamá (2010: 161) referred to the (unattributed) use of *Valerianella coronata* (Valarianaceae).

Fig. 12.327. *Viola sieheana*, 5 May 2019. *Christodoulos Makris*

Fig. 12.328. *Viola odorata*, 27 February 2017. *Christodoulos Makris*

Ecology and behaviour

As discussed under 'Phenology', work carried out by García-Barros (2000) strongly hinted at univoltinism, aestivation and delayed oviposition for *A. pandora*. This study showed that upon dissection of five females caught during July and August, all were found to have immature ovaries, whereas four females captured in early September readily oviposited. These were found to have high fecundity, one female laying 2400 eggs (García-Barros, 2000). In the wild, eggs are laid on the remains of dead plants, often prior to the appearance of fresh aerial parts of the hostplant. *A. pandora* is found in light woodland, clearings and nearby fields, especially those hosting *Centaurea aegialophila*, *Cirsium vulgare* and *Onopordum cyprium*, and is also readily attracted to other nectar-rich sources (e.g. buddleia) in gardens (Makris, 2003: 234).

Description/identification aids

Wingspan: up to 80 mm. A large butterfly with an **upperside** yellow-gold ground colour and postdiscal, submarginal and marginal rows of black marks across fore- and hindwings, typical of the larger fritillaries. The ♂ has two pairs of androconia in the discal area of Cu1 and Cu2. **Undersides** are distinctive, the forewing of both sexes having *a large expanse of red*, partially hidden by the hindwing, and a yellow-green apex. Hindwing has a yellow-green ground colour, ♂ with obscure pale discal, postdiscal and submarginal bands, which are *strongly marked in silvery-white* in the ♀.

Similar species

Other members of the genus are present elsewhere in the Levant, but the much smaller *Issoria lathonia* is the only other fritillary recorded in Cyprus, where it has been reported once only.

Conservation status in Cyprus

Not threatened. Listed by Van Swaay *et al.* (2010a) and Numa *et al.* (2016) as 'LC' in Europe and in the Mediterranean, respectively.

Issoria lathonia (Linnaeus, 1758) **Nymphalidae** (Heliconiinae)

Queen of Spain Fritillary

TL: Europe.

Fig. 12.329. *Issoria lathonia* photographed in Lebanon, 16 July 2016. *Christodoulos Makris*

Fig. 12.330. *Issoria lathonia* nectaring on *Clinopodium vulgare*, 26 May 2016. The distinctive mother-of-pearl underside markings are clearly visible. Courtesy of *Yiannakis Makriyianni*

Fig. 12.331. *Issoria lathonia* photographed in Lebanon, 16 July 2016. *Christodoulos Makris*

Presence in the eastern Mediterranean

Issoria lathonia, as the nominotypical ssp., is present throughout much of the region where, in Mediterranean Turkey, Atahan *et al.* (2018: 61) described it as common in Hatay Province. In countries of the Levant to the south, it is rare in Mediterranean Syria (Mudar Salimeh, Latakia, 2021, personal communication to first author), while Larsen (1974: 122) referred to *I. lathonia* as locally quite common at higher elevations, 'especially in the Antilebanon and the northern Lebanon'. Larsen (1975) further commented on the species as 'undoubtedly a migrant [in Lebanon], although there are no actual observations [of migrant activity]'. The species appears to have a sparse presence elsewhere in the region. Occasional migrants penetrate south into northern Israel (Benyamini, 1993).

Distribution in Cyprus

Altitudinal: data deficient, with just a single record at *ca.* 1400 m.
Geographical: a single specimen was observed and photographed at Pano Amiantos on 26 May 2016 (Yiannakis Makriyianni, Cyprus, 2021, personal communication to first author); there have been no reported sightings since.

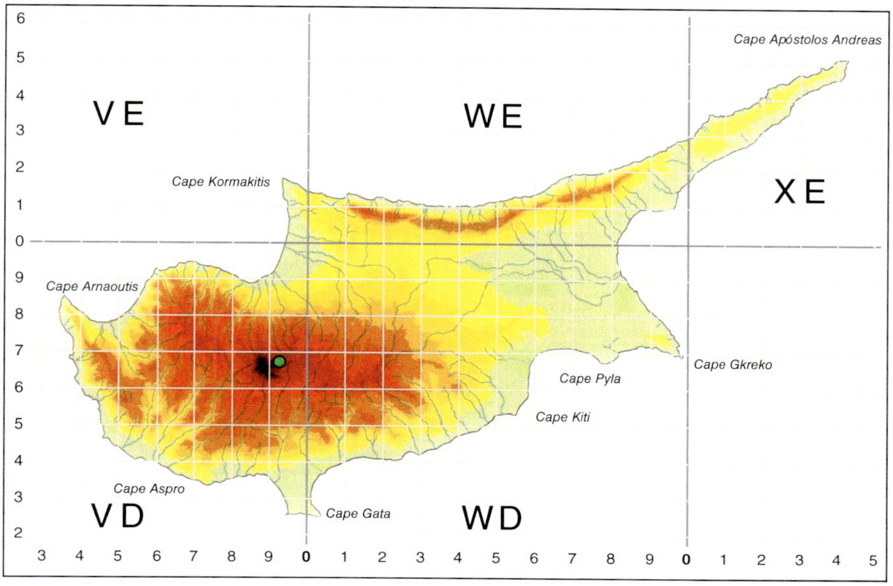

Fig. 12.332. *Issoria lathonia* distribution map showing the location of the single 5 km² observation. © *Eddie John*

Phenology

Generally, polyvoltine in overlapping broods in Mediterranean areas.

Larval hostplants

Violaceae: *Viola* spp. Ovipositing has not been observed in Cyprus, but several species of *Viola* are present (see entry for *A. pandora*).

Ecology and behaviour

A hilltopping and basking species that is strongly migratory (Larsen, 1974: 122 and 1976; Kotiaho *et al.*, 2005), a trait possibly accounting for its presence in Cyprus, as *I. lathonia* is known to be capable of undertaking a sea crossing (Asher *et al.*, 2001: 317). However, the migratory status of the species in Turkey has not been established (Zobar and Genc, 2008), although, in Turkey, the species is reported to disperse into the mountains in summer, in search of cooler conditions (Hesselbarth *et al.*, 1995: 1086). Details of immature stages and life cycle of *I. lathonia* may be found in Zobar and Genc (2008) and Benyamini (2021b: 130).

Description/identification aids

Wingspan: ♂ 34-43 mm, ♀ 40-45 mm (therefore, much smaller than *A. pandora*). Furthermore, *I. lathonia* is clearly distinguished by the presence of extensive and very large *mother-of-pearl markings* on the hindwing underside. Sexes are similar in appearance.

Similar species

Other members of the genus are present elsewhere in the Levant, but *A. pandora* is the only other fritillary recorded in Cyprus, from which *I. lathonia* is easily separated (see above).

Conservation status in Cyprus

Data deficient in Cyprus. Listed by Van Swaay *et al.* (2010a) and Numa *et al.* (2016) as 'LC' in Europe and in the Mediterranean, respectively.

Hipparchia syriaca cypriaca (Staudinger, 1879) Satyridae (Satyrinae)

Eastern Rock Grayling

TL: Syria, Cyprus.

Fig. 12.333. *Hipparchia syriaca cypriaca* (Eastern Rock Grayling) typically rests well camouflaged in shade on trunks of trees (here on alder). Forewings are usually hidden until alerted. 21 June 2012. *Christodoulos Makris*

Fig. 12.334. *Hipparchia syriaca cypriaca* (Eastern Rock Grayling) well camouflaged on *Alnus* (alder), 18 June 2012. *Christodoulos Makris*

Figs 12.335 and 12.336. *Hipparchia syriaca cypriaca* (Eastern Rock Grayling) on *Echinops spinosissimus*, 23 July 2012. The butterfly appears to be nectaring (L), but nectar is not (yet) available from the flower head on the right – see text. (Note the phoretic mite infestation around the head and basal wing areas.) *Christodoulos Makris*

Presence in the eastern Mediterranean

Nominotypical *H. syriaca syriaca* (Staudinger, 1871) is present in northern countries of the Levant, i.e. Turkey, Syria and Lebanon (not known for Israel), whereas Cyprus has the endemic ssp. *H. syriaca cypriaca*, about which Kudrna (1977: 37) stated, ‘Genitalia provide in both sexes decisive diagnostic characters, differing from all related species of the genus.’ Similarly, Olivier (1993: 128, 196), on evidence at the time, had ‘not the slightest doubt about the justification of subspecific separation’ and Hesselbarth *et al.* (1995: 896) also acknowledged the validity of the ssp.

Recorded for Cyprus as ‘Satyrus Hermione L.’ by Lederer (1855), this was later split by Bucknill (1911) into two ‘forms’: ‘Satyrus hermione L. form cypriaca. Stgr. The Cyprian Queen’ and ‘Satyrus hermione L. form nova-cypriaca. Bucknill. The Cyprian King’. The last mentioned is no longer recognized. Turner (1920) referred to ‘Satyrus hermione L. [race cypriaca Stdgr.]’, while Rebel (1939) referred to ‘Satyrus syriaca (Stgr.) cypriaca Stgr.’, adding that the ‘form cypriaca is even darker than syriaca’ (translated from the original German).

Distribution in Cyprus

Altitudinal: from sea level (far less commonly) to the highest elevations.
Geographical: a common butterfly, locally present in western, central and northern parts of the island, but largely absent from the central plain and unrecorded from south-eastern areas (CBRS data). Several *H. syriaca*

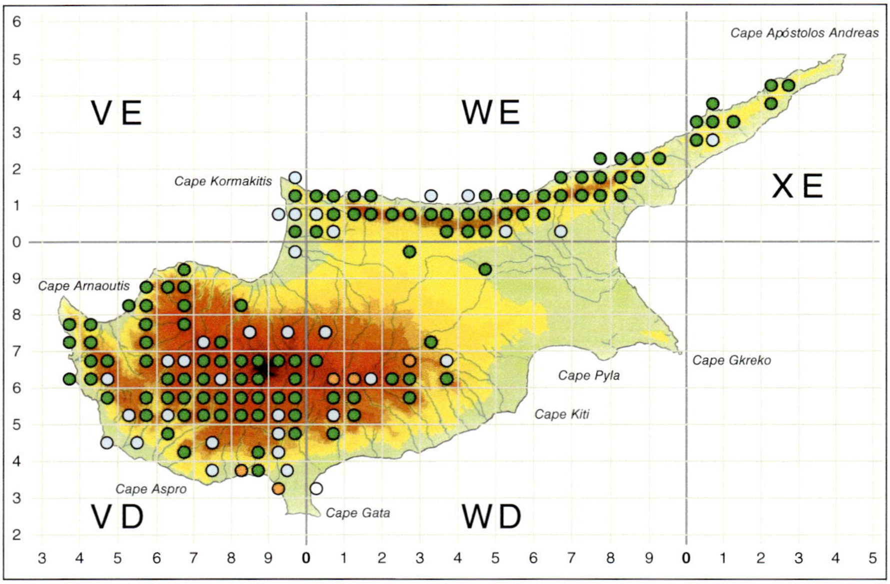

Fig. 12.337. *Hipparchia syriaca* distribution map showing 5 km^2 observations within 10 km^2 UTM squares. © *Eddie John*

appeared from late May 1998 among conifers in the garden (240 m asl) of the first author's rented house in Mosfiloti (WD36 NE), the NE and SE of WD36 seemingly being on the margin of the species' range in the south-east of the island.

Phenology

Univoltine, from May (sometimes late April) to October or later. Overwinters as a larva.

Larval hostplants

Poaceae: including *Piptatherum* spp. and (elsewhere) *Brachypodium* spp. (Bálint *et al.*, 2016).

Ecology and behaviour

The favoured habitat for this species appears to be pine forests, river valleys, parks or gardens where pine, oak, alder or other trees offer suitable shade and resting places. *H. syriaca* enters light aestivation at higher elevations in summer but remains very alert to movement, when disturbed flying to a tree trunk nearby, at which point the forewing quickly closes on alighting and the butterfly becomes well camouflaged against the bark. Often, large aestivating aggregations may form at higher elevations, occasionally in company with *H. cypriensis*. Ellison and Wiltshire (1939) reported that, in Lebanon, 'It may be absent from large parts of a wood and occur in dozens within a few yards.' Hilltopping activity was anticipated by Bálint *et al.* (2016), but this has yet to be confirmed in Cyprus.

H. syriaca is not thought to seek nectar, and the species might be among those adult butterflies that rarely feed. However, the authors have observed *H. syriaca* on white-flowered *Lantana camara*, possibly seeking fermenting nectar (John and Skule, 2016: 317), and on *Echinops spinosissimus* (Figs 12.335 and 12.336). With reference to the latter, the second author has noted that the floral heads of *E. spinosissimus* attract many insects; the apparent nectaring activity might be explained by *H. syriaca* probing for excretions from these insects or for sap resulting from insect damage. Mud-puddling and feeding from fermenting fruit and tree sap has also been reported (Makris, 2003: 240), and possible seawater puddling was discussed in John and Tennent (2012).

Mating has been observed in early summer (7 June 2004) and during autumn; more males appear to survive aestivation than is the case with *H. cypriensis*, although this needs corroborating. As with *H. cypriensis*, it is considered that delayed ovarian maturation results in egg-laying coinciding with the re-growth of grasses following the return of rain in the autumn. An attempt at mating was observed in late September/early October, the sexes alighting on the ground facing head-to-head (Daniel Haines, London, 2016, personal communication to first author). An accompanying series of photographs showed the female adopting an apparent rejection posture by opening wings fully and raising the abdomen; no successful autumn pairings were observed at the time. See Dennis (2020: Fig. D22) for illustrations of *Hipparchia* courtship behaviour.

Description/identification aids

Wingspan: up to 65 mm. **Uppersides** of both sexes are similar, with a very dark brown ground colour, *a pale yellow postdiscal band* (subdued in the male) and a single apical white-pupilled, black ocellus, with a vestigial submarginal ocellus usually visible midway towards the anal angle. Forewing **underside** mirrors the upper, but is lighter overall. Hindwing ground colour is grey-brown, with a lighter, but variable, postdiscal band bordered by irregular black lines. An obscure submarginal small white-pupilled, black ocellus is present near the anal angle.

Similar species

When at rest, the *light yellow apex* of the forewing underside is often visible briefly, helping to separate this species from the similar *Hipparchia cypriensis*, which (below the apical ocellus) has an *orange-yellow* ground colour, markedly different from *H. syriaca*.

Conservation status in Cyprus

Not threatened. Nominotypical *H. s. syriaca* is listed by Van Swaay *et al.* (2010a) and Numa *et al.* (2016) as 'LC' in Europe and in the Mediterranean, respectively.

Hipparchia cypriensis (Holik, 1949) **Satyridae** (Satyrinae)

Cyprus Grayling

TL: Cyprus.

Fig. 12.338. *Hipparchia cypriensis* (Cyprus Grayling) ♂, 1 May 2010. *Christodoulos Makris*

Fig. 12.339. *Hipparchia cypriensis* (Cyprus Grayling) ♂, 7 May 2016. To assist camouflage when at rest, the forewing ocellus and typical orange markings of the forewing underside are often partially or completely concealed. *Eddie John*

Fig. 12.340. *Hipparchia cypriensis* (Cyprus Grayling) ♂ nectaring on *Onopordum cyprium*, 14 June 2011. *Christodoulos Makris*

Fig. 12.341. *Hipparchia cypriensis* (Cyprus Grayling) ♀ nectaring on *Ptilostemon chamaepeuce*, 17 May 2018. *David and Rosalyn Sparrow*

Fig. 12.342. *Hipparchia cypriensis* (Cyprus Grayling), pairing (♀ on left), 5 May 2010. *Christodoulos Makris*

Fig. 12.343. A pair of worn *Hipparchia cypriensis* (Cyprus Grayling), ♀ on right, a late mating on 8 October 2009 post-aestivation, some six months after spring emergence. *Christodoulos Makris*

Fig. 12.344. *Hipparchia cypriensis* (Cyprus Grayling), worn ♂ in typical aestivating position on the trunk of a pine, 26 July 2014. Courtesy of *David and Rosalyn Sparrow*

Presence in the eastern Mediterranean

Hipparchia pellucida (Stauder, 1924), found in Turkey and Syria, was raised to specific rank (from *Satyrus semele* ssp. *pellucida*) by de Lattin (1949), who later wrote 'All *semele* I have at hand from this island [Cyprus] belong to the purely West-Asiatic species *pellucida* Frhst. They form a special subspecies of this, which should be called *cypriensis* Holik' (de Lattin, 1952). Holik (1949) designated '*Satyrus semele* ssp. *cypriensis*' as a valid subspecies, and Wiltshire (1949), too, accepted ssp. status for *cypriensis*, but favoured the name '*Eumenis pellucida* Fruhst. *cypriensis* Holik', in preference to 'the Cyprian species hitherto called "*Satyrus semele mersina* Stgr."' Higgins (1973), in an oblique reference in his account of butterflies collected in Crete, referred to '*semele* from Cyprus (*cypriensis* Holik)'.

Kudrna (1975) similarly considered subspecies status for *cypriensis* likely, hence its inclusion as *Hipparchia pellucida cypriensis* in Kudrna (1977: 93), later being raised to full species rank in Kudrna (1986: 184), an elevation not universally accepted, for example by Olivier (1993: 222), Olivier and Coutsis (1997), Karsholt and Razowski (1996) and Dennis *et al.* (2000). However, species rank was maintained by, among others, Kudrna (2002: 174; 2019: 159), Makris (2003: 242) and by Verovnik and Wiemers (2010), Tshikolovets (2011: 413), Wiemers *et al.* (2018, 2020), Maes *et al.* (2019) and Middleton-Welling *et al.* (2020). Although noting differences in wing morphology, Sbordoni *et al.* (2018: 85) stated of *cypriensis*, 'Male genitalia: in all respects identical to those of nominotypical *pellucida*'. In contrast, investigations on *H. cypriensis* by Jutzeler *et al.* (2019) revealed 'a clear differentiation' of male and female genitalia, as well as differences in wing morphology when compared with nominotypical *H. p. pellucida*. Jutzeler *et al.* (2019) summarized: 'From the results of the present study, it is concluded that *H. cypriensis* (Holik) is unique with regard to several characters and should be therefore considered as a distinct species.'

Furthermore, DNA barcoding on a small number of specimens showed small but consistent differences between *H. cypriensis* and *H. pellucida* (Vazrick Nazari, Ottawa, 2011, personal communication to first author), strengthening the species status of *H. cypriensis*. However, it is recognized that the *Hipparchia* are among a number of genera that 'have notoriously difficult taxonomy' (Dincă *et al.*, 2021).

Possible confusion with *H. syriaca* (by Zach in 1853) might explain the reason for this very common species being a surprising omission from Lederer (1855), not least because Bucknill (1911), who recorded this as '*Satyrus semele. L. form mersina, Stgr. The Mersina Grayling*', stated, 'Occurs all over the Island'. Turner (1920) listed this species as 'Hipparchia semele L. [race mersina Stdgr. ; ab. triocellata Rag.]', while Rebel (1939) gave 'Satyrus semele (L.) mersina Stgr.'

Distribution in Cyprus

Altitudinal: found from sea level to the highest elevations.
Geographical: a very common endemic, well distributed throughout Cyprus, less so on the open Mesaoria Plain. Presence in the more arid and low-lying south-east of the island is sparse.

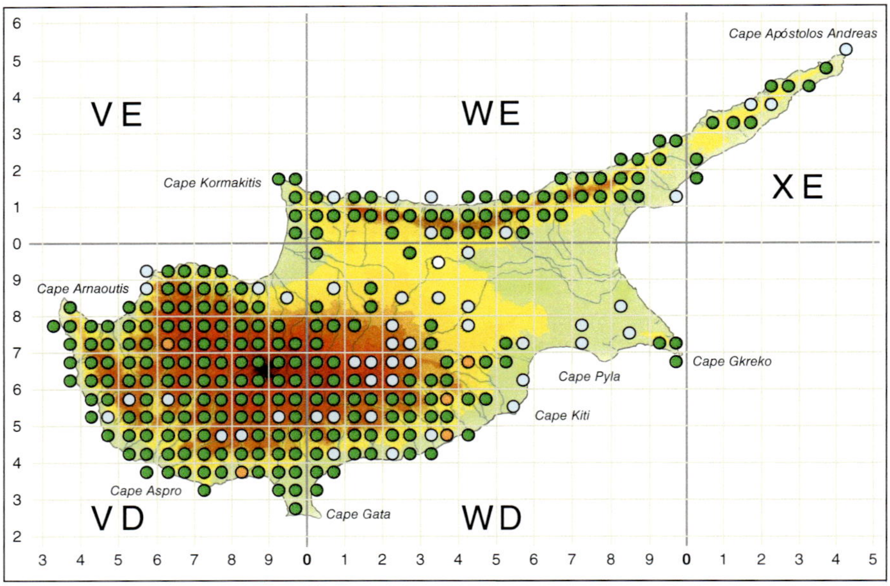

Fig. 12.345. *Hipparchia cypriensis* distribution map showing 5 km² observations within 10 km² UTM squares. © *Eddie John*

Phenology

Univoltine, April to November, earlier in warm springs. Larvae develop slowly throughout the winter.

Larval hostplants

Poaceae spp.: *Festuca ovina* (not listed for Cyprus) was readily accepted by larvae in captivity (Jutzeler *et al.*, 2019). Benyamini (2021: 100) noted acceptance of *Piptatherum miliaceum*, *Poa annua* and *Brachypodium sylvaticum*, the first mentioned being particularly favoured.

Ecology and behaviour

H. cypriensis emerges as early as March in lowland areas, these and intermediate altitudes providing the main breeding sites for this species. As summer temperatures rise, the species disperses into the hills and mountainous regions, where it aestivates in the shade of trees and shrubs along with others of the genus. Mating generally occurs in the spring and it is highly probable that *H. cypriensis* mimics behaviour of *Hipparchia semele* (Linnaeus, 1758) in Spain, where there is early mating and delayed ovarian maturation, followed by ovipositing in autumn (Garcia-Barros, 1988). Figure 12.343 shows a pairing of *H. cypriensis* photographed at Panagia (*ca.* 850 m) on 8 October 2009 (mating has also been observed by the second author on 14 October), indicating that autumn mating occurs on occasions but it is not known if these females had been 'overlooked' during spring or had mated again during autumn. Very few late summer/autumn matings of this species have been observed, but ♀ rejection behaviour, resembling that described under *H. syriaca*, was observed on 27 October 2017 (Ian Bloomer, Cyprus, 2017, personal communication to first author).

Diapause is light, and the butterfly is easily disturbed before quickly settling again, usually on a pine tree or in the shade of a bank beneath trees. Dispersal in numbers to sources of water or mud was investigated by the first author over a 15-minute period near Pedoulas (1270 m) on 8 June 2001, when *ca.* four *H. cypriensis* per minute were seen heading downhill along a track towards mud at a water-spill. Up to 20 individuals at a time were observed mud-puddling before they returned to shade along the same route. Of the 60 or so individuals counted flying downhill, eight were caught in order to determine sex; all were males and upon release some continued to the water source, others returned uphill (John and Parker, 2002).

A seasonally reversed 'migration' to lower elevations occurs in autumn, when shorter day length, rainfall and reducing temperatures bring summer diapause to an end, triggering *H. cypriensis* to return to breeding areas and to conditions favourable for egg-laying (John and Parker, 2002). Such dispersals are rarely reported, as they involve small numbers dispersing over relatively short distances over a period of days.

This is particularly true with early summer dispersals *into* the mountains, which appear to be even thinner in nature, hence less frequently noted or recognized as dispersal. John and Parker (2002) documented one such uphill dispersal (the late Lyndon Thomas, 2001, personal communication to the first author) but, as Kudrna (2013) observed, the study of butterfly dispersal is challenging and is one of the least-known aspects of butterfly ecology.

Nectaring has been observed mainly on flowers of *Onopordum cyprium*, *Ptilostemon chamaepeuce*, *Cistus creticus* and *Plocama calabrica*. In addition to being a frequent mud-puddler, *H. cypriensis* has been observed puddling at damp beach sand (Ken Hearne, Sussex, 2011, personal communication to first author; John and Tennent, 2012).

Description/identification aids

Wingspan: up to 65 mm. A large butterfly, generally more numerous and more widely distributed than *H. syriaca*, and superficially similar to the latter when at rest with wings tightly closed, although ♂ *cypriensis* are smaller. **Uppersides** of both sexes have a dark brown ground colour, marked with orange (not light yellow, as in *H. syriaca*) in the postdiscal area. ♂ and ♀ forewings have two white-pupilled, black forewing ocelli, smaller in the ♂ (where the ocellus between veins Cu1 and Cu2 may be blind). Forewing **underside** has a *large expanse of yellow-orange* in both sexes, brown markings along the costa and two ocelli as on the upperside. Hindwings are grey-brown, the ♂ with a broad, usually discernible, *irregular paler postdiscal band* (Figs 12.338 and 12.339). In both sexes, a small white-pupilled, black ocellus, sometimes faintly encircled in yellow, is present between veins Cu1 and Cu2.

Similar species

See *H. syriaca*.

Conservation status in Cyprus

Although an island endemic, *H. cypriensis* is not threatened. Van Swaay *et al.* (2010a) and Numa *et al.* (2016) listed this endemic species as ‘LC’.

Chazara briseis larnacana (Oberthür, 1909) Satyridae (Satyrinae)

The Hermit

TL: Cyprus.

Fig. 12.346. *Chazara briseis larnacana* (The Hermit) ♂ nectaring on *Limonium sinuatum*, 8 May 2016. Courtesy of *David and Rosalyn Sparrow*

Fig. 12.347. *Chazara briseis larnacana* (The Hermit) ♀ nectaring on *Limonium sinuatum*, 8 May 2016. *Eddie John*

Fig. 12.348. *Chazara briseis larnacana* (The Hermit) ♂ nectaring on *Onopordum cyprium*, 10 May 2020. *Christodoulos Makris*

Fig. 12.349. *Chazara briseis larnacana* (The Hermit) ♀ nectaring on *Limonium sinuatum*, 8 May 2016. *Eddie John*

Fig. 12.350. *Chazara briseis larnacana* (The Hermit) ♀ nectaring on *Onopordum cyprium*, 16 May 2013. *Christodoulos Makris*

Fig. 12.351. *Chazara briseis larnacana* (The Hermit) ♀ 'f' pirata nectaring on *Onopordum cyprium*, 15 May 2013. *Christodoulos Makris*

Presence in the eastern Mediterranean

Elsewhere in the Levant, the species is present only in Turkey (as ssp. *Chazara briseis meridionalis* (Staudinger, 1886)). Although very common in Cyprus, *C. briseis* is rare in Mediterranean coastal regions of Turkey (Hesselbarth *et al.*, 1995: map 282) and with only one known record from Hatay Province (Atahan *et al.*, 2018: 82). Cyprus has the endemic ssp. *Chazara briseis larnacana* (described by Oberthür (1909) as a 'race speciale'). Kudrna *et al.* (2011: 112) stated, 'The appearance of *Chazara briseis larnacana* (OBERTHÜR, 1909) of Cyprus is similar to *Chazara briseis meridionalis* (STAUDINGER, 1886) of Asia Minor and very distinct from any European phenotype of *C. briseis*; it is well worth of thorough examination, both anatomically and molecularly.' In Kudrna *et al.* (2015: 396) the entry was updated to '*C. briseis larnacana* (OBERTHÜR, 1909) of Cyprus appears to be a distinct subspecies', and in Kudrna (2019: 102) further updated to 'appears to be the most distinct geographical form of the species and is well worthy of recognition'.

Earlier publications, for example Lederer (1855) and Turner (1920), referred to *Satyrus briseis*, the latter adding '[race fergana Obthr.; ab. pirata ♀, Esp.]'. Rebel (1939) recognized the ssp.: 'Satyrus briseis (L.) larnacana Obthr.'

Distribution in Cyprus

Altitudinal: all elevations, but most commonly found at lower and intermediate elevations.

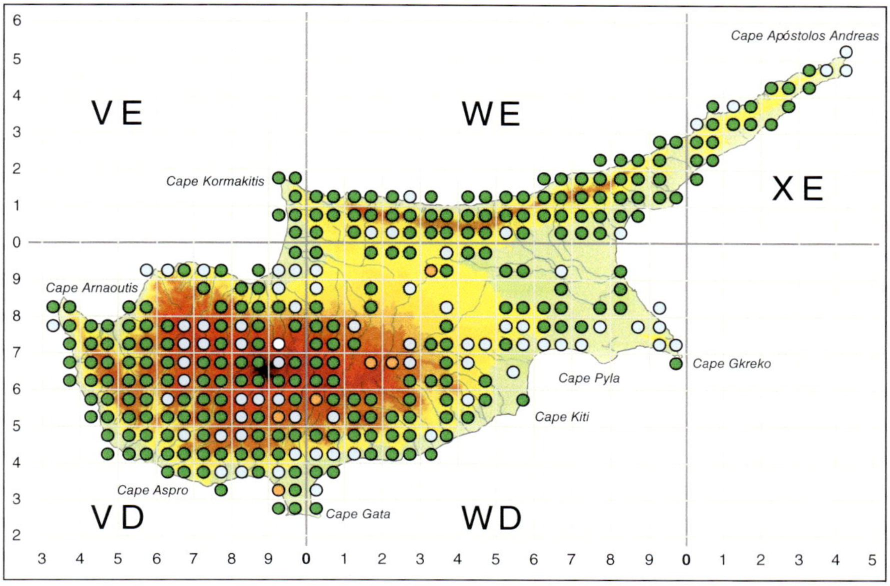

Fig. 12.352. *Chazara briseis* distribution map showing 5 km² observations within 10 km² UTM squares. © *Eddie John*

Geographical: a very common butterfly, present over much of the island, but with fewer records from the Mesaoria Plain, in part a recording artefact attributable to reduced recording effort in an unfavourable environment.

Phenology

Univoltine, late April to early November. Overwinters as a larva.

Larval hostplants

Poaceae: *Poa* spp.

Ecology and behaviour

Found in a range of habitats, commonly along tracks with steep-sided banks on which *C. briseis* rests and from which it is very easily disturbed. As Cyprus represents the southernmost extent of the species' range in the eastern Mediterranean, and is considered rare in nearby provinces of Turkey and 'near threatened' at a European level, it is surprising that *C. briseis* is plentiful on the island. Indeed, within the wider region, Cyprus could be considered a 'hotspot' for the species. At times, the presence of *C. briseis* may not be apparent, as shade is often sought within rocky openings or tunnels beneath roadways that also provide aestivation refuges. For example, over a two-week period in May/June 2013, when early, light aestivation might already have been sought, the first author observed *C. briseis* at just 14 (12%) of 118 locations visited across the island. In contrast, in mid-late October 2017, by which time few males would have survived and losses to predation of both sexes incurred, the species was noted at 45 (34%) of 131 widely scattered locations.

As with *Hipparchia* spp. in Cyprus, egg-laying does not take place until the autumn, indicating delayed ovarian maturation; Makris (2003: 246) referred to oviposition on dry grass in autumn. Post-aestivation females of *C. briseis* were observed puddling at coarse wet sand at eight different sites along the shoreline and at mud in October 2017 (John and Dennis, 2019), complementing the statement by Boggs and Dau (2004) that long-lived females of monandrous species have been observed to puddle when older. Normally the preserve of males, this activity might replenish female sodium reserves that have become depleted during the months of aestivation, thereby benefiting female fecundity. John and Dennis (2019) expressed this as shown in Fig. 12.353.

An avid seeker of nectar, *C. briseis* is often seen at roadsides and field margins nectaring on the thistle species *Onopordum cyprium* (a very common endemic) and *Ptilostemon chamaepeuce* (Makris, 2003: 246). At coastal locations, *Centaurea aegialophila* and *Limonium sinuatum* hold a strong attraction, as does *Thymbra capitata* in the hills; nectaring has also been observed on *Asphodelus ramosus* (first author, personal observation).

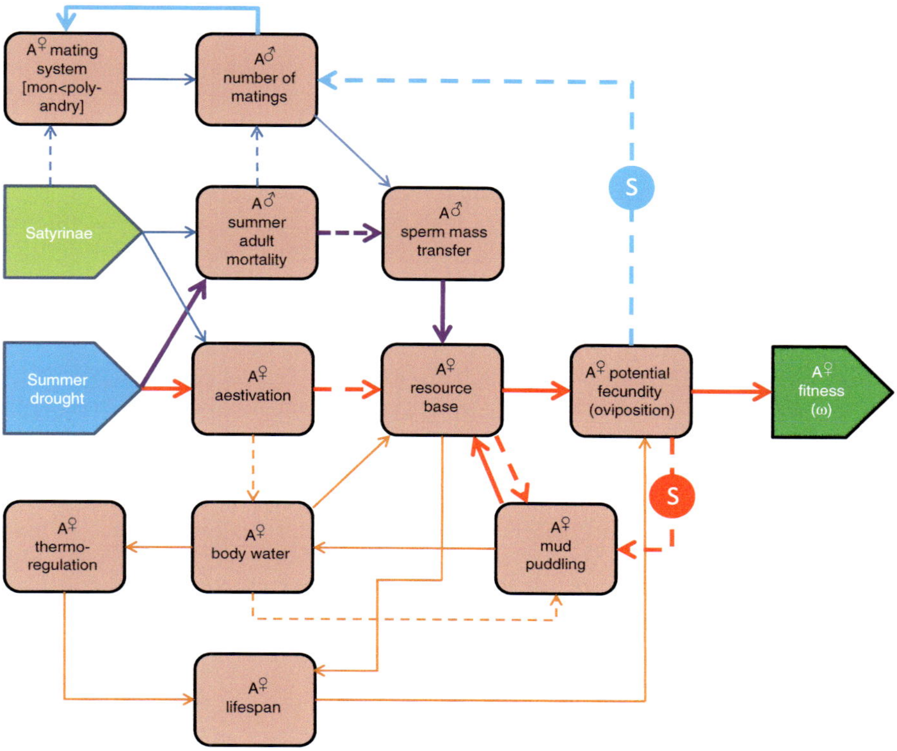

Fig. 12.353. 'A model of potential factors leading to female mud-puddling in Cyprus Satyrinae. The key feature is female aestivation and high levels of pre-summer drought mortality in males. Both female aestivation and male losses should lead to depletion in both water and nutrients in females during a time of drought and limited nectar sources. The males pass nutrients over to females in spermatophores, but this source of nutrients may be limited by monandry and smaller spermatophores in males that have already mated. Nutrient and water resources (hydration) are maintained by mud-puddling. The additional nutrients then add to female maintenance (i.e. water by controlling thermoregulation) and egg laying, which ensures high levels of fecundity and female fitness. Red arrows: mud-puddling links related to aestivation; orange arrows: links with hydration; purple arrows: links associated with male contributions; thin blue arrows are associated with the mating system; thick pale blue arrows indicate a possible evolutionary pathway to polyandry, triggered by the need for additional contributions from males; A, adult; S, potential selection pathway; continuous arrows, positive associations; pecked arrows, negative associations.' Diagram and text by Roger Dennis, originally published in *Entomologist's Gazette*, reproduced here courtesy of Pemberley Books (Publishing).

Description/identification aids

Wingspan: generally up to 65 mm, but occasionally larger; Bretherton (1954) referred to having in his collection a ♀ with a wingspan of 88 mm. **Uppersides** of both sexes are dark brown with a wide, cream discal/postdiscal band (*very noticeable in flight*) that extends *obliquely* across both fore- and hindwings. Two white-pupilled, black ocelli are present on the forewing, less distinct in the ♂. Forewing **underside** pattern is complex, with a broad irregular postdiscal band, cream-white at the inner margin and becoming orange at the first of two large ocelli, that nearest the grey wing apex *always* visible when nectaring. Hindwing undersides have a characteristic patterning of irregular brown banding on a grey-brown (♂) or light brown (♀) ground colour. Females are much larger and have a commonly seen dimorphic form, f. 'pirata', *where the cream-white bands on upper- and undersurfaces are replaced by orange.* Parker (1983) estimated f. 'pirata' to comprise *ca.* 40% of the female population; we would place this figure at 20–30%. Bucknill (1911) noted that f. 'pirata' was not common – an opinion shared by Wilson (cited in Turner, 1920), who stated 'but is rare. I saw one on Troödos in 1918'. These remarks, from experienced people who spent much time in Cyprus, appear to suggest that f. 'pirata' has become considerably more common than hitherto.

Similar species

While uppersides are markedly different, underside markings of *C. briseis* bear a resemblance to those of *C. persephone* (see next species).

Conservation status in Cyprus

Not threatened locally, although a species of European conservation concern where it is accorded a status of 'NT' (Van Swaay *et al.*, 2010a, 2011). At the Mediterranean level, *C. briseis* is listed as 'LC' (Numa *et al.*, 2016).

Chazara persephone (Hübner, [1805]) Satyridae (Satyrinae)

Great Steppe Grayling/Dark Rockbrown

TL: 'Russland, bei Sarepta' (Russia).

Fig. 12.354. *Chazara persephone* (Great Steppe Grayling), photographed in Turkey. Courtesy of *Ahmet Baytaş*

Fig. 12.355. *Chazara persephone* (Great Steppe Grayling) Kantara Mountains (500 m), 27 June 2010. Courtesy of *Daniel and Hilary Haines*

Fig. 12.356. *Chazara persephone* (Great Steppe Grayling) ♀, photographed in Turkey, 23 June 2014. Courtesy of *Onat Başbay*

Fig. 12.357. *Chazara persephone* (Great Steppe Grayling) ♀ (upperside top, underside below) ex Latakia Mountains, Syria, 10 July 2020. Courtesy of *Mudar Salimeh*

Presence in the eastern Mediterranean

Found, usually sparingly, throughout much of the Levant, including Cyprus, as the ssp. *Chazara persephone transiens* Zerny, 1932, TL: N. Lebanon. *C. persephone* is shown as being present in Mersin Province, Turkey, by Hesselbarth *et al.* (1995: Map 285), although is not listed for Hatay (Turkey) by Atahan *et al.* (2018). Zarikian and Ghrejyan (2018) reported numerous sightings from the Syrian Anti-Lebanon, but the species is scarce in the Latakia mountain range (Mudar Salimeh, Latakia, 2021, personal communication to first author). Whereas it was once considered common in Lebanon, Larsen (1974: 134) found the species to be scarce during his time in the country from 1970 to 1974. Presence in Israel is restricted to an area along the border with Lebanon (Benyamini, 2002a: 142).

Not listed by Lederer (1855) but reportedly taken as *Satyrus anthe* (= *Chazara persephone*) in the Troodos Mountains in 1911 and 1918 (Bucknill, 1911; Turner, 1920). Bucknill (1911) stated, 'We have only met with it very rarely on Troödos in August.' Wilson (in Turner, 1920) further informed: 'A very rare species met with on Troödos at 5000–6000 ft.' Rebel (1939) added, 'Although only found 2 decades ago in Cyprus, probably the species is one of the oldest faunal elements on the island' (translated from the original German). Parker (1983) listed '*Chazara anthe* Ochsenheimer, 1807 (Giant Steppe Grayling)' among doubtful records but, in questioning these, astutely noted, 'but have to concede that it is possible, since *anthe* is found both in Turkey and Lebanon. Also I have established that both Bucknill and Wilson were able to differentiate between their "*anthe*" and both *C. briseis and P. anthelea*'. See John *et al.* (2011) for further details.

Distribution in Cyprus

Altitudinal: there is insufficient information for Cyprus, where (in recent years) *C. persephone* has been found only once, at *ca.* 500 m; in Lebanon, noted historically from sea level to 2500 m (Larsen, 1974: 134).
Geographical: in recent years, known only from the mountains near Kantara – and then only once.

Phenology

Univoltine, although phenology in Cyprus is unknown. If resident, likely to be from May/June into the autumn. Overwinters as a larva elsewhere.

Larval hostplants

Poaceae: *Poa* spp.

Ecology and behaviour

This species had remained unrecorded in Cyprus since the early 1900s, at which time it was found only rarely in the Troodos Mountains (see above). On 27 June 2010, when in the Kantara Mountains in northern

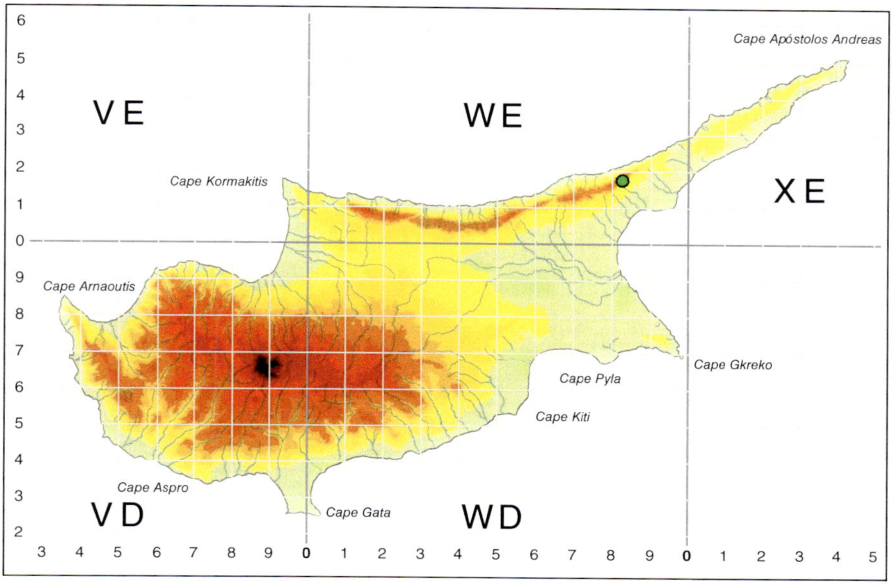

Fig. 12.358. *Chazara persephone* distribution map showing the single 5 km² location. © *Eddie John*

Cyprus, Drs Daniel and Hilary Haines observed an unrecognized satyrine that had alighted on a vertical bank recently cleared of vegetation. An image (Fig. 12.355) was sent to the first author, and following consultation with others, a determination of *Chazara persephone transiens* was accepted, based on the wing morphology described below. The reason for the species' apparent absence for almost 100 years is uncertain. It is tempting to consider a westerly migration from the Syrian Anti-Lebanon, but *C. persephone* is not known as an obligate migrant, although Larsen and Nakamura (1983) speculated on the occasional dispersal of the species from Lebanon 'into Jordan and Palestine ... in years of exceptional abundance'. Although unlikely, it is possible that *C. persephone* is simply exceedingly rare in Cyprus, and overlooked in the presence of far more numerous members of other satyrine genera.

Hesselbarth *et al.* (1995: 932) stated that for much of the time *C. persephone* remains on rock faces or steep embankments, a description of behaviour fitting perfectly with that of the observation by D and H Haines in Cyprus in 2010. The known cryptic nature of *C. persephone* also mirrors the experience in Israel, where it is rarely seen other than on Mt Hermon (Dubi Benyamini, Beit-Arye, 2010, personal communication to first author). The species aestivates during the hottest months (Tshikolovets, 2011: 394).

Description/identification aids

Wingspan: up to 65 mm. A large butterfly, the male noticeably smaller. *C. persephone* differs from *C. briseis* as follows: '...wing morphology, especially the pronounced wing veining, is unlike that of other satyrids [satyrines] found in Cyprus. The single visible postdiscal ocellus on the forewing underside is blind, also ... the pale yellow band extending either side of the underside forewing ocellus and reaching to the costa, together with the pale hindwing underside postdiscal and submarginal bands are typical for *Chazara persephone transiens*' (John *et al.*, 2011). A differently marked form of the female (f. 'hanifa') has the white banding replaced by orange-brown.

Similar species

There is a superficial underside similarity to *C. briseis*; however, this lacks *the very prominent white wing veining*, characteristic of *C. persephone*.

Conservation status in Cyprus

Data deficient. Listed by Van Swaay *et al.* (2010a) and Numa *et al.* (2016) as 'LC' in Europe and at the Mediterranean level, respectively.

Pseudochazara anthelea (Hübner, [1824]) Satyridae (Satyrinae)

White-banded Grayling

TL: Not stated (Asia Minor).

Fig. 12.359. *Pseudochazara anthelea* (White-banded Grayling) ♂, 25 June 2011. *Christodoulos Makris*

Fig. 12.360. *Pseudochazara anthelea* (White-banded Grayling) ♂, 26 June 2011. *Christodoulos Makris*

Fig. 12.361. *Pseudochazara anthelea* (White-banded Grayling) ♀, 11 June 2013. *Christodoulos Makris*

Fig. 12.362. *Pseudochazara anthelea* (White-banded Grayling) ♀ nectaring on *Ptilostemon chamaepeuce*, 17 June 2006. *Christodoulos Makris*

Fig. 12.363. *Pseudochazara anthelea* (White-banded Grayling) ♀ (L), ♂ (R), 11 June 2013. *Christodoulos Makris*

Fig. 12.364. *Pseudochazara anthelea* (White-banded Grayling) ♀ (L), ♂ (R), 11 June 2013. *Christodoulos Makris*

Presence in the eastern Mediterranean

Within the Levant, the species is known only from Turkey and Cyprus, probably as nominotypical *P. anthelea anthelea*, with just one record given for nearby Hatay Province, Turkey (Atahan *et al.*, 2018: 83). Cyprus represents the southernmost extent of the species' range in the Mediterranean, where Lederer (1855) first recorded *P. anthelea* as '*Satyrus Anthelea* Hb.'. Bucknill (1911) referred to '*Satyrus anthelea. Hbn. The Mountain Flower*', Turner (1920) to 'Hipparchia anthelea Hbn.'

Later works (e.g. Parker, 1983; Manil, 1990) agreed with the designation by Rebel (1916) of the endemic status of ssp. *P. anthelea acamanthis* (described by Rebel as '*Satyrus anthelea* (Hb.) *acamanthis* n. subsp.'). However, based on wing patternation, the ssp. was later synonymized with *P. anthelea anthelea* by Olivier (1996). Further detailed work by Wakeham-Dawson *et al.* (2003) revealed a general similarity of genitalia and androconia, and supported conspecificity of *P. anthelea acamanthis* with *P. anthelea anthelea*, but suggested further investigation. *P. anthelea acamanthis* was not referred to as a valid spp. in their work on the genus *Pseudochazara* by Wakeham-Dawson and Dennis (2001) or by Verovnik and Wiemers (2016). Pyrcz *et al.* (2015) proposed the synonymy of *Pseudochazara* as a subjective junior synonym of *Chazara*, but the generic name of *Pseudochazara* was maintained by Wiemers *et al.* (2018, 2020).

Distribution in Cyprus

Altitudinal: usually found above 1000 m, rarely at submontane elevations (see below).

Geographical: a local, generally uncommon butterfly, well distributed over the Troodos mountain range (where it may be locally common), but far less frequently reported from the Pentadaktylos range, where it is known only from western areas. *P. anthelea* is absent from eastern areas of

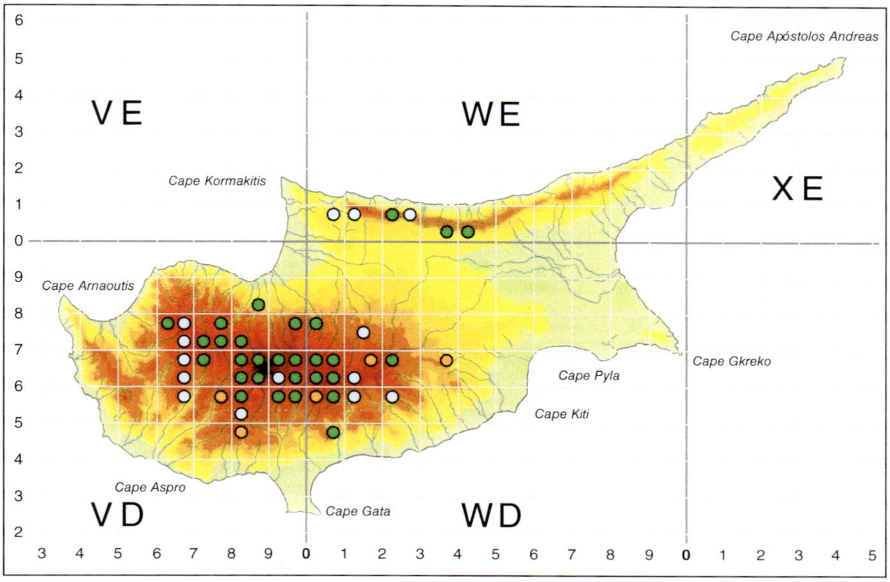

Fig. 12.365. *Pseudochazara anthelea* distribution map showing 5 km² observations within 10 km² UTM squares. © *Eddie John*

the island, coastal regions and the Mesaoria Plain. Population declines are possible, since Bucknill (1911), although encountering the species only on Troodos, referred to *P. anthelea* being common there in July, August and September.

Phenology

Univoltine, late April to September. As with other univoltine satyrines in Cyprus, *P. anthelea* probably aestivates prior to egg-laying in the autumn upon the return of seasonal rainfall.

Larval hostplants

Poaceae: *Poa* spp.

Ecology and behaviour

Generally inhabits steep, rocky, mainly south-facing hillsides above 1000 m and in Cyprus is very rarely recorded below this elevation. The altitudinal range accords well with an analysis of almost 300 observations of *P. anthelea anthelea* listed for the Turkish mainland by Hesselbarth *et al.* (1995: 962). Wakeham-Dawson *et al.* (2003) cited this work as follows: 'On the Turkish mainland, only three specimens (1% of sightings) are listed as being noted below an altitude of 250 m while, in sharp contrast, 282 (96%) were found above 500 m (including 237 records (81%)

observed at 1000 m or higher).' Of 23 specimens collected in Cyprus for the Wakeham-Dawson *et al.* (2003) study, one was taken at 800 m, the remainder from 1120 m to 1450 m. A single observation of *P. anthelea* nectaring on *Lantana* at 250 m (Daniel Haines, London, 2010, personal communication to first author) represents the only low elevation record known in Cyprus.

Males are territorial, usually with a low fast-flying habit, frequently alighting on rocks and often returning to the same perching spot. Figures 12.363 and 12.364 show a courtship display by the male, associated with probable release of pheromones. *P. anthelea* is attracted to flowers of *Centaurea aegialophila* (Makris, 2003: 250) and males have been observed mud-puddling in company with *H. cypriensis*.

Description/identification aids

Wingspan: up to 55 mm. The ♀ is a large butterfly, but sexes show *very marked* dimorphism: males are smaller, appearing *black and white* in flight. ♂ **uppersides** (not seen at rest) are almost black, with a wide, *white postdiscal band* on the fore- and hindwings, the forewing having two white-pupilled black ocelli and a line of androconia indistinct against the ground colour. Small white vestigial ocelli are present on all wings. ♂ **underside** is similar to the upper, but with more extensive forewing white. ♀♀ are larger, uppersides marked much as the ♂, but with a brown ground colour and *orange-brown replacing the white.* Ocelli are larger and more prominent than in the male. ♀ forewing underside lacks the brown ground colour of the upperside and is replaced by *orange-brown*; hindwing underside is grey and *largely unmarked* except for vestigial ocelli.

Similar species

Chazara persephone – see above. In having twin forewing ocelli and somewhat similar colouring, ♀♀ of *P. anthelea* bear a slight resemblance to *C. briseis.*

Conservation status in Cyprus

Not threatened. Listed by Van Swaay *et al.* (2010a) and Numa *et al.* (2016) as 'LC' in Europe and in the Mediterranean, respectively.

Maniola cypricola (Graves, 1928) Satyridae (Satyrinae)

Cyprus Meadow Brown

TL: "Patria. – Cyprus".

Fig. 12.366. *Maniola cypricola* (Cyprus Meadow Brown) ♂ showing the prominent forewing scimitar-shaped androconia, 2 May 2014. Courtesy of *Gerrit Ros and Jeannette Dewaerheijt*

Fig. 12.367. *Maniola cypricola* (Cyprus Meadow Brown) ♀, 5 June 2011. *Christodoulos Makris*

Fig. 12.368. *Maniola cypricola* (Cyprus Meadow Brown) ♀ post-aestivation, nectaring on *Dittrichia viscosa*, 19 October 2017. (Note the relatively unblemished appearance, at least seven months after the spring emergence.) *Eddie John*

Fig. 12.369. *Maniola cypricola* (Cyprus Meadow Brown) ♂ nectaring on *Centaurea hyalolepis*, 1 May 2008. *Christodoulos Makris*

Fig. 12.370. *Maniola cypricola* (Cyprus Meadow Brown) ♀ nectaring on *Stachys cretica*, 8 May 2012. *Christodoulos Makris*

Fig. 12.371. *Maniola cypricola* (Cyprus Meadow Brown) pairing, ♂ left, ♀ right, 25 April 2014. *Eddie John*

Presence in the eastern Mediterranean

An endemic species to Cyprus, originally described as *Epinephele cypricola* Graves, 1928. As discussed in Chapter 9, molecular work by Kreuzinger *et al.* (2015) has raised the question of 'taxonomic "oversplitting"', with *M. cypricola* belonging to a 'superspecies' – *M. jurtina.* This conflicts with earlier held views by, for example, Thomson (1976, 1987: Fig 185, 2011: Fig 185) and the late Otakar Kudrna (2019, personal communication to first author) who, in their work on various *Maniola* spp., noted differences in both male and female genitalia of *cypricola* when compared with other members of the genus. Thomson (1992) also noted differences in egg surface morphology. Olivier (1993: 172) stated, 'While the female of *M. cypricola* bears a close resemblance to that of *M. telmessia* in external features, the males of both taxa can be distinguished at a first glance (*M. cypricola* males look more similar to those of *M. chia and M. halicarnassus*).' Moreover, Grill *et al.* (2004) stated, 'On basis of the valvae, however, *M. chia* and *M. cypricola* can be clearly distinguished from the other *Maniola*.' Tshikolovets (2011: 340) considered *cypricola* to be a ssp. of *M. telmessia*, whereas Olivier (1993: 145), commenting on wing morphology, stated 'The ♂ of *M. cypricola* can easily be separated from the ♂ of *M. telmessia*'. Vlad Dincă (Oulu, 2019, personal communication to first author) informed that analyses based on mitochondrial DNA (DNA barcoding) have led to inconclusive results regarding the relationships within European *Maniola*, and additional molecular data are required before any conclusions can be drawn. Wiemers *et al.* (2018) continued to list *M. cypricola* as a valid species, as did Maes *et al.* (2019), Middleton-Welling *et al.* (2020) and Wiemers *et al.* (2020).

Inexplicably, in view of his arrival on the island in early May, *M. cypricola* was overlooked by Zach during his 1853 visit (see Lederer, 1855), and this species was probably first reported for Cyprus by Bucknill (1911), as '*Epinephele jurtina. L. form telmessia Z. The Cyprian Meadow Brown*'. Turner (1920) referred to 'Epinephele telmessia Zell. (E. jurtina v. telmessia

Zell.) [race kurdistana Rühl. (?)]'. Rebel (1939), clearly aware of Graves' (1928) work, but appearing not to accept his elevation of *cypricola* to species status, listed the taxon as 'Epinephele jurtina (L.) cypricola Grav.'. Olivier (1993: 172) designated BMNH specimens (both collected by G.A. Mavromoustakis in 1926) as the lectotype and paralectotype of *Maniola cypricola*.

Distribution in Cyprus

Altitudinal: found at all elevations.
Geographical: a very common endemic species, found throughout the island, but with fewer records from the Mesaoria Plain, perhaps due to under-recording in an arid region (John and Skule, 2016: 321). It is only matter of time before the species is recorded in all 5 km squares.

Phenology

Univoltine, usually April to October, but may be seen earlier or later. Overwinters in the larval stage.

Larval hostplants

Poaceae: *Poa* spp.

Ecology and behaviour

M. cypricola, a very common endemic species, is lethargic in flight and largely sedentary in nature. Found in all types of habitat, including

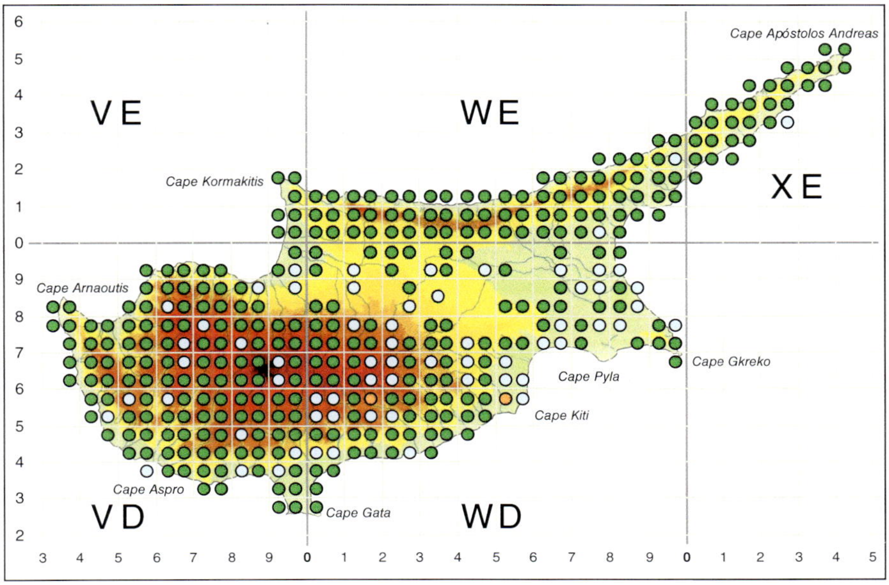

Fig. 12.372. *Maniola cypricola* distribution map showing 5 km² observations within 10 km² UTM squares. © *Eddie John*

urban areas, but less commonly on the central plain where cereals are extensively grown. Although apparently more tolerant of heat than most of Cyprus's satyrines, *M. cypricola* is usually found near bushes, into which the species readily retreats to seek shade, especially during the height of summer, but unlike the *Hipparchia* spp. in Cyprus, *M. cypricola* does not appear to retreat to higher, cooler elevations at which to aestivate. In summer, populations at low elevations become much less active and enter light aestivation, while those present in the cooler conditions of the mountains are more active and are commonly encountered.

Following mating in spring, delayed ovarian maturation ensures that oviposition is timed to coincide with the re-growth of grasses upon the return of rain in the autumn (Makris, 2003: 252), a trait in common with *Maniola* spp. elsewhere in the Mediterranean (e.g. Grill *et al.*, 2013; Haeler *et al.*, 2014). In spring, *M. cypricola* is especially attracted to the flowers of *Cistus creticus*, on which it nectars gregariously with others of the same species. *Centaurea aegialophila*, *C. hyalolepis*, *Echium angustifolium*, *Limonium sinuatum*, *Thymbra capitata* and *Rubus sanctus* are among other favoured nectar plants. Females of the species have been observed mud-puddling in October, an intake that possibly replenishes depleted sodium reserves and thereby enhances fecundity (John and Dennis, 2019) – see also Fig. 12.353 and entry for *Chazara briseis*.

Description/identification aids

Wingspan: up to 50 mm. Superficially resembles European *M. jurtina*, ♂ **upperside** is dark brown, relieved by partial orange postdiscal scaling, a white-pupilled, black apical ocellus surrounded by orange and a *wide curved black androconial band* on the forewing (Fig. 12.366). ♀ fore- and hindwing uppersides are brighter, with a predominance of brown and orange markings and a larger apical ocellus. **Undersides** are dull brown, with a lighter postdiscal band, *clearly delineated* in the female. The *forewing underside of both sexes is orange*, with an apical ocellus. Up to five ocelli of *varying size* may be present on the hindwing underside.

Similar species

In Cyprus, only *Hyponephele lupina* – see next species.

Conservation status in Cyprus

Not threatened. Although described by Grill *et al.* (2006) as one of Europe's rarest butterflies, their assessment of *M. cypricola* should be viewed in a Europe-wide context, as this is a very common species in Cyprus. For example, during a late-season visit to Cyprus by the first author in 2017 (12–25 October), *M. cypricola* was recorded at 64 (49%) of 131 locations visited, at a time when male survival rate would be low and natural losses would have taken a toll on females. Van Swaay *et al.* (2010a) and Numa *et al.* (2016) listed *M. cypricola* as 'LC', i.e. a species of least concern.

Hyponephele lupina cypriaca Riley, 1921 — Satyridae (Satyrinae)

Oriental Meadow Brown

TL: Bosco di Guagnano nelle foreste di Terra d'Otranto (Puglia, Italy).

Fig. 12.373. The rarely seen upperside of *Hyponephele lupina* (Oriental Meadow Brown) ♀ nectaring on *Cardopatium spinosum*, 8 May 2016. Courtesy of *David and Rosalyn Sparrow*

Fig. 12.374. *Hyponephele lupina* (Oriental Meadow Brown) ♂, 12 May 2015. *Eddie John*

Fig. 12.375. *Hyponephele lupina* (Oriental Meadow Brown) ♂, 12 August 2011. *Christodoulos Makris*

Fig. 12.376. *Hyponephele lupina* (Oriental Meadow Brown) ♀ nectaring on *Cistus creticus*, 17 June 2012. *Christodoulos Makris*

Fig. 12.377. *Hyponephele lupina* (Oriental Meadow Brown) ♀, 30 April 2016. Note the smaller, lower blind ocellus, not present in *M. cypricola*. *Eddie John*

Presence in the eastern Mediterranean

H. lupina is well represented in northern countries of the Levant, but is rather local in Mediterranean Syria, Lebanon and Israel (Larsen, 1974: 139; Benyamini, 2002a: 146), although Benyamini (2021b: 58) noted an extension of range into the northern Negev Desert. *Hyponephele lupina intermedia* Staudinger, (1886) was identified as a subspecies for Turkey and Cyprus, a ssp. about which Olivier (1993: 179) wrote, 'I somewhat reluctantly maintain the name *intermedia* for the Eastern Aegean, Turkish and remaining Asiatic populations', and who then further expressed the view that populations from much of the wider region might finally be 'lumped under nominotypical *H. lupina*'. Hesselbarth (1995: 844) similarly listed ssp. *intermedia* for Turkey, as did Samodurow *et al.* (2001) for countries of the former USSR in their extensive overview of the genus *Hyponephele*. Ssp. *Hyponephele lupina centralis* Riley, 1921 had been listed for some countries of the Middle East, including Lebanon and Israel, but has since been synonymized with ssp. *intermedia* by Eckweiler and Bozano (2011: 49) and with the nominotypical ssp. by Lukhtanov and Pazhenkova (2021).

H. lupina was another of the species overlooked in Cyprus by Zach during his 1853 visit (see Lederer, 1855). Bucknill (1911) listed both '*Epinephele lycaon. Rott.*' and '*Epinephele lycaon. Rott. form lupinus*' for Cyprus, while Gibbs (cited in Turner, 1920), stated, 'I am inclined to think one may call all the *lycaon*, var. *lupinus*.' Riley (1921), noting the 'extreme in depth of coloration in both sexes above and below', described '*Epinephele lupinus cypriaca*' as a new subspecies, designating Bucknill's specimens from 1909 as the types. Olivier (1993: 178) disagreed: 'Riley, 1921 described distinct subspecies from Irak/Iran, Turkey and Kípros [Cyprus], but these seem superfluous', a view supported by Hesselbarth *et al.* (1995: 845), who synonymized '*Epinephele lupinus cypriaca*, subsp. n.' with *Hyponephele lupina intermedia*. However, Eckweiler and Bozano

(2011: 49), in stating 'both sexes ups and uns with deep coloration; female upf yellow markings of a very deep shade', supported a subspecies status for *cypriaca*, as did Lukhtanov and Pazhenkova (2021), who also regarded the taxon as morphologically distinct: 'Therefore the subspecies status of *H. lupina cypriaca* Riley, 1921 should be preserved.' However, it remains to be seen if future molecular work supports Riley's designation of *cypriaca* as a subspecies. The species is regarded by some authors, e.g. Eckweiler and Bozano (2011: 48), as *Hyponephele lupinus*, but we follow Wiemers *et al.* (2018, 2020).

Distribution in Cyprus

Altitudinal: may be encountered at any elevation.
Geographical: *H. lupina* has a similar distribution across the island to *M. cypricola*; however, the former is far less commonly reported, possibly because of close similarity to (i.e. confusion with) the more familiar *M. cypricola*.

Phenology

Univoltine, late April to October; generally appearing on the wing four to five weeks later than *M. cypricola* and not persisting as late into the autumn.

Larval hostplants

Poaceae: *Poa* spp. Overwinters in the larval stage.

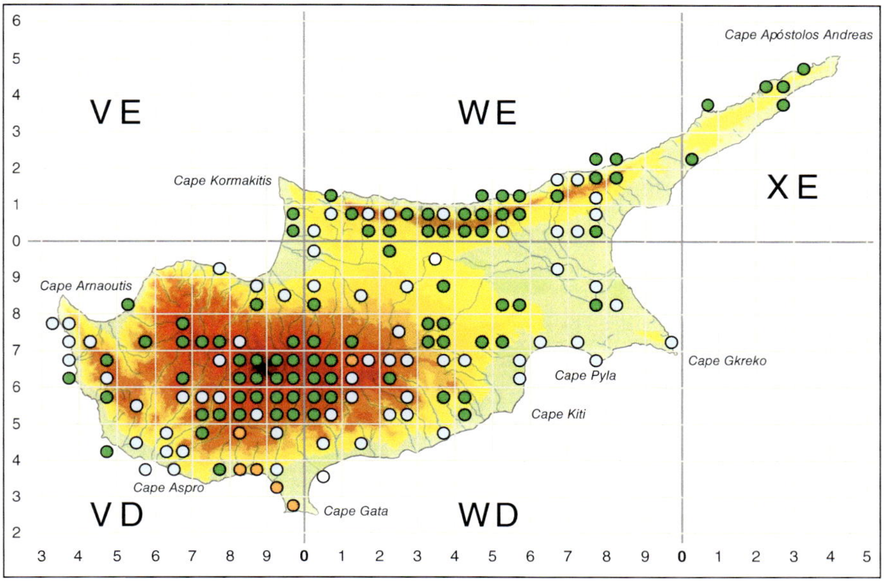

Fig. 12.378. *Hyponephele lupina* distribution map showing 5 km^2 observations within 10 km^2 UTM squares. © *Eddie John*

Ecology and behaviour

Resembles *M. cypricola* in flight, with which it shares similar habitats – low-lying fields, garigue, maquis and mountainous areas – but *H. lupina* is usually, although not always, outnumbered by *M. cypricola*. *H. lupina* perhaps shows a greater tendency to inhabit shadier locations than *M. cypricola*, especially as summer approaches; it is similarly attracted to the flowers of *Cistus creticus* and *Rubus sanctus*, and is also a mud-puddler. As with *M. cypricola*, there is delayed ovarian maturation followed by oviposition in the autumn (Jutzeler and Lafranchis, 2005).

Description/identification aids

Wingspan: up to 50 mm. Very similar to *M. cypricola*, but differs in the following ways.

- Uppersides of both sexes (rarely displayed) are dark brown (slightly lighter in the ♀), *with no orange* other than a small amount on the forewing of the ♀.
- The ♂ sex brand is much narrower and not scimitar-shaped as in *M. cypricola*.
- The clearest distinguishing character is in the ♀, where *two* large submarginal forewing ocelli (non-pupilled and encircled in light orange on the upperside – see Fig. 12.373) are present *on upper- and undersides* (the lower, blind forewing underside ocellus (Fig. 12.377) is *sometimes* visible when at rest). *M. cypricola* has a single forewing ocellus.
- The postdiscal band on the ♀ hindwing upper- and underside is less well defined than in *M. cypricola*.
- Both sexes have *noticeably serrated* hindwing margins (*cf.* the less sharply dentate hindwing margin of *M. cypricola*).

Similar species

In Cyprus, *M. cypricola*; elsewhere in the Levant, *Hyponephele lycaon* (Kühn, 1774).

Conservation status in Cyprus

Not threatened. Listed by Van Swaay *et al.* (2010a) and Numa *et al.* (2016) as 'LC' in Europe and in the Mediterranean region, respectively.

Pararge aegeria (Linnaeus, 1758) **Satyridae** (Satyrinae)

Speckled Wood

TL: S Europe and North Africa.

Fig. 12.379. *Pararge aegeria* (Speckled Wood) ♂ nectaring on *Allium trifoliatum*, 19 April 2010. Courtesy of *Alison McArthur*

Fig. 12.380. *Pararge aegeria* (Speckled Wood) ♂, 7 February 2015. *Christodoulos Makris*

Fig. 12.381. *Pararge aegeria* (Speckled Wood) nectaring on *Sonchus oleraceus*, 14 February 2013. *Christodoulos Makris*

Fig. 12.382. *Pararge aegeria* (Speckled Wood) nectaring on *Sinapis alba*, 14 February 2013. *Christodoulos Makris*

Presence in the eastern Mediterranean

Absent from Jordan and Sinai, but present elsewhere in the Levant as nominotypical *P. aegeria aegeria.* Overlooked by Zach in 1853 and hence unrecorded by Lederer (1855), this species was documented for Cyprus by Bucknill (1911) as '*Pararge ægeria*' and under the familiar name by subsequent authors.

Distribution in Cyprus

Altitudinal: from sea level to the highest elevations.
Geographical: a moderately common butterfly, but local in distribution: recorded over much of the Troodos range and foothills and from coastal regions of the west. Sparsely recorded in the north, unrecorded from much of the Mesaoria Plain and absent from the more arid south-east.

Phenology

Probably trivoltine in Cyprus, January to December in overlapping broods (CBRS data). Elsewhere, *P. aegeria* overwinters as a third instar larva or as a pupa; in Cyprus, possibly as a slowly developing larva or in the pupal stage.

Larval hostplants

Poaceae: *Poa* spp. A preference for perennial grasses such as *Brachypodium pinnatum* that grows in damp localities was reported by Makris (2003: 258).

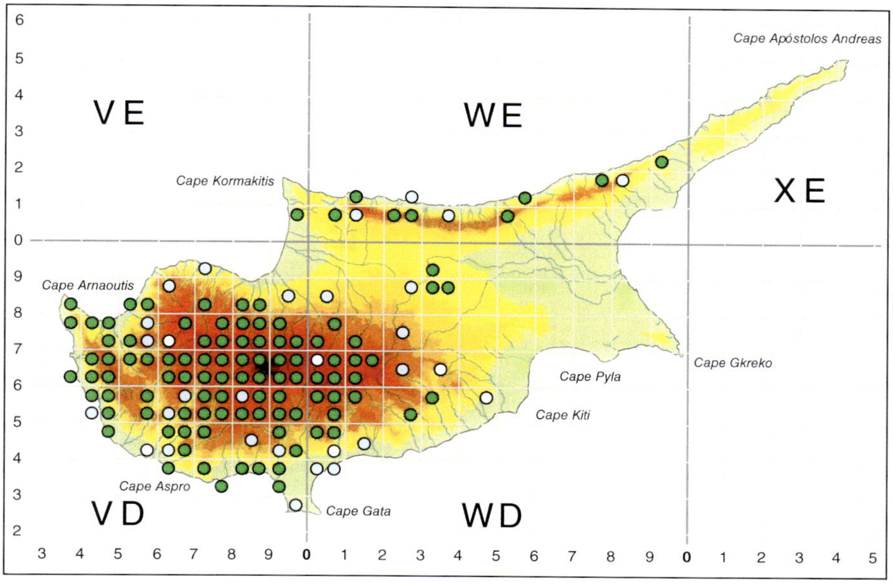

Fig. 12.383. *Pararge aegeria* distribution map showing 5 km² observations within 10 km² UTM squares. © *Eddie John*

Ecology and behaviour

Found in light riverine woodland, watercourses with overhanging vegetation, e.g. *Arundo donax* (giant reed), and trees that provide light shade, as well as other types of habitat, usually near water. Dappled sunlight is a prerequisite, where *P. aegeria* males perch in sunshine on low vegetation or on the ground, awaiting the arrival of females. In such circumstances, males become aggressively territorial, repeatedly returning to the same sunlit perch after chasing away intruders, often in a spiralling flight. In fragmented agricultural landscapes, Merckx and Van Dyck (2005) noted a greater tendency towards patrolling, a mate location behaviour less commonly seen in woodland. Bálint *et al.* (2016) reported the taking of a singleton at light in Lebanon. Nectaring is rarely noted, as *P. aegeria* is more attracted to honeydew, but has been observed at *Allium trifoliatum*, *Rubus sanctus*, *Sonchus oleraceus* and *Sinapis alba*.

Description/identification aids

Wingspan: up to 45 mm. In Cyprus, *P. aegeria* is present as the nominotypical (yellow-orange) *P. aegeria aegeria*, not as the darker, white-spotted ssp. *P. a. tircis* found in much of northern Europe and western Asia. **Uppersides** are characteristically marked in yellow-orange and brown, the ♂ darker, the ♀ with more yellow-orange markings. Both sexes have a forewing apical ocellus and three distinct white-pupilled, black ocelli on the hindwing.

A fourth, vestigial ocellus may be present. ♂ and ♀ forewing **undersides** are similar to those of the upperside but less heavily marked, giving a brighter appearance. Hindwing undersides are similar in both sexes, lightly patterned yellow-brown-grey, with three indistinct white-pupilled ocelli in the submarginal band.

Similar species

Lasiommata megera and *L. maera* uppersides are superficially similar, but these are generally found in different habitats. Forewing uppersides of both *Lasiommata* spp. have a *larger and brighter apical ocellus*, while hindwing undersides have a series of *clearly defined ocelli*.

Conservation status in Cyprus

Not threatened. Listed by Van Swaay *et al.* (2010a) and Numa *et al.* (2016) as 'LC' in Europe and in the Mediterranean, respectively.

Lasiommata megera (Linnaeus, 1767) Satyridae (Satyrinae)

Wall Brown

TL: "Austria, Dania".

Fig. 12.384. *Lasiommata megera* (Wall Brown) ♂, 3 May 2015. *Eddie John*

Fig. 12.385. *Lasiommata megera* (Wall Brown) ♂ nectaring on *Dittrichia viscosa*, 15 October 2016. *Christodoulos Makris*

Fig. 12.386. *Lasiommata megera* (Wall Brown) ♀, 17 February 2013. *Christodoulos Makris*

Fig. 12.387. *Lasiommata megera* (Wall Brown), 17 February 2013. *Christodoulos Makris*

Fig. 12.388. *Lasiommata megera* (Wall Brown), 18 May 2007. *Christodoulos Makris*

Fig. 12.389. *Lasiommata megera* (Wall Brown), 22 June 2012. Courtesy of *David and Rosalyn Sparrow*

Presence in the eastern Mediterranean

There is uncertainty about the naming of the ssp. in the wider region: Larsen (1974: 142) listed *Lasiommata megera emilyssa* (Verity, 1919) for Lebanon, while *L. m. lyssa* (Hübner, 1823) was given for Cyprus by Parker (1983) and Makris (2003: 260). Olivier (1993: 203) listed nominotypical *L. m. megera* for S Turkey and Cyprus, while Larsen and Nakamura (1983) and Benyamini (2002a: 148, 2021b: 48) gave *L. m. emilyssa* for Jordan and Israel, respectively. Hesselbarth *et al.* (1995: 783) considered *emilyssa* synonymous with yet another ssp., *L. m. transcaspica* (Staudinger, 1901), in Turkey. Because of 'great phenotypic variability', Olivier (1993: 184) argued that 'no subspecies should … be recognized for *L. megera* in the West-Palearctic, despite the recent argumentation of de Freina (1991), who ascribes the East Mediterranean populations to *L. megera lyssa* (Hübner, [1829]).'

The species was first recorded for Cyprus as '*Pararga Megaera*' in Lederer (1855). Bucknill (1911) stated, 'A form called Pararge megæra. L. Var. lyssa Boisd. in which the underside is very pale occurs on Troödos fairly commonly.' Turner (1920) referred to 'Pararge megera L. [race lyssa Hb.]', while Rebel (1939) listed 'Pararge megaera (L.) lyssa Hb'.

Distribution in Cyprus

Altitudinal: all elevations, from sea level to the highest peaks.
Geographical: a fairly common but localized butterfly. Well recorded from most western and northern parts of the island but appears to be largely absent from the Mesaoria Plain and is sparsely reported in the far south-east, but may have a slightly greater presence than is shown on the distribution map.

Phenology

Trivoltine to polyvoltine, generally February to November, but occasionally earlier or later. Overwinters as a larva, pupating earlier in warm areas.

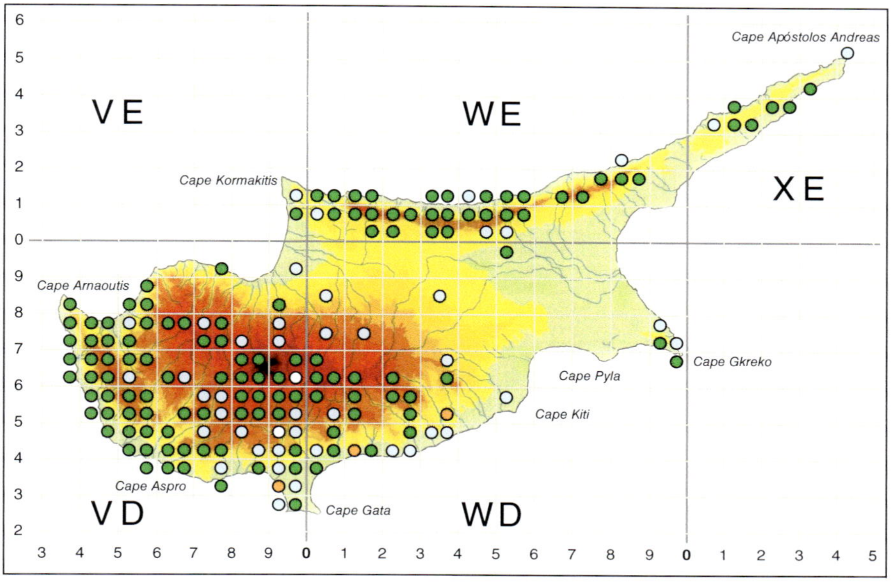

Fig. 12.390. *Lasiommata megera* distribution map showing 5 km² observations within 10 km² UTM squares. © *Eddie John*

Larval hostplants

Poaceae: *Poa* spp.

Ecology and behaviour

A known hilltopper throughout its range (Dennis, 2020: 246), at which hilltop areas large aggregations may occasionally form as they await the arrival of females (Makris, 2003: 260). *L. megera* is found in most types of habitat, including urban locations and even sand dunes, but is usually associated with well-vegetated rocky areas, especially vertical rock faces – a habitat typical also for *Lasiommata maera*. Males may perch or patrol in search of females (Thomas and Lewington, 1991: 166).

Description/identification aids

Wingspan: up to 45 mm. Forewing **upperside** ground colour is orange-yellow with dark brown markings and a white-pupilled, black apical ocellus. The ♂ has an *extensive androconial patch* in the discal area which, together with a more *heavily marked hindwing upperside* (basal and discal areas are *brown*), readily separates this from the lighter ♀. **Undersides** of both sexes are very similar: forewing with orange-yellow ground colour lightly marked in brown and with a large white-pupilled, black ocellus encircled with yellow and a narrow brown ring. Hindwing undersides are intricately patterned in brown on a greyish ground colour. As on the upperside, the hindwing underside has a *well-defined series of postdiscal ocelli*, white-pupilled on black and with brown and orange rings.

Similar species

At rest, with wings tightly closed, the species may be difficult to separate from *Lasiommata maera*. Note that on the forewing underside (in both sexes) the postdiscal wavy line in *L. megera* is *well defined* in brown, but is *very faint* in *L. maera* – see this species for remarks on hindwing underside ocelli.

Conservation status in Cyprus

Not threatened. Listed by Van Swaay *et al.* (2010a) and Numa *et al.* (2016) as 'LC' in Europe and in the Mediterranean, respectively.

Lasiommata maera (Linnaeus, 1758) Satyridae (Satyrinae)

Large Wall Brown

TL: Sweden.

Fig. 12.391. *Lasiommata maera* (Large Wall Brown) ♂, 5 February 2013. *Christodoulos Makris*

Fig. 12.392. *Lasiommata maera* (Large Wall Brown) ♂ nectaring on *Dittrichia viscosa*, 22 October 2013. *Christodoulos Makris*

Fig. 12.393. *Lasiommata maera* (Large Wall Brown) ♀, 4 November 2010. Courtesy of *Alison McArthur*

Fig. 12.394. *Lasiommata maera* (Large Wall Brown) ♀ nectaring on *Dittrichia viscosa*, 15 October 2015. *Christodoulos Makris*

Fig. 12.395. *Lasiommata maera* (Large Wall Brown) ♂ nectaring on *Dittrichia viscosa*, 22 October 2013. *Christodoulos Makris*

Fig. 12.396. *Lasiommata maera* (Large Wall Brown) ♀ nectaring on *Dittrichia viscosa*, 15 October 2016. *Christodoulos Makris*

Presence in the eastern Mediterranean

The ssp. *Lasiommata maera orientalis* Heyne, 1894 was listed by Larsen (1974: 141) and Larsen and Nakamura (1983) for Lebanon, with the accompanying comment in the former, 'does not differ much from nominotypical material'. The same subspecies was given by Parker (1983) for Cyprus and by Benyamini (2002a: 150) for Israel. However, Olivier (1993: 188, 201) treated this as nominotypical *L. maera* for Turkey and Cyprus, adding: '*L. maera* has no distinct subspecies in Europe and Turkey, but that its phenotype is very sensitive to environmental influences'. Hesselbarth *et al.* (1995: 788) concurred for Turkey.

Lederer (1855) referred to the Cyprus species as '*Pararga Maera* L. Var. *adrasta* Esp.', Bucknill (1911) as '*Pararge mæra. L. The Dull Wall*', whereas Gibbs (cited in Turner, 1920) commented 'I think *P. maera* are Staudinger's var. *orientalis* rather than true *adrasta*'; Turner agreed, using 'Pararge maera L. [race orientalis Stdgr.]', as did Rebel (1939) with 'Pararge maera (L.) orientalis Stgr.'

Distribution in Cyprus

Altitudinal: from sea level to the highest elevations.
Geographical: as with *L. megera*, a local butterfly, with a similar distribution in the west and north, absent from the Mesaoria Plain and with one record only from the far south-east, where it may have a greater presence.

Phenology

Trivoltine to polyvoltine, February to November, but occasional specimens may be seen earlier or later. There may be a partial diapause in August/early September.

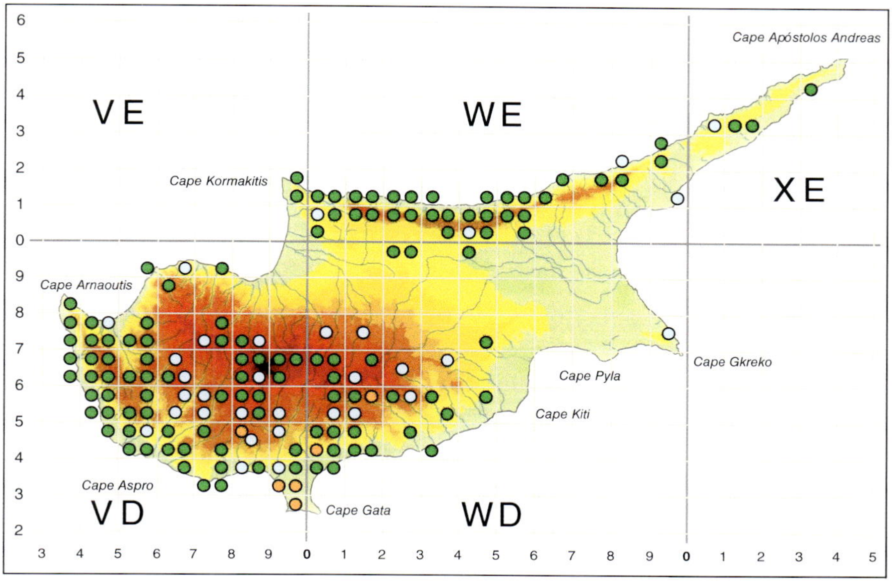

Fig. 12.397. *Lasiommata maera* distribution map showing 5 km² observations within 10 km² UTM squares. © *Eddie John*

Larval hostplants

Poaceae: *Poa* spp.

Ecology and behaviour

Most likely to be found in hilly/mountainous areas, flying around walls of ravines (often hilltopping) and vertical rock faces alongside tracks. Occasionally present in small groups in such locations, rather than as individuals. Dwarf forms are known. *Dittrichia viscosa* provides a much-favoured source of nectar late in the year.

Description/identification aids

Wingspan: up to 55 mm. ♂ and ♀ **uppersides** are very similar to *L. megera*, with an orange-yellow ground colour, but with brown markings *subdued* and *less well defined*, especially in the larger ♀, where fore- and hindwings are *much lighter* than in the sister species, the hindwing upperside in particular being largely devoid of markings, other than the presence of postdiscal ocelli. In both sexes, forewing **undersides** are predominantly yellow-orange, largely unbroken by dark brown lines (*cf. L. megera*), with a grey-brown margin, apex and costa and a large white-pupilled, black ocellus encircled in yellow and ringed in brown. The hindwing underside has a *well-defined series of postdiscal ocelli*, white-pupilled on black and with brown and orange rings.

Similar species

L. megera (see previous species). In ♀ *L. maera*, the forewing apical ocellus is *generally* larger than in *L. megera*, as are the hindwing ocelli, but sufficiently variable in size from specimen to specimen that ocelli size alone should not be used to separate *maera* from *megera*. Although usually larger than *L. megera*, overall size is an insufficiently reliable character to enable separation of the Cyprus species.

Conservation status in Cyprus

Not threatened. Listed by Van Swaay *et al.* (2010a) and Numa *et al.* (2016) as 'LC' in Europe and in the Mediterranean, respectively.

Kirinia roxelana (Cramer, 1777)

Satyridae (Satyrinae)

Lattice Brown

TL: İstanbul, İzmir.

Fig. 12.398. *Kirinia roxelana* (Lattice Brown) ♂, 27 May 2014. *Christodoulos Makris*

Fig. 12.399. *Kirinia roxelana* (Lattice Brown) ♂, May 2016. *Christodoulos Makris*

Fig. 12.400. *Kirinia roxelana* (Lattice Brown), ♂ at tree sap, 22 May 2010. Courtesy of *Sami Tamson*

Fig. 12.401. *Kirinia roxelana* (Lattice Brown) ♀, 17 May 2018. Courtesy of *David and Rosalyn Sparrow*

Presence in the eastern Mediterranean

Nominotypical *K. roxelana* is present in all coastal countries of the wider region except Sinai. Formerly included in the genus *Pararge*, this was first reported for Cyprus (as '*Pararga Roxellana* Fab.') by Lederer (1855) and as '*Pararge roxelana. Cr. The Balkan Beauty*', by Bucknill (1911). Turner (1920) noted much uniformity in size and markings of the short series he received and ascribed no 'form' names to nominotypical 'Pararge roxelana Cram.'. Recent molecular studies prompted the restoration of *K. roxelana* to '*Esperarge roxelana* (Cramer, [1777]) comb. rev.' (Spitsyn *et al.*, 2019), but until this combination becomes more widely accepted, we follow Wiemers *et al.* (2018, 2020).

Distribution in Cyprus

Altitudinal: from 100 m, but generally found at intermediate to higher elevations.
Geographical: generally uncommon, but may occasionally be locally numerous, especially in the northern mountain ranges. Confined to the Troodos and Pentadaktylos Mountains and foothills, absent from the Mesaoria Plain and south-east of the island.

Phenology

Univoltine, April to September, rarely earlier or later, with light aestivation in this long-lived species. Larsen (1974: 143) stated, 'It is just possible that

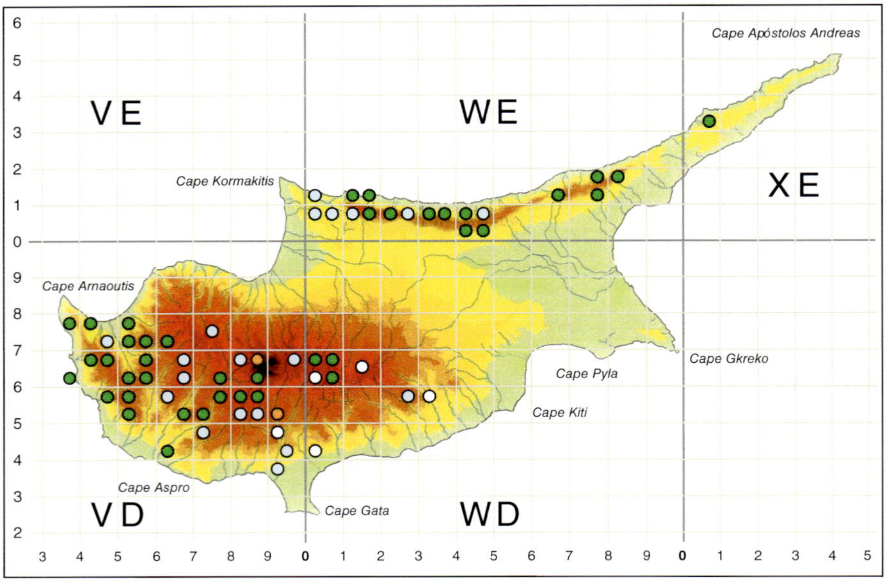

Fig. 12.402. *Kirinia roxelana* distribution map showing 5 km² observations within 10 km² UTM squares. © *Eddie John*

there are two broods'; Pamperis (2009: 596) indicated 'one and two generations' for Greece. Overwinters in the larval stage.

Larval hostplants

Poaceae: *Poa* spp: *P. annua*, *P. pratensis*, *P. trivialis* and other members of the family.

Ecology and behaviour

Found in hilly or mountainous areas and often favouring shade, *K. roxelana* is usually associated with hedges, thickets or small trees, where the species' typical behaviour of weaving in a laboured fashion between branches before alighting sets it apart from other satyrines. Both Bucknill and Wilson (in Turner, 1920) noted the species' attraction to carob (*Ceratonia siliqua*) trees. *K. roxelana* is an alert, reclusive species that, when disturbed, quickly seeks shelter in the depth of bushes. In urban areas, occasionally roosts on walls (first author, personal observation) or under eaves (Daniel Haines, London, 2007, personal communication to first author).

Mating occurs soon after emergence, females then aestivating in shade in high summer, to reappear in autumn upon the return of rain and consequent regrowth of grasses, when eggs are attached to grass blades, or may be laid in bark crevices on trunks of trees (Hesselbarth *et al.*, 1995: 772; Lafranchis, 2019). Dennis (1991) reported egg-laying beneath the bark of an olive tree in late July in Corfu, reducing the risk of egg desiccation and indicating that, in some females at least, egg maturation occurs earlier than autumn. Males may not survive beyond mid- to late summer.

Although not widely known to seek nectar, *ca.* 30 individuals were observed on 30 July 2006 on *Echinops spinosissimus* (see earlier entry on *H. syriaca*) at various roadside locations in northern Cyprus (Daniel Haines, London, 2006, personal communication to first author), an activity also reported in Jordan by Eid *et al.* (2009). Hesselbarth *et al.* (1995: 772) observed the species visiting *Onopordum* sp. and similar Asteraceae, while Jutzeler *et al.* (2005) reported *K. roxelana* on *Lilium martagon*. The last-mentioned authors also reported an attraction to the dry inflorescences of *Centaurea* sp., *Cirsium* sp. and *Onopordum* sp. However, the butterfly is more frequently encountered at tree sap and is attracted to aphid 'honeydew', fermenting liquids or fruit (John and Skule, 2016: 326). On occasions, an attraction to light has been reported.

Description/identification aids

Wingspan: up to 60 mm. An easily recognized species, with hindwings that appear *disproportionally large* in the ♂. Forewing **upperside** ground colour is orange, the ♂ with a *heavy black mark running near the costa* and *conspicuous black lines of androconia*, the ♀ with an *irregularly shaped black mark near the costa* (where it is bordered by yellow) and extending inwards into the discal area. A black forewing apical ocellus is present in both sexes, with a pale yellow spot, smaller in the male, above and below. Hindwing upperside is brown, largely unmarked other than for a series of black ocelli ringed with orange, and a strong black line crossing the basal and discal areas in the male. Forewing **undersides** are similar in both sexes, with an orange ground colour and brown markings (fewer in the male). The ♀ has a small series of lighter markings running from the forewing costa into the postdiscal area. Ocelli mirror those on the upperside, which, in the ♀ are on a *lighter ground colour*. Hindwing underside is highly patterned, *deeply serrated* (emphasized by a pale submarginal line) and with a complete series of *highly characteristic white-pupilled, black ocelli encircled by yellow and dark brown.*

Similar species

None. *Kirinia climene* (Esper, [1783]), the Lesser Lattice Brown, although present in Turkey, is not known for countries of the Levant.

Conservation status in Cyprus

Not threatened. Listed by Van Swaay *et al.* (2010a) and Numa *et al.* (2016) as 'LC' in Europe and in the Mediterranean, respectively.

Ypthima asterope (Klug, 1832)

Satyridae (Satyrinae)

African Ringlet

TL: "Syria, Arabia félici et deserta".

Fig. 12.403. *Ypthima asterope* (African Ringlet) ♂ on *Cardopatium spinosum*, 27 May 2018. *Christodoulos Makris*

Fig. 12.404. *Ypthima asterope* (African Ringlet) ♀ (reared). Courtesy of *Martin Gascoigne-Pees*

Fig. 12.405. *Ypthima asterope* (African Ringlet) ♂ nectaring on *Cardopatium spinosum*, 27 May 2018. *Christodoulos Makris*

Fig. 12.406. *Ypthima asterope* (African Ringlet) ♂ (reared). Courtesy of *Martin Gascoigne-Pees*

Fig. 12.407. *Ypthima asterope* (African Ringlet) pairing, 1 April 2016, the smaller ♂ on the left. *Christodoulos Makris*

Presence in the eastern Mediterranean

Kattulas and Koutsaftikis (1977) described an endemic subspecies *Y. asterope marlenii* from Kastellórizo (Megisti), off the south-west coast of Turkey, but Olivier (1990: 182) synonymized this with nominotypical *Y. a. asterope*, which is present throughout the region, other than Sinai. Hesselbarth *et al.* (1995: 793) concurred. Larsen (1986) remarked, 'Its presence also on Cyprus suggests a certain antiquity. If very recent, then its presence in the Mediterranean makes it the only successful coloniser that is not migratory.'

Initially described as 'Hipparchia Asterope' Klug (1832), this species was not listed by Lederer (1855), but was noted as '*Ypthima asterope. Klug. The Eye-starred Grey*' by Bucknill (1911). Turner (1920) stated that 'In the Asiatic dry-season form, to which the Cyprian race belongs, the eye-spots are very much suppressed', and applied the name 'ab. *inocellata*' to a specimen 'quite destitute of eye-spots'; Rebel referred to 'Yphtima [*sic*] asterope Klug'.

Distribution in Cyprus

Altitudinal: presently known from sea level to 400–500 m in Cyprus, although Gibbs (in Turner, 1920) referred to specimens taken from the Pentadaktylos range only 'between 1200 ft [366 m] and 3000 ft [914 m]'.

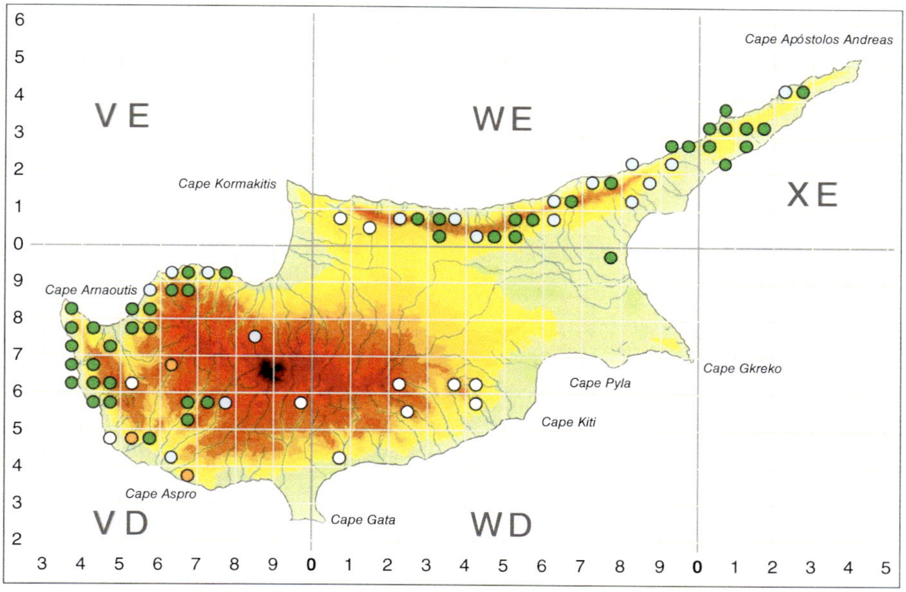

Fig. 12.408. *Ypthima asterope* distribution map showing 5 km² observations within 10 km² UTM squares. © *Eddie John*

Geographical: a generally rare species, although occasionally it may be very common locally. CBRS data shows the species to be distributed along coastal and some inland areas of the west, and particularly in some hilly and coastal areas of the north, especially Karpasia. Pre-1990 records from southern central areas (see map) are fully credible, but are not supported by data from the past 20 years, suggesting that *Y. asterope* has disappeared from these areas or retains a minimal presence.

Phenology

Bivoltine to trivoltine, mid-March to June/July and September/October to November. There is likely to be a partial summer brood in some years (Makris, 2003: 268). This view is supported by reports of singletons on 3 August 2000 (a fresh specimen) by Peter Flint (Kent, 2000, personal communication to first author), and on 7 August 2018 by David and Rosalyn Sparrow (Cyprus, 2018, personal communication to first author), both observations at known sites for this species. Hesselbarth (1983) reported on the captive rearing of *Y. asterope* from Hatay Province, Turkey in late April/early May, noting that development time from egg-laying to adult eclosion occupied about two months, coinciding with the appearance of the next generation in Hatay. Hesselbarth *et al.* (1995: 793) considered *Y. asterope* to be polyvoltine in Turkey, a view on voltinism shared by Larsen (1974: 140) and Olivier (1993:183) and, for Jordan, by Katbeh-Bader *et al.* (2003, 2004). Benyamini (2021b: 124) also considered *Y. asterope* polyvoltine in Israel, with observations throughout the year in warmer areas. There is no known evidence supporting aestivation. Probably overwinters in the larval stage.

Larval hostplants

Poaceae: *Poa* spp., probably including *Hyparrhenia hirta* in Cyprus (John *et al.*, 2010).

Ecology and behaviour

The smallest and least conspicuous of the Cyprus satyrines, *Y. asterope* is a highly sedentary, tropical species (Larsen, 1986) that has adapted to eastern and northern Mediterranean habitats west to the Dodecanese. In Cyprus, the species tends to fly along dry gullies, dry riverbeds, roadside drainage channels, or along the base of steep rock faces, but also frequents open habitats such as grassy slopes. Pittaway (1985) referred to *Y. asterope* in Saudi Arabia favouring deep shade, but this is not apparent in Cyprus, although, as with other satyrines in Cyprus, the shade of bushes might be sought in high temperatures, an observation noted also by Hesselbarth *et al.* (1995: 793). Indeed, adults of the summer brood in Cyprus might well seek refuge in this way.

Larsen (2015) gave the species' presence in the Levant as 'locally common at low elevations from southern Turkey to Israel and Jordan', which is supported by the statement in Atahan *et al.* (2018: 102): 'Common at sea shores.' Emergence of the autumn generation appears to be triggered by rainfall, synchronizing eclosion of the adult with regrowth of grasses (John *et al.*, 2010).

Flight is variable, usually gentle and bobbing, occasionally more direct rendering the species difficult to follow in its usual rocky habitats. *Y. asterope* has been observed to mud-puddle, in addition to taking nectar from *Heliotropium hirsutissimum* and *Dittrichia viscosa* (authors' personal observations).

Description/identification aids

Wingspan: up to 38 mm (females are noticeably larger than males, although similarly marked). **Upperside** ground colour of both sexes is a uniform brown with a greyish sheen. The forewing upperside has a *large black apical ocellus* (larger in the ♀), with *two metallic-blue pupils*. Encircling the ocellus are rings of yellow and brown, within a brownish-grey area slightly lighter than the ground colour. The hindwing has a yellow-ringed single small black ocellus, with one metallic-blue pupil. **Underside** ground colour is greyish-brown, densely irrorated with darker brown. There is a *large black ocellus at the forewing apex* and *three* (perhaps four) *submarginal small, blue-pupilled, black ocelli ringed with yellow* on the hindwing. Size and number of ocelli *may vary considerably*, as illustrated in Figs 28 and 29 in John *et al.* (2010) and discussed further under 'Natural variation' in the cited paper.

Similar species

None.

Conservation status in Cyprus

Uncertain, but considered vulnerable. It seems unlikely that the apparent loss of *Y. asterope* from central and southern areas is entirely attributable to urbanization but the reason for the loss is unexplained.

Not assessed by Van Swaay *et al.* (2010a), while Numa *et al.* (2016) listed *Y*. asterope as 'LC' at the Mediterranean level.

Family: HESPERIIDAE

Lederer (1855) listed no fewer than eight species of hesperiid for Cyprus, only two of which, *Carcharodus alceae* and *Thymelicus acteon*, remain confirmed for the island, although *Gegenes nostrodamus* is considered probable (see Chapter 13).

Carcharodus alceae (Esper, 1780)	Mallow Skipper	Pyrginae
Thymelicus acteon (Rottemburg, 1775)	Lulworth Skipper	Hesperiinae
Gegenes pumilio (Hoffmansegg, 1804)	Pigmy Skipper	
Pelopidas thrax (Hübner, [1821])	Millet Skipper	

Carcharodus alceae (Esper, 1780) Hesperiidae (Pyrginae)

Mallow Skipper

TL: Erlangen, southern Germany.

Fig. 12.409. *Carcharodus alceae* (Mallow Skipper) ♂, 4 May 2015. *Eddie John*

Fig. 12.410. *Carcharodus alceae* (Mallow Skipper) ♀, 10 September 2011. Courtesy of *Colin Baker*

Fig. 12.411. *Carcharodus alceae* (Mallow Skipper), 1 July 2006. *Eddie John*

Fig. 12.412. *Carcharodus* alceae (Mallow Skipper) ♂, 2 May 2016. *Eddie John*

Presence in the eastern Mediterranean

Present in all countries of the Levant as nominotypical *C. alceae alceae*. Recorded as '*Hesperia Malvarum* O.' by Lederer (1855), this is one of numerous historical synonyms used for *C. alceae*, including that used by Bucknill (1911): '*Spilothyrus alceæ. Esp. The Common Skipper*'. Mistakenly, Bucknill (1911) appears to have considered Lederer's '*Hesperia Malvarum*' to be '*Hesperia malvae. L. The Grizzled Skipper*', now known as *Pyrgus malvae* (Linnaeus, 1758), a species not known for Cyprus and not observed by Bucknill. Turner (1920) and Rebel (1939) used the currently accepted name *C. alceae*, the latter adding '*(malvarum)*'. Parker (1983) noted that earlier authors had variously misidentified this species and added, 'The 33 specimens in the BM(NH) [British Museum of Natural History, London] are all *alceae*.'

Distribution in Cyprus

Altitudinal: predominantly a species of low and intermediate elevations, but may be encountered anywhere in Cyprus.
Geographical: a common butterfly, found throughout the island.

Phenology

Polyvoltine, mainly February to November, but recorded in all months of the year in Cyprus, with numbers increasing during the autumn. Overwinters as a small larva (Makris, 2003: 272).

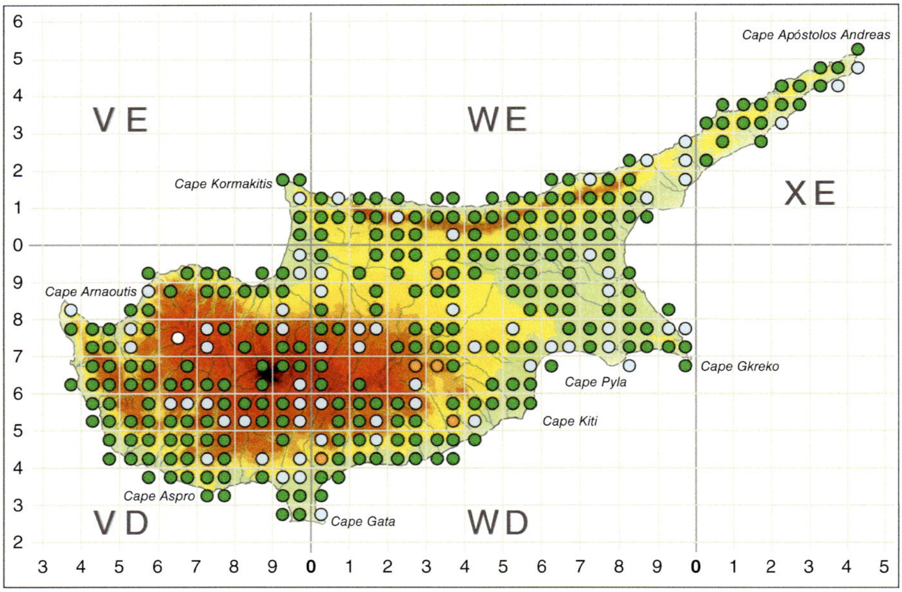

Fig. 12.413. *Carcharodus alceae* distribution map showing 5 km² observations within 10 km² UTM squares. © *Eddie John*

Larval hostplants

Principally Malvaceae: *Malva sylvestris* (common mallow) (Fig. 12.414), *Hibiscus* spp. In Cyprus, the second author confirmed the use there of *Chrozophora tinctoria* (turnsole) (Euphorbiaceae, subf. Acalyphoideae) first reported for Israel by Benyamini (2005).

Ecology and behaviour

C. alceae often frequents field margins and sparsely vegetated waste ground, particularly in coastal and urban areas. The species is usually found singly (males are territorial) or in small groups, but may occasionally be present in larger numbers. In the absence of similarly marked species such as *Muschampia orientalis*[4] (Reverdin, 1913) Oriental Marbled Skipper, *C. alceae* is readily recognizable with its whirring flight and will usually be seen at rest with wings fully or partially open. Flight is rapid, close to the ground and with frequent pauses in order to perch or nectar, particularly favouring the near-prostrate *Heliotropium hirsutissimum* or, in the autumn, *Dittrichia viscosa*.

Fig. 12.414. *Malva sylvestris*. Courtesy of *Yiannis Christofides*

[4]New combination, formerly in the genus *Carcharodus* (Zhang *et al*., 2020).

Description/identification aids

Wingspan: up to 30 mm. ♂ and ♀ **upperside** ground colour is grey-brown with large blotches of darker brown forming irregular discal and post-discal bands across fore- and hindwings. There is white spotting on the forewing, the most noticeable being a *line of three white spots near the apex*. Hindwing margins present a *serrated appearance*, becoming more apparent as wings show signs of wear. **Undersides** are straw-coloured, lighter in the ♀, and with white markings mirroring those on the upper surfaces.

Similar species

Other *Carcharodus/Muschampia* spp. are found in nearby mainland countries, but *C. alceae* is the only member of the genus known in Cyprus.

Conservation status in Cyprus

Not threatened. Listed by Van Swaay *et al.* (2010a) and Numa *et al.* (2016) as ‘LC’ in Europe and in the Mediterranean, respectively.

***Thymelicus acteon* (Rottemburg, 1775)** **Hesperiidae (Hesperiinae)**

Lulworth Skipper

TL: Landsberg an der Warthe, Poland.

Fig. 12.415. *Thymelicus acteon* (Lulworth Skipper) ♂, 2 May 2011. Courtesy of *David and Rosalyn Sparrow*

Fig. 12.416. *Thymelicus acteon* (Lulworth Skipper) ♂, 20 April 2014. *Eddie John*

Fig. 12.417. *Thymelicus acteon* (Lulworth Skipper) ♀ nectaring on *Convolvulus arvensis*, 23 April 2014. *Eddie John*

Fig. 12.418. *Thymelicus acteon* (Lulworth Skipper) ♀ on *Echium angustifolium*, 30 April 2016. *Eddie John*

Fig. 12.419. *Thymelicus acteon* (Lulworth Skipper) ♂ nectaring on *Echium angustifolium*, 23 April 2014. *Eddie John*

Fig. 12.420. *Thymelicus acteon* (Lulworth Skipper) ♂ (L), ♀ (R), 23 April 2013. *Christodoulos Makris*

Fig. 12.421. *Thymelicus acteon* (Lulworth Skipper) pairing, 22 April 2016. *Christodoulos Makris*

Fig. 12.422. Roosting *Thymelicus acteon* (Lulworth Skipper), ♀ above, ♂ below, 8 May 2015. *Eddie John*

Presence in the eastern Mediterranean

Earlier authors referred to various ssp. or forms, including ssp. *Thymelicus acteon phoenix* Graves, 1925 for Levantine populations (e.g. Larsen, 1974: 210; Larsen and Nakamura, 1983), but this is likely to be synonymous with *T. a. acteon* (Hesselbarth *et al.*, 1995: 173). Parker (1983) referred to *Thymelicus acteon heydeni* Plötz, 1884 for Cyprus, a ssp. also regarded as invalid by Hesselbarth *et al.* (1995: 173). Olivier (1993: 36, 202) listed nominotypical *T. acteon* for Cyprus and Turkey and stated, 'It seems impossible to distinguish any subspecies of *T. acteon* in Europe and probably also in [the] whole [of] Turkey and the Near East.' *T. acteon* is described as local, but widespread in Turkey (Hesselbarth *et al.*, 1995: 173), although Atahan *et al.* (2018: 175) referred to just a single record – from Antakya (Hatay Province). It is similarly scarce in southern countries of the Levant and is considered endangered in Israel (Benyamini and John, 2020: 166).

Listed for Cyprus by Lederer (1855) as '*Hesperia Actaeon* Esp.', this became *Pamphila acteon* in Bucknill (1911), and was referred to as 'Thymelicus (Adopaea) acteon Rott. [race obsoleta Tutt; clara Tutt]' in Turner (1920), 'Adopaea actaeon Rott.' in Rebel (1939) and similarly named in Bretherton (1954).

Distribution in Cyprus

Altitudinal: coastal regions to above 1000 m.

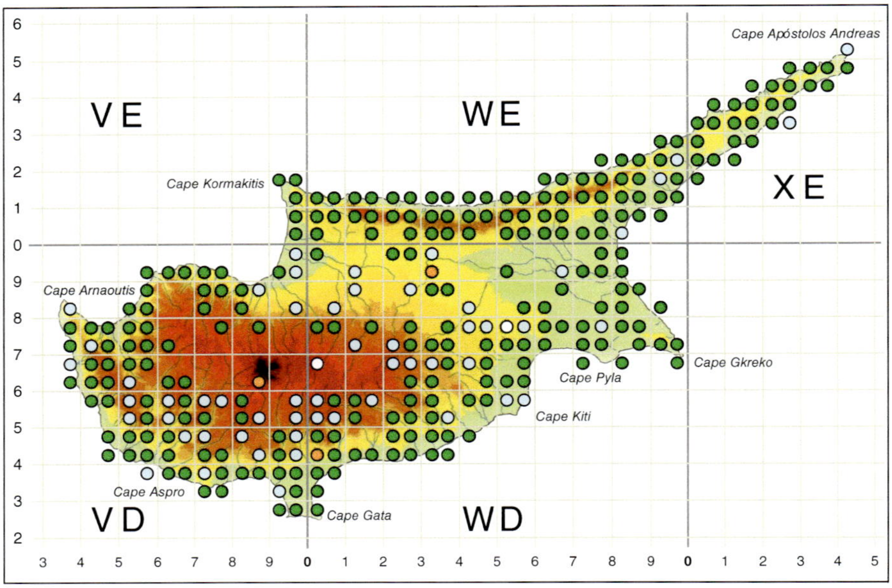

Fig. 12.423. *Thymelicus acteon* distribution map showing 5 km² observations within 10 km² UTM squares. © *Eddie John*

Geographical: a very common butterfly, absent from the highest mountain regions, but widely represented from the far west to the tips of all peninsulas. Although Bucknill (1911) described the species as 'common in the plains', and Bretherton (1954) stated *T. acteon* to be locally abundant near Keryneia, the species was considered as generally uncommon until very recently. Parker (1983), for example, described it as 'fairly rare but widely distributed', showing the species' occupancy in only 11 × 10 km² UTM squares (*cf.* current distribution map opposite). Manil (1990) appears to have quoted Parker: 'Espèce rare à Chypre mais assez dispersée', while John (2000) incorrectly described it as 'uncommon'. However, extensive recording effort over the past 20 or so years has shown that at the peak of its flight period it is now easily the most common of Cyprus's skippers, supporting the view held by Makris (2003: 274). This was further confirmed on 20 May 2003 when hundreds were seen at Agios Kendeas Monastery (WD77 NE) and again during a late April/early May visit to Cyprus in 2016, when the first author (in company with David and Rosalyn Sparrow) recorded *T. acteon* at 80 (54%) of 148 locations visited across much of the island.

Phenology

Univoltine, generally April to June, occasionally earlier or later.

Larval hostplants

Poaceae: *Poa* spp. Overwinters as a young larva within a cocoon formed inside a dry grass sheath into which the female had inserted the eggs in May/June. Larvae probably emerge from the cocoon to feed on mild winter days.

Ecology and behaviour

Most frequently found in areas of long grass in a variety of habitats such as roadside verges, waste ground and field margins, where *T. acteon* is very attracted to the flowers of *Echium angustifolium*, *Onopordum cyprium* and *Phagnalon rupestre*. Occasionally found away from such

Fig. 12.424. Larva of reared *Thymelicus acteon* (Lulworth Skipper) on *Poa* sp. *Eddie John*

habitats in search of nectar sources such as *Thymbra capitata*. The species is a known mud-puddler; furthermore, John and Tennent (2012), in a review of reports of *marine* puddling, reported the species imbibing seawater from a wave-splashed pebble in north-western Cyprus.

Description/identification aids

Wingspan: up to 30 mm. This is the only member of the 'golden' skippers to be found in Cyprus, so is easily identified. NB it is *much brighter* than *T. acteon* found in, for example, UK specimens. **Uppersides** of both sexes are orange-brown with dark wing margins. The ♂ has a long narrow, but distinctive, forewing sex brand, the ♀ an *indistinct part-circle of lighter marks* in the cell and postdiscal areas near the costa, creating the typical 'ray' characteristic associated with this species. **Undersides** of both sexes are golden orange. When nectaring, *T. acteon* frequently adopts the typical posture of many skippers, with the hindwings fully open and the forewings held at an angle of 45°.

Similar species

Several other *Thymelicus* species are present elsewhere in the Levant, but *T. acteon* is the sole member of the genus known for Cyprus.

Conservation status in Cyprus

Considered 'NT' (near threatened) by Van Swaay *et al.* (2010a, 2011) at the European level, but *T. acteon* is not threatened in Cyprus. Numa *et al.* (2016) listed *T. acteon* as 'LC' at the Mediterranean level.

Gegenes pumilio (Hoffmansegg, 1804) Hesperiidae (Hesperiinae)

Pigmy Skipper

TL: Königreich Neapel.

Fig. 12.425. *Gegenes pumilio* (Pigmy Skipper) nectaring on *Dittrichia viscosa*, 23 October 2007. Courtesy of *Martin Gascoigne-Pees*

Fig. 12.426 (top). *Gegenes pumilio* (Pigmy Skipper) ♂ nectaring on *Dittrichia viscosa*, 31 October 2015. *Christodoulos Makris*

Fig. 12.427 (bottom). *Gegenes pumilio* (Pigmy Skipper), typically alighting on a hot surface, 14 May 2007. *Eddie John*

Fig. 12.428. *Gegenes pumilio* (Pigmy Skipper) nectaring on *Heliotropium hirsutissimum*, 5 August 2012. Courtesy of *Clive Burrows*

Fig. 12.429. *Gegenes pumilio* (Pigmy Skipper) nectaring on *Dittrichia viscosa*, 31 October 2012. *Christodoulos Makris*

Fig. 12.430. *Gegenes pumilio* (Pigmy Skipper) pairing, 31 October 2012. *Christodoulos Makris*

Presence in the eastern Mediterranean

Chromosome analysis comparing populations of western Mediterranean *G.* '*pumilio*' with those of the eastern Mediterranean has revealed the presence of two cryptic species (de Lesse, 1960, 1967; Larsen, 1982b, 1983; Saitoh, 1984). Larsen (1982b) wrote of western populations having 'a haploid chromosome number of n = 24', whereas eastern populations 'have n = 41'.

He further stated, 'The presence of the taxon with n = 41 in both the eastern Mediterranean and in southwestern Arabia makes it almost certain that the African, Arabian and Asian populations belong to one taxon [i.e. either *G. pumilio* or *G. gambica* (Mabille, 1878)] ... we do not know to which of the two species the name *pumilio* applies'. Moreover, populations cannot be separated either by molecular means or by differences in morphology, rendering the systematic status unresolved. Until this is clarified, we remain with the conventional naming of *G. pumilio* for the eastern Mediterranean populations. The contact zone between the (possible) two taxa is also unknown. For further information, see Hesselbarth *et al.* (1995: 181), de Jong and Coutsis (2017) and Benyamini and John (2020: 174).

This species appears not to have been collected by Zach, and hence was not reported in Lederer (1855). Instead, Lederer (1855) listed '*Hesperia Nostradamus* Fab. (= *Gegenes nostrodamus*)', a species also included, as '*Pamphila nostradamus Fabr.*', by Bucknill (1911). Turner (1920) listed 'Gegenes nostrodamus Fab.', but with no mention of *G. pumilio*. However, Rebel (1939), seemingly on the advice of P.P. Graves, gave the Cyprus *Gegenes* as 'Gegenes lefebvrei Rbr. ... *(nostrodamus)*' = *G. pumilio*.

Distribution in Cyprus

Altitudinal: found from sea level to intermediate elevations.
Geographical: a less than common butterfly, *G. pumilio* is widely distributed across the island, although usually seen singly. The relative lack of records from the Mesaoria Plain might reflect reduced recording effort.

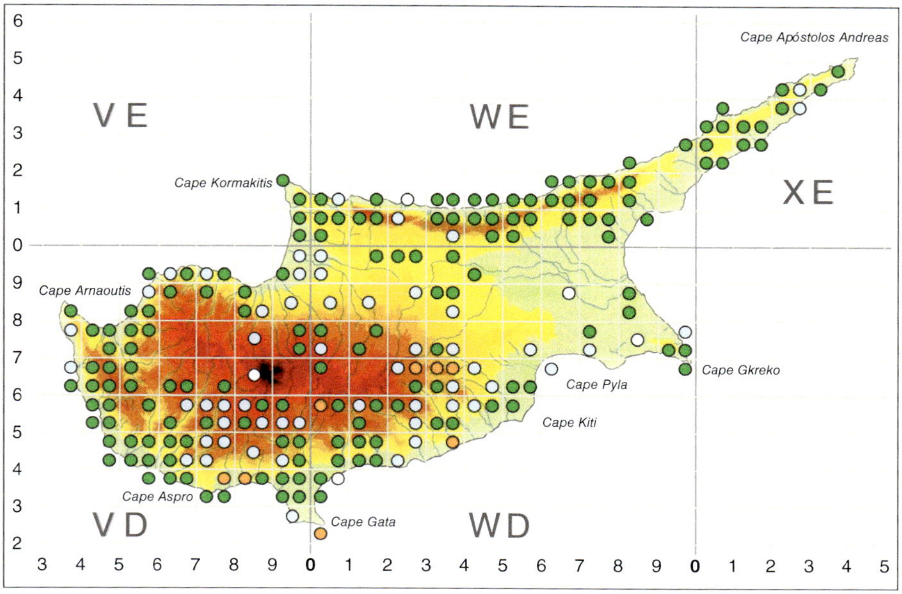

Fig. 12.431. *Gegenes pumilio* distribution map showing 5 km^2 observations within 10 km^2 UTM squares. © *Eddie John*

Phenology

Trivoltine, possibly polyvoltine, generally late March to early December (CBRS data), occasionally earlier; singletons of *G. pumilio* were reported on 3 January 2013 and 2 February 2018 by Mark Easterbrook and Francis Plowman, respectively (personal communications to first author). Photographs showed both to be fresh specimens. Numbers build up in late summer and early autumn. Reported to overwinter as a slowly growing larva in the Levant (Benyamini and John, 2020: 174), but occasional CBRS records early in the year also suggest overwintering at some point in the pupal stage.

Larval hostplants

Poaceae: *Poa* spp. including *Hyparrhenia hirta* and *Setaria verticillata*. Lafranchis (2005, 2019) noted the additional use of *Phragmites australis*, *Imperata cylindrica* and *Sorghum halepense* in Greece; all are present in Cyprus.

Ecology and behaviour

G. pumilio is attracted to hot, dry areas, principally stony tracks, dry riverbeds and sparsely vegetated areas in lowland areas, where its rapid, darting flight makes it difficult to follow. Males are territorial, easily disturbed, but repeatedly return to one of a few selected perches – a blade of grass, twig or stone. Flowers of the very common *Heliotropium hirsutissimum* and *Dittrichia viscosa* provide favoured nectaring opportunities (Makris, 2003: 278). *G. pumilio* is also known to mud-puddle (first author, personal observation).

Description/identification aids

Wingspan: up to 30 mm. A small, nondescript species, ♂ **uppersides** are a uniform drab dark brown, the ♀ is similar in colour, but with fairly *subdued* (*cf. Pelopidas thrax*) off-white postdiscal markings on the forewing. **Undersides** of both sexes are similarly coloured a lighter grey-brown, with fore- and hindwing postdiscal markings (sometimes *indistinct*) present on both ♂ and ♀.

Similar species

The very similar *Gegenes nostrodamus* Fabricius, 1793 (Mediterranean Skipper), found elsewhere in the Levant, is now considered (on photographic evidence) to be present on Cyprus but we require specimens for genitalic examination before confirming this as a new species for Cyprus (see Chapter 13 for description). *Pelopidas thrax* is similar in shape, but is *much* larger and with *brighter* hyaline markings.

Conservation status in Cyprus

Not threatened. Listed by Van Swaay *et al.* (2010a) and Numa *et al.* (2016) as 'LC' in Europe and in the Mediterranean, respectively.

Pelopidas thrax (Hübner, [1821]) Hesperiidae (Hesperiinae)

Millet Skipper

TL: No location (Evans, 1937).

Fig. 12.432. *Pelopidas thrax* (Millet Skipper), ♂ with wings held in typical skipper fashion, 20 October 2017. *Eddie John*

Fig. 12.433. *Pelopidas thrax* (Millet Skipper) ♀ nectaring on *Dittrichia viscosa*, 31 October 2015. *Christodoulos Makris*

Fig. 12.434. *Pelopidas thrax* (Millet Skipper) ♂ nectaring on *Dittrichia viscosa*, 20 October 2012. *Christodoulos Makris*

Fig. 12.435. *Pelopidas thrax* (Millet Skipper) ♀ nectaring on *Dittrichia viscosa*, 20 October 2012. *Christodoulos Makris*

Fig. 12.436. *Pelopidas thrax* (Millet Skipper) ♂, 20 October 2017. The hooked antennal tip is characteristic of *P. thrax* but not of *G. pumilio* or *G. nostrodamus*. *Eddie John*

Fig. 12.437. *Pelopidas thrax* (Millet Skipper) pairing, 23 September 2021. Courtesy of *Antonis Konis*

Presence in the eastern Mediterranean

An eremic species, present throughout the Levant as nominotypical *Pelopidas thrax thrax*. This hesperiid was not recorded by Lederer (1855) and appears to have been misidentified by Bucknill (1911) as *Parnasa* (*sic*) *mathias*, while Turner (1920) and Rebel (1916, 1939), using Bucknill's data, referred to *Parnara mathias*, a similar species now known as *Pelopidas mathias* (Fabricius, 1798), the Lesser Millet Skipper.

In addition, Turner (1920) included an unconfirmed report of *Parnara zelleri* (now known as *Borbo borbonica* (Boisduval, 1833)) Zeller's Skipper, which was repeated by Rebel (1916, 1939) and perpetuated by Wiltshire (1948) as '*Pelopidas borbonica*, Boisd., subsp. *zelleri*, Led.'. Although *Borbo borbonica* is known as a very infrequent migrant to the southern Levant (Larsen, 2015; Benyamini and John, 2020: 180), it has not been confirmed for Cyprus, suggesting that these earlier reports relate to observations of *P. thrax*.

Parker (1983), using the correct nomenclature at the time, referred to *Parnara thrax*, but this should not be confused with *Erionota thrax* (Linnaeus, 1767), otherwise known as the Banana Skipper (TL: Indonesia, "Java"), a hesperiid species principally found in south-east Asia and listed erroneously, as '*Erienota* (*sic*) *thrax* L. (*mathias* F.)', for Cyprus by Bretherton (1974).

Distribution in Cyprus

Altitudinal: mainly found at coastal locations, less frequently to 500 m and very rarely at high elevations.

Geographical: a locally common species especially in areas with reed beds, now widely distributed across the island. Data contributed by the first author in December 2012 to Martin and Russell (2013) gave the presence of *P. thrax* in 119 × 5 km squares, whereas it is now known from 164 × 5 km squares (37% of the total). The increase is considered explained by greater recording effort, rather than by range extension.

Phenology

Polyvoltine in some countries of the Levant, but probably trivoltine in Cyprus; late March to November. Numbers build up in late summer and early autumn. Phenology would suggest that *P. thrax* probably overwinters as a slowly growing larva in Cyprus.

Larval hostplants

Poaceae: *Piptatherum miliaceum* (on which ovipositing has been observed by the second author), but principal hostplants are *Phragmites australis* and possibly *Arundo donax*, as thought likely by Martin and Russell (2013) and with which member of the Poaceae *P. thrax* is frequently associated in Cyprus.

Ecology and behaviour

P. thrax, the largest of the Cyprus hesperiids, is known as a vigorous migrant, and is found throughout Africa and tropical Arabia, extending to

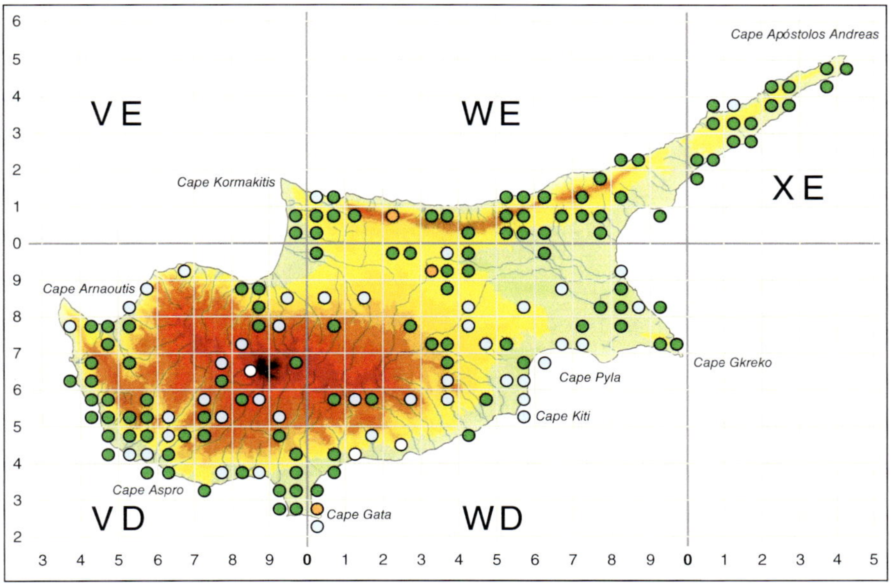

Fig. 12.438. *Pelopidas thrax* distribution map showing 5 km^2 observations within 10 km^2 UTM squares. © *Eddie John*

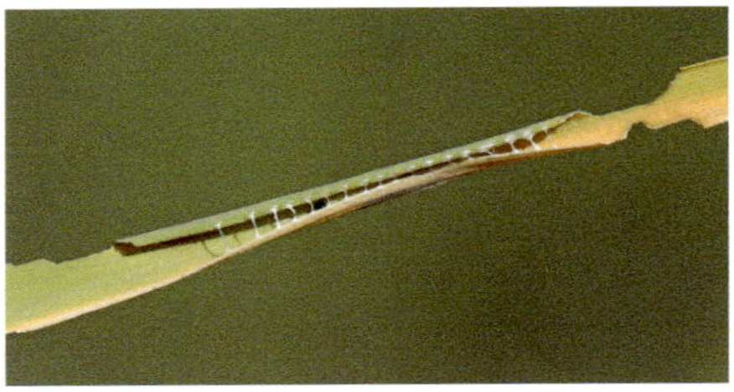

Fig. 12.439. *Pelopidas thrax* (Millet Skipper) larva within a folded leaf of *Phragmites australis*, 1 October 2017. *Christodoulos Makris*

India (Larsen, 1975, 2015; Benyamini and John, 2020: 178). The presence of a singleton in a well-glazed dining room of a hotel in coastal Bogazi in northern Cyprus on 26 May 2013 hinted at this being a migrant specimen, since no nectar plants to attract it to the area were present below the large, open hotel windows. However, the absence of any published evidence of immigration into Cyprus, together with the consistency with which sightings of (non-migrating) individuals have been reported to the CBRS in recent years, promote a growing belief that this species is now firmly established in Cyprus. Indeed, it may no longer need to rely on immigration from the south to sustain local populations, as was once thought a requirement for Cyprus and other countries of the Levant (Larsen, 1974: 215; Larsen and Nakamura, 1983; Parker, 1983).

A robust, fast-flying species, *P. thrax* is usually found near watercourses or drainage ditches where reed (*Phragmites australis* and *Arundo donax*) dominates; it alights in typical skipper fashion, with forewings held at a 45° angle (Fig. 12.432), usually perching on foliage and tending to favour shady areas. As with other hesperiids (and some lycaenids) in Cyprus, numbers increase towards the autumn. Favoured nectar sources include *Heliotropium hirsutissimum*, *Rubus sanctus* and, in the autumn, *Dittrichia viscosa*.

Description/identification aids

Wingspan: up to 45 mm. **Uppersides** of both sexes are brown, with highly characteristic discal and postdiscal *bright hyaline markings*, which are especially large in the ♀. The ♂ has a *distinctive forewing androconial streak*. **Undersides** are similar in both sexes, but with the *hyaline markings larger in the* ♀. Unlike the antennal tips of *G. pumilio* and *G. nostrodamus*, those of *P. thrax* are *noticeably* hooked (Fig. 12.436).

Similar species

See entry under *G. pumilio*.

Conservation status in Cyprus

Not threatened. Not assessed by Van Swaay *et al.* (2010a), but listed by Numa *et al.* (2016) as 'LC' at the Mediterranean level.

13 Potential Changes to the Cyprus List

Adventive Species and Range Expansion/Retraction

Due to human-mediated activities, accidental introduction of 'alien' species may occur occasionally: the term adventive is applied to such species. While there is no evidence to date of any adventive butterfly species having gained a foothold in Cyprus, several alien invasive moth species have been discovered in recent years, possibly resulting from accidental introduction. The leafminer moth *Phyllocnistis citrella* (Stainton, 1856) (Gracillariidae) has been reported on citrus (Iordanou and Charalambous, 1998), the stem borer *Chilo partellus* (Swinhoe, 1884) (Pyraloidea: Crambidae) caught in a light trap (Barton *et al*., 2013) and the *Ficus* leaf miner *Choreutis sexfasciella*, taken at three sites in 2021 (Jakovos Demetriou, Cyprus, 2021, personal communication to authors).

An adventive butterfly species *has* been observed and others are likely to make an appearance. In 2013, linked to release at a wedding, *Danaus plexippus* (Monarch) was reported in Cyprus. Investigations by the first author revealed that specimens, captive-bred in Spain and shipped to an 'event organizer' in Cyprus, were released in Pafos where a specimen was subsequently seen nectaring on an invasive shrub, *Lycium ferocissimum* (boxthorn), and reported to the first author. A similar experience was reported from the Balearic Islands, the wedding organizer there having received stock from the same Spanish breeder/supplier; full details may be found in John *et al*. (2015b). Not least because of a paucity of milkweed hostplants in Cyprus, the butterfly failed to become established and, due to intentional release, we do not consider this species worthy of inclusion in Chapter 12. Note the similarity between the alien *D. plexippus* and the related *D. chrysippus*, the latter a well-known migrant to Cyprus; the wing veining on *D. chrysippus* is much subdued (Figs 13.1 and 13.2).

DOI: 10.1079/9781800621275.0013

Fig. 13.1. *Danaus plexippus* (Monarch) released at a wedding and later photographed in Pafos, 19 October 2013. Courtesy of *David Taylor*

Fig. 13.2. The related and very similar *Danaus chrysippus* (Plain Tiger), a regular migrant to Cyprus. Pafos District, 28 November 2016. *Eddie John*

In recent decades, a South African lycaenid, *Cacyreus marshalli* (Geranium Bronze) appears to have been introduced into the Mediterranean on imported geranium plants (the larval hostplant for this species). Since first observed, this highly opportunistic species has been found extensively throughout Europe and from west to east in the Mediterranean basin, and can be expected to appear in Cyprus soon, if not already present. Further information on this and potential migrant species is given below.

A warming climate has already influenced the composition of butterflies and other invertebrate faunas in the eastern Mediterranean (Benyamini and John, 2020: 15) and will continue to encourage ingress of species into Cyprus, especially of migrants from regions to the south and south-east. As a possible consequence of temperature change, a degree of adaptation by butterflies may be necessary if hostplant and butterfly synchrony is not to be lost, i.e. adoption of normally less-favoured hostplants.

Although not contiguous with other countries, the proximity of Cyprus to nearby mainland countries, where butterfly diversity is significantly higher, adds to the prospect of a changing species count for Cyprus in the future. Whether by northerly/north-westerly migration (most likely) or by wind-assisted passage from countries such as Turkey or Syria, the species described below are candidates to become additions to the Cyprus list. Other species, already with a limited distribution and a possible tenuous hold (e.g. *Libythea celtis*), may be lost to the Cyprus list due to climate-driven range retraction.

In terms of range expansion, as described earlier, Cyprus has recently (2021) experienced the arrival of the spectacular swallowtail *Papilio demoleus*. It seems inevitable that there will be further gains and losses in the decades ahead. Anthropogenic spread, as with the lycaenid *Cacyreus marshalli* throughout much of the Mediterranean region, will also have an effect in the future.

Anticipated Arrivals to the Island

Cacyreus marshalli Butler, 1898 — Lycaenidae: Lycaeninae

Geranium Bronze/Pelargonium Butterfly

Fig. 13.3. *Cacyreus marshalli* (Geranium Bronze), Cala Galdana, Menorca, 12 June 1996. *Eddie John*

Fig. 13.4. *Cacyreus marshalli* (Geranium Bronze), Dordogne, France, 16 September 2011. Courtesy of *David Tomlinson*

Following probable accidental introduction on imported *Geranium* plants from South Africa, the first known Mediterranean observation of this small brown lycaenid was in Mallorca in 1987 (Waring and Thomas, 1994; Parker and John, 2016; John *et al.*, 2018b), following which the Mediterranean's first published record was by Eitschberger and Stamer (1990). Since then, *C. marshalli* has spread rapidly throughout much of the Mediterranean basin. In the east of the Mediterranean in recent years, *C. marshalli* has been reported from Athens (Anastassíu *et al.*, 2010), western Turkey (Başbay, 2013; Soyhan *et al.*, 2013; Ambarlı and Başbay, 2014; Başbay and John, 2021), the Aegean Sea and Rhodes (Galanos, 2014), Egypt (Abu-Ghonem *et al.*, 2017) and Israel (Benyamini and Kurman, 2020). If not already present in Cyprus and south-eastern coastal provinces of Turkey, where it is already known from Antalya Province (Onat Başbay, Ankara, 2021, personal communication to first author), it is only a matter of time before *C. marshalli* arrives, possibly via infested cuttings harbouring the larvae. The absence (to date) of observations of *C. marshalli* from Mediterranean coastal areas to the north of Israel seems to point to an accidental introduction in Israel, as no links with populations in south-western Mediterranean Turkey are apparent.

Despite its small size (wingspan up to *ca.* 23 mm), C. *marshalli* is a mobile species and has been recorded far from habitation and even at elevations up to 2400 m in north-western Italy (Paradiso *et al.*, 2019). In Cyprus, any small member of the family of blues (note that both sexes are predominantly brown) seen flying around cultivated geraniums should be examined critically.

Gegenes nostrodamus (Fabricius, 1793) **Hesperiidae:** Hesperiinae

Mediterranean Skipper

Among seven other species/'forms' of hesperiid, *G. nostrodamus* was first listed for Cyprus (as '*Hesperia nostradamus* Fab.') by Lederer (1855), although no details were provided and there is no mention of *Gegenes pumilio* having been taken, raising the question of a possible misidentification or of the specimens having been captured in Lebanon (see Chapter 2). Bucknill (1911), in a section on hesperiids, noted that '*Pamphila nostradamus*' (= *G. nostrodamus*) was 'recorded by Lederer and from Nicosia in October'. Rebel (1916) included '*Parnara nostrodamus*' (= *G. nostrodamus*), but in his 1939 list excluded this in favour of *Gegenes lefebvrei* (= *G. pumilio*). Turner (1920) wrote, 'Two or three specimens received are quite typical [of *G. nostrodamus*] and indistinguishable from a specimen I have from Central Italy.' However, Parker (1983) noted that early collectors 'recorded *nostrodamus* in place of *pumilio*; they are not easily separated. The BM(NH) has plenty of *pumilio*, but no *nostrodamus* from Cyprus.' Distribution maps in Hesselbarth *et al.* (1995: Map 12) and Atahan *et al.* (2018: 170) show *G. nostrodamus* to be rare in Turkish locations to the

Fig. 13.5. A probable *Gegenes nostrodamus* (Mediterranean Skipper) showing costal 'hair' tufts (circled), photographed at Protaras, October 2015. Courtesy of *Charles Cuthbert*

Fig. 13.6. A probable *Gegenes nostrodamus* (Mediterranean Skipper) showing costal 'hair' tufts (circled), photographed near Polis, 22 November 2015. Courtesy of *Richard Somers Cocks*

Fig. 13.7. A possible *Gegenes nostrodamus* (Mediterranean Skipper) showing a plain underside, lacking the typical markings of *G. pumilio*, but also without obvious costal 'hair' tufts. Photographed near Polis, 19 November 2015. Courtesy of *Richard Somers Cocks*

Fig. 13.8. A possible *Gegenes nostrodamus* (Mediterranean Skipper) mud-puddling and simultaneously excreting surplus water, Ezousa riverbed, 9 April 2016. (The Papilionidae have been observed taking in water and excreting at the same time, but it is rare (in Cyprus) to see skippers engaged in this activity.) Courtesy of *David and Rosalyn Sparrow*

north and north-east of Cyprus, with the latter publication noting that the species was not recorded from Hatay until May 2009. Baytaş (2007: 190) similarly stated, 'Very local and rare.' Recent records from Lebanon indicate a continued presence in coastal areas to the south-east of Cyprus (Seizmair, 2013). The species is present in the Levant as nominotypical *G. nostrodamus nostrodamus*.

Both *Gegenes* species are of similar size, *nostrodamus* perhaps being slightly the larger and with somewhat larger and brighter upperside markings. The late Torben Larsen (2012, personal communication to first author) wrote: '*Gegenes pumilio* and *G. nostrodamus* are very similar species. The literature is full of misidentifications. Normally *G. pumilio* is slightly smaller than *G. nostrodamus*, with less pointed forewings. *G. pumilio* is generally dark blackish-brown with relatively little contrast between the base and the margin of the forewing. In *G. nostrodamus*, the marginal area of the forewing is usually much lighter than the basal area and has a distinct cinnamon tone: specimens from arid localities such as Iraq and Mauritania may be very light. Lighter-than-normal specimens of *G. pumilio* are seen but normally lack the hint of cinnamon.'

In recent years, the receipt of several photographs of possible *G. nostrodamus* from Cyprus led to much discussion initially involving Rob Parker, David Lees at the BMNH and the first author, but without arriving at a firm conclusion. However, on viewing Figs 13.5–13.7, John Coutsis and Rienk de Jong (Athens and Leiden, 2021, personal communications to first author) have expressed the opinion that these specimens from Cyprus are *G. nostrodamus*. Wing undersides of the specimens illustrated are lighter in colour than *pumilio* (a plain grey/brown) and, importantly, the hindwings show the characteristic long tuft of costal 'hairs' (circled in Figs 13.5 and 13.6) associated with that species (Lorković, 1971; Lafranchis, 2004: 23; de Jong and Coutsis, 2017).

Larsen (1982b), when noting the ecological preferences of the two species, commented that 'It is quite possible to develop a "feel" for *pumilio* habitats' and further added that 'It should be noted that *Gegenes nostrodamus* ... is basically an oasis insect or an insect of the shade. It is rare to find the two in the same locality.' Similarly, de Jong and Coutsis (2017) wrote: '*Gegenes pumilio* and *G. nostrodamus* have different ecological preferences, the first being mainly found in open Mediterranean garrigue (a type of low, soft-leaved scrubland) and dried out river beds, often near the coast, while *G. nostrodamus* is more shade-loving, frequenting areas with scattered trees, and may occur far inland, as in Spain ...'. Larsen and Nakamura (1983) also raised the possibility that while both species use grasses as hostplants, different species (of grasses) might be used, 'as the two [*Gegenes* spp.] show decided differences in habitat choice'. However, their views on habitat would appear to conflict with the findings of Wiemers (1983) in Greece and Hesselbarth *et al.* (1995: 182) in Turkey, who reported the finding of *G. nostrodamus* in hot locations, typical of those inhabited by *G. pumilio*.

In summary, *Gegenes* specimens lacking underside markings and displaying hindwing costal 'hairs' would benefit from molecular analysis, or genitalic examination (♂ or ♀), as the two species can easily be separated in this way (Coutsis, 2012). Meanwhile, until such work has been undertaken, this species remains unconfirmed for Cyprus, but the degree of confidence in its presence is high. North-westerly migration from southern countries of the Levant should be considered, as *G. nostrodamus* is known to reach coastal areas of the region (Benyamini and John, 2020: 176). Note also that the individuals shown in Figs 13.5–13.7 were photographed in autumn of the same year at points in the far west and far east of Cyprus, perhaps indicative of a migration.

Other Possible Future Arrivals

Iphiclides podalirius (Linnaeus, 1758) — Papilionidae: Papilioninae

Scarce Swallowtail

A large swallowtail with a kite-like shape, easily separable in flight from *Papilio machaon* and present throughout much of the Levant, but yet to be recorded in Cyprus and not predicted to appear there by Dennis *et al.* (2001: Appendix 2). Nevertheless, its absence is surprising, since ecological conditions, especially at intermediate elevations, would appear to favour its survival. *I. podalirius* was included for Cyprus in a checklist of species for Asia Minor by Staudinger (1881), but lacked substantive data. In an earlier account, Staudinger (1878) stated that 'This species occurs almost everywhere in Asia Minor' (translated from the original German) but the inclusion for Cyprus in his 1881 summary appears to have been erroneous. Rebel (1916) included *I. podalirius* on the 'strength' of Staudinger's listing, but excluded the species from his 1939 paper, reasoning that it was Staudinger's error.

I. podalirius was described as common in Hatay by Atahan *et al.* (2018: 19) and there were recent records from the Syrian Anti-Lebanon by Zarikian and Ghrejyan (2018). Larsen (1975) described *I. podalirius* as 'common and widespread' in Lebanon, noting a potential for large-scale population build-up, although falling short of describing the species as a migrant. Larsen (1997) included *I. podalirius* in a list of only four species of butterfly noted during a brief late-October visit to Cyprus; however, this 'slip of the pen', as it was described, was corrected (to *P. machaon*) by Larsen during a stay with the first author in 2006. *I. podalirius* was described by Stefanescu *et al.* (2003, 2006) as a highly mobile species and there are occasional reports of migration in Europe (e.g. de Vos, 1999). The species' presence on the Mediterranean islands of, for example, Crete and Rhodes (Sala and Bollino, 1997) would seem to indicate that it is capable of a sea crossing. Various Rosaceae, including species of *Prunus*, *Malus* and *Crataegus*, are used as hostplants (Geier and Schurian, 2014; Benyamini and John, 2020: 97), all available in Cyprus.

Madais (*Colotis*) *fausta* (Olivier, [1804]) Pieridae: Pierinae

Large Salmon Arab

This is a highly distinctive salmon-buff coloured pierid, approximately the size of *Pieris rapae* and a regular and predictable migrant to mainland Levant countries northwards to southern Turkey. By late summer, numbers of this Oriental species may become prolific before dying out completely by the onset of winter (Larsen, 1976; Benyamini and John, 2020: 98). *M. fausta* is one of the most likely of the tropical/subtropical species to appear in Cyprus; indeed, Larsen (2015) expressed surprise at the absence of occasional records from the island. While not resident in Israel, such is the species' migratory capability that Comay *et al.* (2021) reported *M. fausta*, along with *P. brassicae* and *P. rapae*, to be the butterfly species next in order of abundance to *V. cardui* in Israel, based on its presence in 44 of 47 transects surveyed over a 5-year period. Current winter conditions in the Levant prevent permanent populations from becoming established (Benyamini and John, 2020: 98) and in the view of Larsen (1975) the species might be 'a permanent nomad with regular migration'. In Lebanon, Beale and Sprenger (2006: 2) considered *M. fausta* as 'being exclusively an autumn migrant to the Aammiq area'. Hostplants include *Capparis spinosa* (Pittaway, 1981, 1985), a common plant in Cyprus.

Belenois (Anaphaeis) aurota Fabricius, 1793

Caper White

A medium-sized white butterfly heavily marked in grey and black, with yellow on the hindwing underside. *A. aurota* is another vigorous (Palaeotropical) migrant but, compared with *M. fausta*, appearances in the Levant are random and unpredictable; Comay *et al.* (2021) noted its presence in 33 of 47 transects in Israel. *A. aurota* was reported by Larsen (1974: 94) to reach its northern (Levant) limit of distribution in Lebanon, but, possibly because of a warming climate, it is now known to reach Syria and Turkey (John *et al.*, 2020). Nevertheless, the species is unable to survive the winter and maintain a presence in the Levant (Benyamini and John, 2020: 94). Larsen (1990: 33) considered that no subspecies should be recognized. Hostplants include various members of the Capparaceae and Brassicaceae (Benyamini and John, 2020: 97), both families being abundant in Cyprus.

Pontia glauconome Klug, 1829

Desert White/Desert Bath White

Superficially similar in size and appearance to both known *Pontia* species in Cyprus, *P. glauconome*, an eremic species, is less likely to reach Cyprus, but is a known migrant to the Levant, where Comay *et al.* (2021) noted its presence in just five of 47 transects surveyed over a 5-year period in Israel. *P. glauconome* has recently arrived and established colonies in south-eastern Turkey (John *et al.*, 2018a; John *et al.*, 2020), although the permanency of these has yet to be determined. Hindwing undersides of early and late forms have much dark green scaling and *pronounced* yellow veining; in the summer form, the green scaling is largely lost. *P. glauconome* has a remarkable ability to survive adverse (desert) conditions by remaining in the pupal stage for up to 5 years, pending the return of rain to stimulate growth of its larval hostplants (Larsen, 1984a: 33); these include members of the Brassicaceae, Resedaceae and, less frequently, Capparaceae.

Callophrys rubi (Linnaeus, 1758) — Lycaenidae: Theclinae

Green Hairstreak

Although predicted (> 50% probability) as a possible species for Cyprus by Dennis *et al.* (2001: Appendix 2), this very distinctive green lycaenid has yet to be reported. The nominotypical subspecies is present in Mediterranean mainland coastal areas of Turkey (Atahan *et al.*, 2018: 103, described it as common in Hatay), Syria and Lebanon to the east of Cyprus, but *C. rubi* is not known as a migrant in the region. *C. rubi* larvae are highly polyphagous (Fabaceae and Rhamnaceae included), so there would be no lack of available hostplants in Cyprus.

Chilades (Luthrodes) pandava **(Horsfield, 1829)** **Lycaenidae:** Polyommatinae

Cycad Blue, Plains Cupid

This invasive lycaenid was first recorded as new to the Mediterranean region following observations in 2012, when its appearance in the Nile Delta was linked to probable importation of infested ornamental *Cycas revoluta* (Sago Palm) (Fric *et al.*, 2014). In the summer of 2014, *C. pandava* was reported in Dubai (United Arab Emirates), also for the first time (Feulner *et al.*, 2014). In the Mediterranean basin, *C. pandava* was next reported from Alexandria on the Mediterranean coast in the same year (Abu-Shall *et al.*, 2014) and, in 2019, it appeared near Tripoli in northern Lebanon (Lepidoptera Libanotica, 2019). All observations were associated with locations in which *Cycas* spp. were present and on which plant *C. pandava* is regarded as an invasive pest species (Wu *et al.*, 2010: 12). Wet and dry season forms are known elsewhere, e.g. Tiple *et al.* (2009). The species is known to disperse but does not appear to migrate, so any future appearance in Cyprus will be the result of accidental introduction, as is likely elsewhere in the Mediterranean.

Junonia orithya **(Linnaeus, 1758)** **Nymphalidae:** Nymphalinae

Blue Pansy / Eyed Pansy

This small (size of *P. rapae*) migratory nymphalid has unmistakable uppersides (♂ with vivid blue hindwings, ♀ mainly brown) and two distinctive, pupilled orange fore- and hindwing ocelli. *J. orithya* was formally recorded for the first time in Turkey (Diyarbakır Province) on 30 October 2010, in the south-east of the country (Biricik, 2011), and later photographed in Hatay Province on 3 September 2011 (Trakel, undated). The species has appeared regularly in Turkey's south-eastern provinces to the east of Hatay since 2011 (AdaMerOs, 2021; Trakel, undated), and appeared in Osmaniye Province, adjoining Hatay to the north, in October 2016 (Trakel, undated). It is a surprising omission from Atahan *et al.* (2018).

Larsen (2015) described *J. orithya* as a regular local and occasional long-distance migrant, 'common throughout Africa and tropical Arabia and penetrates well into the non-tropical parts of Arabia.' *J. orithya* is also known from Jordan (Larsen, 1984b; Bozano, 1990) and Israel (Benyamini, 2002a: 110). DNA barcoding of a specimen taken in Israel indicated the likelihood of an Asian (Arabian) source of this migrant to the Middle East (Lukhtanov *et al.*, 2016). However, winter survival in the northern Levant seems unlikely (Larsen, 1984b). Hostplants of possible relevance in Cyprus include members of the families Convolvulaceae, Scrophulariaceae, Plantaginaceae and Lamiaceae (Labiatae).

Polygonia egea **(Cramer, 1775)**

Southern Comma

Another species that is widely distributed in the Levant, as well as on other Mediterranean islands, but is surprisingly absent from Cyprus, although recorded there historically as '*Vanessa Triangulum* Fab.' by Lederer (1855) and '*Vanessa Egea* Cr. (*Triangulum* F.)' by Staudinger (1878). However, this was not found by Bucknill (1911), who noted, 'It is common enough near Beyrout', providing perhaps another indication that specimens collected by Zach in 1853 from Cyprus and Lebanon had become mixed (see Chapter 2). However, in referring to Lederer's (1855) inclusion for Cyprus, Rebel (1939) offered the opinion, 'like all information provided by Zach [it] is to be regarded as reliable' (translated from the original German) and he considered that the species had disappeared from the island in the meantime. *P. egea* is well distributed throughout Turkey's southern Mediterranean coastal provinces (Hesselbarth *et al.*, 1995: Map 307; AdaMerOs, 2021), but is described as rare in Hatay (Atahan *et al.*, 2018: 54). A widespread presence in mainland countries to the east of Cyprus makes its absence from Cyprus even more surprising, yet *P. egea* was not predicted to appear in Cyprus by Dennis *et al.* (2001: Appendix 2). Hostplants are of the family Urticaceae: *Parietaria judaica* and others of the genus.

Melitaea trivia **([Denis & Schiffermüller], 1775)**

Lesser Spotted Fritillary

Melitaea syriaca **Rebel, 1905**

Syrian Fritillary

M. trivia was predicted, with a > 50% probability, as a possible species for Cyprus by Dennis *et al.* (2001: Appendix 2) and appears to be the most common of the *Melitaea* fritillaries found in coastal areas of Mediterranean Turkey. It is listed there as nominotypical *Melitaea trivia trivia* by Hesselbarth *et al.* (1995: Map 321) and by Atahan *et al.* (2018: 68).

Taxonomy is uncertain, however, with the elevation of ssp. *M. trivia syriaca* to species' status by van Oorschot and Coutsis (2014), an elevation not universally recognized (e.g. Numa *et al.*, 2016; Kudrna, 2019: 198), but accepted by Benyamini (2021: 212). Hostplants include Scrophulariaceae.

Spialia orbifer (Hübner, [1823]) Hesperiidae: Pyrginae

Orbed Red-underwing Skipper

Predicted (> 50% probability) as a possible species for Cyprus by Dennis *et al.* (2001: Appendix 2). Lederer (1855) listed both '*Hesperia Eucrate*' (= *Spialia sertorius*, Red Underwing Skipper) and '*Hesperia Eucrate* Var. *orbifer*' (= *S. orbifer*) for Cyprus, as did Bucknill (1911) (using the name '*Hesperia orbifer*'), who noted that the former ('*Hesperia Eucrate*') was recorded 'In June and July on the Kyrenia hills'. *Spialia sertorius* is, in fact, a species of the western Mediterranean (e.g. Tshikolovets, 2011: 64), whereas Hesselbarth *et al.* (1995: Map 23) and Atahan *et al.* (2018: 167) show *S. orbifer* to be a common species in Mediterranean Turkey; it is also common elsewhere in the Levant, south into Israel and Jordan. Various members of the Rosaceae are used as hostplants (Benyamini and John, 2020: 142).

Doubtful Species

With the publication of more reliable species accounts and the advent of electronic communication, bringing with it easier access to such data, it has been increasingly evident in recent decades that significant misidentifications, for example *Glaucopsyche melanops* (Boisduval, 1828) or *Cyaniris semiargus* (Rottemburg, 1775) for *G. paphos*, and the uncorroborated records of Ayberk and Akkuzu (2007), as discussed in Chapter 2, have become a rarity. Many of the questionable species listed in early publications have been discussed in the preceding pages and were dealt with in impressive detail by Parker (1983). There is little to be gained by repeating Parker's work, the more so because, other than *Chazara persephone* and potentially *Gegenes nostrodamus*, the questionable species listed by Parker remain unconfirmed. However, some may yet appear in Cyprus, as discussed earlier.

The recent arrival of *P. demoleus* and the likely presence of *C. marshalli* and *G. nostrodamus* is a reminder that claims of a new species should be investigated with an open mind, especially in view of Cyprus's positioning on the fringes of Asia, coupled with the influences of a warming climate. The appearance of highly unlikely species such as *D. plexippus* should continue to be viewed with scepticism and not be accepted as a natural phenomenon. However, other than for organized 'ceremonial releases', the accidental or intentional release into the wild of captive-reared species is far less likely to occur in Cyprus than in countries such as the UK or perhaps America.

14 Status of Butterflies in Cyprus

In common with some other countries of Europe, the absence, to date, of a sustainable Butterfly Monitoring Scheme in Cyprus hampers accurate assessment of population changes and hence the opportunity of presenting such data here; however, information on a recently implemented (2021) monitoring scheme follows in the next chapter. Nevertheless, despite the lack of a formal assessment, there is ample everyday evidence to indicate that, as elsewhere in Europe, habitat loss has adversely affected butterfly numbers overall.

In 1999, the lack of either a monitoring or a recording scheme at the time contributed to some erroneous data for Cyprus being accepted for Butterfly Conservation Europe's Red Data Book of European Butterflies (Van Swaay and Warren, 1999), although inaccuracies in the island's butterfly checklist were quickly corrected (Van Swaay *et al.*, 2010a). In preparation for the 2010 update, an assessment by the authors of this book listed two Cyprus species as giving some cause for concern. While not (yet) viewed as seriously threatened in Cyprus, *Apharitis acamas* (Levantine Leopard) and *Luthrodes galba* (Small Desert Blue) were regarded as species of concern due to the vulnerability of their respective habitats. Of the 43 species of butterfly assessed for Cyprus (migrants or species of marginal occurrence were excluded), 91% were classified as of 'Least Concern', while 9% were listed at the European Level as 'Near Threatened' in the IUCN Red List (2013).

Considering *L. galba* as an example of a species giving some cause for concern, this comparatively weak-flying species, although reasonably widespread in Cyprus, is restricted in its distribution by the limited geographical availability of its larval hostplant (*Prosopis farcta*). Absent from much of Cyprus, *L. galba* may be locally abundant at some sites, but always in close association with its larval hostplant that is found only at low elevations, often in areas most at risk to urbanization. Unfortunately,

DOI: 10.1079/9781800621275.0014

the attraction of such sites to developers is increasingly depleting areas of the hostplant, *P. farcta*, confining the butterfly to smaller, disjunct territories, sometimes lacking suitable 'corridors' (links) through which even this sedentary species might otherwise be able to connect with more suitable breeding areas. This is noticeable, for example, along the eastern and northern boundaries of the Larnaka salt lake, where land clearance for house-building and associated infrastructure has destroyed once valuable biotopes for *L. galba* and other species of butterfly, especially *Zizeeria karsandra* (Dark Grass Blue) and *Lycaena thersamon* (Lesser Fiery Copper), which share similar habitats. On the western fringes of the salt lake, habitat has been severely damaged due to over-grazing by goats.

Because of their marginal occurrence within Europe, neither *A. acamas* nor *L. galba* is accorded a 'threat' status by Van Swaay *et al.* (2010a), earning instead the classification 'NA', i.e. not assessed. Also similarly non-classified at the European level are 45 other species, including, for Cyprus, *Z. karsandra*, *Ypthima asterope* (African Ringlet) and *Pelopidas thrax* (Millet Skipper). To be able to assess these five species in like manner to that adopted for the other 529 species classified as 'true European' would require a wider study of their status in other countries of the eastern Mediterranean; at present, they are regarded as data deficient. Where appropriate, further remarks appear for each of these five species within their respective species' descriptions.

Libythea celtis (Nettle-tree Butterfly) gives cause for concern because of the species' tenuous presence at the southernmost extent of its range in the eastern Mediterranean, added to which the hostplant trees *Celtis tournefortii* and *C. australis* are uncommon in Cyprus. For the reasons stated in the text for each species, *Aporia crataegi* (Black-veined White), *Pontia chloridice* (Small Bath White) and *Ypthima asterope* are also considered vulnerable in Cyprus.

It should be noted that the endemic species, *Maniola cypricola* (Cyprus Meadow Brown) and *Glaucopsyche paphos* (Paphos Blue), while being shown in the 1999 study (Van Swaay and Warren, 1999: Table 7) as species of conservation concern, are not considered threatened at present (see species entries for details). That is not to say that their habitat requirements do not require protection (see Özden and Hodgson, 2011); they are so listed simply because of their restricted European distribution, i.e. (where these taxa are concerned) found only in Cyprus. Somewhat misleadingly, the terminology used in the 1999 Red Data Book (Van Swaay and Warren, 1999), allowed such endemics to be designated as 'threatened' at the European level even when that endemic is abundant within its territory (Chris van Swaay, Wageningen, 2008, personal communication to first author). In the updated Red Data Book, and based on data provided by the authors of this book, Van Swaay *et al.* (2010a: Appendix 2) and Maes *et al.* (2019) concur with the IUCN Mediterranean Red List (Numa *et al.*, 2016), wherein *M. cypricola* and *G. paphos* are considered 'LC', i.e. species of least concern (not threatened). This category is also applied to the third endemic species, *Hipparchia cypriensis* (Cyprus Grayling), omitted from Van Swaay and Warren (1999) but now included in the 2010 Red Data Book list for Cyprus (Van Swaay *et al.*, 2010a).

However, Kudrna *et al.* (2015: 545), writing of European endemics in general, stated, 'These species – mostly island isolates – are potentially exceptionally vulnerable for various reasons related in the first place to their miniature ranges highly susceptible to land use change [*cf. L. galba* and the Larnaka Salt Lake] and their apparent inability to spread outside their range since the end of the last glacial period.' Fernández-Palacios *et al.* (2021) expressed similar concerns in stating: 'Nevertheless, the majority of these endemic species are inherently vulnerable due to genetic and demographic factors linked with the way islands are colonized.'

Notwithstanding, and in common with Van Swaay *et al.* (2010a), the three Cyprus endemics mentioned above are (rightly) not included in the list of rare endemic European species given in Kudrna *et al.* (2015: 545; 2019: 289) and all are listed as 'LC' (Least Concern) in Numa *et al.* (2016). Because of concern at a *European* level, *Chazara briseis* (The Hermit) is listed as NT ('Near Threatened') by Kudrna *et al.* (2011: 112, 2015: 396) and Kudrna (2019: 102), but the Cyprus population is plentiful (see species entry).

Similarly, *Thymelicus acteon* (Lulworth Skipper) and *Pseudophilotes vicrama* (Eastern Baton Blue), although previously listed as vulnerable in Europe (Van Swaay and Warren, 1999) and listed as 'NT' in Europe as a whole (Van Swaay *et al.*, 2010a), are not considered threatened in Cyprus. Populations of *T. acteon* in Cyprus (where it is a very common species) appear under no imminent threat, but assessment of *P. vicrama* numbers is problematic due to its disjunct distribution, smaller numbers overall and superficial similarity to *Polyommatus icarus* (Common Blue), potentially leading to this species being under-recorded by inexperienced observers.

Climate Change

As in other areas of the Levant, climate-driven change seems likely to impact upon the butterfly biodiversity of Cyprus, by encouraging some tropical species, e.g. *Catopsilia florella* (African Migrant) (John *et al.*, 2019b), to extend their ranges northwards. In response to climate change, aestivation, a trait adopted by a number of species in Cyprus, might extend to others, as reported for *Brintesia circe* (Fabricius, 1775) (Great Banded Grayling) in central Europe (Birch *et al.*, 2021), potentially leading to fewer broods per year for some satyrines, for example *Lasiommata* spp.

A warming climate might also see a change in the altitudinal distribution of some resident species such as *A. crataegi*, forcing movement to higher, cooler elevations (Merrill *et al.*, 2008; Rödder *et al.*, 2021). Spatiotemporal responses to climate change have been reported elsewhere in the Mediterranean (e.g. Colom *et al.*, 2021) and are of relevance to Cyprus, particularly in respect of phenology. A negative consequence of this might include phenological asynchrony between butterfly and hostplant interactions (Donoso *et al.*, 2016; Zografou *et al.*, 2020), particularly in satyrines that lay eggs on dry grasses in autumn, as increasingly arid conditions could further delay regrowth of hostplants and consequent absence of edible plants for autumn-emerging larvae, unless adult aestivation periods are extended. Other Cyprus species – *Libythea celtis* could be one example – might retreat northwards towards zones that are more temperate.

Collecting

In recognition of the need to exercise some degree of control, a permit is necessary to allow the collection of invertebrates in Cyprus. The Department of Environment of the Ministry of Agriculture, Rural Development and Environment recognizes the value of records contributed by early collectors and observers over many decades, so the introduction of the need for a permit is not intended to impose unnecessarily strict sanctions on those wishing to continue with their studies.

Collecting Code

Cyprus is a small island, with correspondingly small populations of many species and high rates of endemism. Some populations of invertebrates are confined to particular areas of the island, either because of hostplant/habitat interrelationships or because the sedentary nature of the species, in combination with topographical barriers or the loss of natural 'corridors' due to urbanization, prevents movement from one part of the island to another. Consequently, the over-collecting of species from one such isolated area may mean that others of the same species are physically unable to occupy these biotopes, with the result that the species may go extinct locally. If collecting specimens:

1. Please adhere to the guidelines set out below.
2. To aid future conservation projects, please provide the Department of Environment with detailed records of your observations or specimens taken (species, numbers, location and date).

A Code of Practice for the Collection of Lepidoptera and other Invertebrates in Cyprus

It is strictly prohibited to collect species under Annex II of the Law on the Protection and Management of Nature and Wildlife [No. 153(I)/2003]. Although very few of the invertebrates protected by Annex II have been found in Cyprus (and these include no species of butterfly), the following are known to occur on the island and are thus prohibited by law from collection in any of their stages:

- **Coleoptera:** *Propomacrus cypriacus*
- **Lepidoptera:** *Euplagia quadripunctaria* (Jersey Tiger)

In addition to the species listed above, *Libythea celtis* (Nettle-tree Butterfly) is considered worthy of inclusion in the list of protected species in Cyprus and should not be collected in any of its stages.

15 Conservation of the Island's Butterflies

Conservation

Conservation is comprehensively examined in Dennis (2010: 256), in which it is stated: 'Greater emphasis must be given to plant and animal communities; even when the objective is conservation of a sole species, attention needs to be given to whole landscapes and not just patches, and management for change not stasis.' Hadjisterkotis (2008) gave a similar message: 'Therefore, when Cypriots consider biodiversity conservation and research they should have in mind the unique characteristics of their island, and the need to study and protect not only the habitats and the species which are designated by the European directives, but also the unique Cypriot species, subspecies, varieties, cultivars, and habitats.' Dennis and Hardy (2018: 159) expressed the view: 'Above all, there is a need to understand butterfly habitats better and the extent to which resources and environmental conditions other than the simple presence of the larval host plant(s) affect population parameters, thus population persistence.' While the passion to conserve a single, perhaps favoured, species is an understandable motivation for action, the requirements of other species sharing the same habitat must also be taken into consideration to ensure that any conservation activity does not have a detrimental influence on these species (ICN News, 2011).

Although the expansion of agricultural practices has, in general in Europe, had a detrimental effect resulting in substantial biodiversity loss, the effects of further intensification can be mitigated to some extent by the inclusion within agricultural belts of mosaic habitats, where fields of crops are interspersed with semi-natural or even, with adequate advance planning, natural habitat. General principles of habitat management and practical guidance in respect of the Habitats and Species' Directive and

DOI: 10.1079/9781800621275.0015

the Natura 2000 network are also dealt with by Van Swaay *et al.* (2010b). The experience in Cyprus, however, is rather one of abandonment of agriculture.

Among a number of initiatives in Cyprus was the establishment in 2013 of the Episkopi Paphos Environmental Information Centre (http://epeicentre.com/?lang=en, accessed February 2022), funded by the European Union and the Republic of Cyprus. Included in its facilities is an exhibition hall with displays of local fauna, information on habitat types and an environmental education room.

Where required, expert guidance is available from Butterfly Conservation Europe www.bc-europe.eu (accessed February 2022), with which organization the Cyprus Butterfly Study Group, founded in 2012, is a 'Network Partner'.

Cyprus Butterfly Recording Scheme (CBRS) and Cyprus Butterfly Study Group (CBSG)

The first author established the Cyprus Butterfly Recording Scheme in 2000 and in 2012 he co-founded the Cyprus Butterfly Study Group, which has since been followed by the Cyprus Dragonfly Study Group, run by colleagues David and Rosalyn Sparrow. The mapping system for butterflies is based on a format established by Parker (1983), with updates in Manil (1990), John (2000) and, by the first author, in Makris (2003). Improved and more detailed maps are now maintained at the greater resolution of 5 km^2 UTM squares, and are published here for the first time. Included are records submitted by CBSG members, records received from visitors to the island and those sourced from published accounts, all supplemented with many records contributed by the authors of this book.

The term 'recording artefact' has appeared on occasion throughout. As in any country, particularly where comprehensive fixed transect recording is not practised, there will be favoured areas, for example those with pleasing landscapes, those offering proximity to a home base or where there is an opportunity to see the greatest number of species within a given period. Of course, these are often favoured for transect work, too. Clearly, without an army of recorders, there will inevitably be a degree of bias shown in distribution maps, as demonstrated even in a well-surveyed country such as the UK (Dennis and Thomas, 2000). Consequently, some areas receive little attention. In Cyprus, the Mesaoria Plain is a prime example and remains under-recorded, although it must also be emphasized that this agricultural heartland does not offer much in the way of suitable habitat for many species and is largely uninspiring when visited; nevertheless, this has not been ignored by the authors. To date, the first author's emphasis, through extensive 'square-bashing', has been on building knowledge of the butterflies' distribution, with well over 100 locations checked on each of many visits to Cyprus; for example, 596 species entries were recorded at 183 different sites over a two-week period in the second half of June 2006. The low average count per site reflects less rewarding visits to areas such as the Mesaoria Plain, where few if any species were recorded at some sites in what had become an arid landscape by late June.

As an indicator of an approach to recording taken elsewhere (and its consequential benefits towards the conservation effort) the UK Butterfly Monitoring Scheme began operating in 1976 (Pollard and Yates, 1993), resulting in over 10.5 million butterflies being counted in a 30-year period (Greatorex-Davies *et al.*, 2007); so Cyprus has some ground to make up. However, as an extension of the efforts to date by the Cyprus Butterfly Study Group, a more formal monitoring scheme is the next aim, as proposed by Tzirkalli *et al.* (2019) and explained below.

Butterfly Monitoring in Cyprus

[This section has been contributed by Elli Tzirkalli, BSc MSc (Joint Chair, Cyprus Butterfly Study Group)]

The high biodiversity of Cyprus is reflected in the rich insect fauna observed on the island. The number of insect species is estimated to be more than 7000, of which about 10% are endemic. Butterflies form one of the most familiar and beautiful of the insect groups but, although well known, nevertheless benefit from ongoing study.

Butterflies are well documented, easy to observe and occur in a variety of habitats such as shrublands, forests, river valleys, grasslands and agricultural areas. They have been widely used as indicators in climate change studies, as well as indicators of a healthy ecosystem. It is therefore necessary to collect data not only on their distribution, but also on their population size, which will provide us with useful information for their effective protection and conservation. In Cyprus, butterflies are threatened due to the ongoing urbanization, land-use change and excessive use of insecticides. Additionally, rising temperatures and long periods of drought associated with climate change are expected to cause butterfly populations to decrease due to habitat loss and fragmentation.

To promote the study and conservation of butterflies on the island, the Cyprus Butterfly Study Group (CBSG), formed in February 2012, works towards the protection of Cyprus's butterflies and is a 'Network Partner' of Butterfly Conservation Europe. CBSG activities include the Butterfly Recording Scheme (established by Eddie John in 2000 to collate, record and publish butterfly distribution records), the production of butterfly phenograms and newsletters, attendance at meetings and workshops in Europe and publishing research work by its members.

Although butterfly sightings across the island are important, the need for collecting robust and systematic data, such as the established schemes operating in various Butterfly Monitoring Schemes (BMS) in Europe, is urgent. The UK was the first country to start transect butterfly monitoring in 1976, while the Netherlands was the second, beginning in 1990. Since then, many European countries such as Belgium, Spain, Germany, Finland, etc., have adopted similar schemes. In 2016, the European Butterfly Monitoring Scheme (eBMS) was formed by Butterfly Conservation Europe to bring together data gathered by BMS from different countries into a single database.

To help develop BMS across Europe, a two-year project funded by the EU started in 2018, named Assessing ButterfLies in Europe (ABLE). The ABLE project forms a partnership between Butterfly Conservation Europe, the Centre for Ecology and Hydrology (UK), the Helmholtz Centre for Environmental Research (UFZ, Germany), Dutch Butterfly Conservation and Butterfly Conservation (UK). The CBSG participated as a volunteer partner organization in this project and, with the cooperation of island-wide butterfly experts, established a preliminary BMS in 2019 after a successful butterfly workshop organized under the auspices of the ABLE project (Fig. 15.1).

The methodology of butterfly monitoring follows the eBMS and includes two methods: the transect count and the 15-min count. A transect is a fixed route (walk) established at a site where butterflies are recorded over a period of several years. All butterfly species occurring within a band of 2.5 m on each side of the transect, as well as up to 5 m ahead, are counted while walking a constant pace. The length of the transects is typically about 1 km but can be less, depending on the country and the region (southern European countries have more mountainous terrain). Each transect is divided into sections (e.g. 50 m) that broadly coincide with distinct habitat types within the wider area. Butterflies are counted during their flight period (mainly during spring and summer). The recorder should ideally visit the transect every week; if a weekly walk is not possible, every 10 days or 2 weeks is acceptable. Transect walking is done at a constant pace, between 9 am and 4 pm, in good weather (sunny and warm, with no rain or strong wind). The transect walker can choose which route to follow or ask for advice from the country BMS coordinator.

The second method is the 15-min count, which is a simple method that allows the recording of butterflies on a path, in a given area (meadow or a garden) and at a fixed point. During the 15-minute period, the observer records all the butterflies inside an imaginary box (5 × 5 × 5 m). This more flexible methodology is used to collect butterfly data from various places (gathering data from countries with few or no transects) and follows the guidelines of transect recording. The ButterflyCount app was developed by eBMS to perform this methodology and records the track, while the observer adds butterfly species.

The newly formed Cyprus BMS is currently using both methods and has helped with the creation of a 'field guide' (an illustrated ABLE-funded pamphlet, see Fig. 15.2) to help volunteers identify the less familiar species that can be found on the island. In order to develop the full potential of the Cyprus BMS, more volunteers are needed to join the scheme. It is important to set up more transects and collect long time-series butterfly monitoring data to inform and develop EU (e.g. European Grassland Butterfly Indicator) and national policies.

Fig. 15.1. Participants of the 1st BMS Workshop Cyprus, 2019. Courtesy of *Elli Tzirkalli*

Fig. 15.2. ABLE Cyprus butterfly guide, illustrating 48 of the most common of Cyprus's butterflies. Courtesy of *Cristina Sevilleja*, De Vlinderstichting/Dutch Butterfly Conservation

To find out more information about butterfly monitoring and the submission of butterfly data, visit the Cyprus BMS website: https://butterfly-monitoring.net/cyprus-bms (accessed February 2022). To participate in monitoring butterflies in Cyprus, contact the coordinators: Elli Tzirkalli (ellitzirkalli@gmail.com) or Özge Özden (ozgefuller@gmail.com).

Also, join our CBSG group by sending an email to cbsg2012@gmail.com or Facebook Page 'Cyprus Butterfly Study Group'.

Possible Areas for Future Research

On occasions throughout this book, we have indicated gaps in our knowledge that would benefit from further study, especially in the field of butterfly behaviour. The first author, in particular, regrets not spending more time simply observing butterfly activity when the priority over the years has largely concentrated on distributional recording. Of course, much can be gleaned by so doing, but there is so much more to be understood.

Among subjects in need of investigation are the following.

- Establish the validity of the endemic species/ssp.:
  Allancastria cerisyi cypria
  Hipparchia cypriensis
  Maniola cypricola
  Hyponephele lupina cypriaca

  Confirm the subspecies of:
  Apharitis acamas
  Lycaena phlaeas
  Lycaena thersamon
  Favonius quercus
  Celastrina argiolus
  Pseudophilotes vicrama

Aricia agestis
Polyommatus icarus
Limenitis reducta
Pseudochazara anthelea
Lasiommata megera
Lasiommata maera
Thymelicus acteon

- Voltinism of some of the island's butterfly species. For example, pursuing further the discussions in the species texts on the number of annual broods of the endemic *Glaucopsyche paphos* and those of *Gonepteryx cleopatra*, *Argynnis pandora*, *Kirinia roxelana* and *Ypthima asterope*.
- Determining what accounts for fluctuations in *Pontia chloridice* numbers and whether, in times of drought, the species undergoes pupal diapause. In parallel with this, an investigation into the germination requirements of the species' *Cleome* hostplants.
- Aestivation involving several of the island's satyrine species – establishing ratio of aestivating females to males.
- Post-aestivation puddling by female satyrines – how common is this and is it confined to *Chazara briseis* and *Maniola cypricola*?
- Identifying the hostplant(s) used by *Aricia agestis* in Cyprus.
- *Gegenes pumilio*? Or is it *Gegenes gambica*? (See entry for *G. pumilio*.)
- *Gegenes nostrodamus* – confirmation of presence by means of genitalic or molecular examination remains to be determined.
- An analysis comparable to that reported by Vodă *et al.* (2016) that examines the genetic differentiation of the butterfly populations of Cyprus compared with those of mainland countries of the eastern Mediterranean.
- A study to increase our knowledge of parasitoids in Cyprus. For example, are they implicated in the occasional reduction of *Leptotes pirithous* numbers, or is this winter weather-related?

In conclusion, we express our hope that the contents of this book, together with data gathered from the newly implemented Cyprus Butterfly Monitoring Scheme, will help stimulate additional research, thereby increasing further our knowledge of the island's fascinating butterfly fauna. Volunteers willing to assist with either form of recording as detailed above will receive a warm welcome.

Abbreviations, Acronyms and Glossary

aberration (ab.): a deviation in appearance from the normal.
ABLE: Assessing Butterflies in Europe.
adventive: non-native – an organism accidentally introduced to an area.
aestivation (vb: **aestivate):** a means of surviving through the summer in a quiescent (dormant) state during very hot or dry weather; summer diapause.
androconia: specialized wing scales ('scent' scales) in the male that release pheromones.
aphytophagous: not feeding on plants.
aposematic: displaying warning coloration/markings.
asl: above sea level.
bivoltine: having two broods per year.
BMNH: British Museum of Natural History.
BMS: Butterfly Monitoring Scheme.
ca.: about (from Latin *circa*).
CBRS: Cyprus Butterfly Recording Scheme (website: www.cyprusbutterflies.co.uk).
CBSG: Cyprus Butterfly Study Group.
cf.: compare (from Latin *confer*).
cilia (sing. **cilium):** hair-like structures bordering the wing edges.
clade: a group of organisms believed to comprise all the evolutionary descendants of a common ancestor.
conspecific: belonging to the same species.
cryptic: displaying camouflaged coloration/markings.
cuticle: outer covering (of the pupal case, in this context).
DD: data deficient.
dimorphic (n. **dimorphism):** having two distinct forms, characterized by size, shape or coloration. See also: sexual dimorphism.

eBMS: European Butterfly Monitoring Scheme.
eclosion (vb. **eclose**): the act or process of emerging from the egg or pupal case.
endemic: indigenous, confined to a region.
eremic: relating to deserts or sandy areas.
f. (= form): used here to indicate representatives of a species differing in appearance from the 'normal', e.g. *Colias croceus* f. 'helice'.
facultative myrmecophily: a commonly seen association in a number of lycaenid species, whereby larvae are 'groomed' by ants to encourage the release of nutritious secretions. Most of these interactions appear mutualistic, but are not essential for survival of either ant or butterfly (see Pierce *et al.* 2002).
herbivory (adj. **herbivorous**): feeding on plant material.
hilltopping: a mate-location behaviour whereby males and females may fly to higher areas (summits, hilltops, ridges) in order to seek mating opportunities.
holotype: the single specimen designated by the author of a species as its type specimen.
honeydew: a sticky, sugar-rich substance secreted by aphids as a by-product of digestion.
imago (pl. **imagines**): the adult form.
in cop.: in copula, i.e. in the act of mating.
in coll.: in the collection of.
instar: stage between periods of moults.
LC: Least Concern.
lectotype: a specimen selected in the absence of a holotype and designated through publication to serve as the type.
migrant: seasonal determined and predictable movement of a species from one place to another.
monandrous: where females mate once only.
monophagous: specializing on a single host species.
morph: see polymorphism.
mud-puddling: the imbibing of minerals and salts from various substrates such as mud, gravel at stream edges, water puddles, carrion and excreta.
MYA: million years ago.
myrmecophilous: having a beneficial, close association with ants.
NA: Not Assessed.
NT: Near Threatened.
nominotypical: when a species is split into subspecies, the originally described population is retained as the 'nominotypical (or 'nominate') subspecies' and the species name is repeated, e.g. *Papilio machaon machaon*. A subspecies cannot be recognized in isolation: a species will either be recognized as having no subspecies at all, or two or more (including any that are extinct), never just one.
obligate myrmecophily: in contrast to facultative myrmecophily, a single species of lycaenid in Cyprus (*Apharitis acamas*) has an obligate interaction with ants, which is both mutualistic and parasitic, and depends on this association for survival.

oviposition (vb. **oviposit**)**:** egg-laying.
paralectotype: any of the remaining members of a type series after the selection of a lectotype.
pharmacophagous: in this context, the behaviour of danaines to sequester pyrrolizidine alkaloids direct from suitable plants or other sources.
phenotype: the physical appearance of an individual, which may be influenced due to interaction with the environment.
phylogenetic analysis: the study of evolutionary relationships among biological groups.
phytophagous: herbivorous; feeding on plants.
polymorphic: having many forms.
polyphagous: accepting a range of hostplants belonging to more than one plant genus.
polyvoltine: having two or more (perhaps unknown or unspecified) broods per year (widely interpreted as 'multi-brooded').
riverine: associated with a river or river bank.
sexual dimorphism: difference in appearance (e.g. shape, markings, colour) between male and female members of the same species.
***sic*:** inserted into text to indicate the correct transcription of an author's error.
sp.: species (sing.).
spp. species (pl.).
ssp.: subspecies.
submontane: situated in the foothills of a mountain range.
syn.: synonym.
taxon (pl. **taxa**)**:** A group of organisms sufficiently differing from others to allow the group to be considered unique from the perspective of classification or taxonomy. A single species (or a family, etc.) can be considered a taxon.
taxonomy: a scheme of classification of organisms [Gr. *taxis,* arrangement; *nomos,* law].
trivoltine: having three broods per year.
Type Locality (TL): the geographical location where the name-bearing taxon was captured.
UFZ: Helmholtz Centre for Environmental Research.
univoltine: having one brood per year.
uns: undersides.
upf: upperside forewing.
ups: uppersides.
UTM: Universal Transverse Mercator coordinate system.
vagrant: strayed or blown from its usual range.
var.: variety.
♂ (♂♂): male(s).
♀ (♀♀): female(s).
>: greater than.
<: less than.
≥: greater than or equal to.

Bibliography

Abbes, K., Zouba, A., Harbi, A., Ghrissi, N., Ksantini, M. and Chermiti, B. (2020) The pomegranate butterfly *Deudorix livia* (Lepidoptera: Lycaenidae): an emerging pest on dates in Tunisia. *Bulletin OEPP/EPPO Bulletin* 50(1), 191–196.

Abu-Ghonem, M.A., Ramadan, H.M., Abu-Shall, A.M., Amany, M. and Tawfeek, M.E. (2017) First recorded [*sic*] of the pelargonium butterfly *Cacyreus marshalli* Butler, 1898 (Lepidoptera: Lycaenidae: Polyommatinae: Polyommatini): in Alexandria, Egypt. *Alexandria Journal of Agricultural Sciences* 62(2), 219–222.

Abu-Shall, A., Ramadan, H.M. and Abu-Ghonem, M.A. (2014) Immature stages of *Chilades pandava* (Lepidoptera: Lycaenidae), a new pest of *Cycas* spp. in Egypt. *Alexandria Journal of Agricultural Sciences* 59(3), 197–204.

Ackery, P.R., Smith, C.R. and Vane-Wright, R.I. (eds) (1995) *Carcasson's African Butterflies: An annotated Catalogue of the Papilionoidea and Hesperioidea of the Afrotropical Region*. CSIRO, Victoria, Australia.

Ackery, P.R., de Jong, R. and Vane-Wright, R.I. (1999) The butterflies: Hedyloidea, Hesperioidea and Papilionoidea. In: Kristensen, N.P. (ed.) *Lepidoptera, Moths and Butterflies, Vol. 1 Evolution, Systematics, and Biogeography*. Part 35 in *Handbook of Zoology* series, Vol. IV *Arthropoda: Insecta*. Walter de Gruyter, Berlin and New York, pp. 399–436.

AdaMerOs-Butterflies of Turkey (2021) AdaMerOs Butterflies Monitoring and Photography Society of Turkey. Available at: http://www.adamerkelebek.org/default.asp (accessed 27 February 2021).

Ambarlı, D. and Başbay, O. (2014) Kelebek Gözlemciliği Yeni Kayıtlar, Yeni Türler (Butterfly Observations: New Records, New Species). *Bilim ve Teknik Dergisi* 47(558), 72–75 (in Turkish).

Anastassíu, H.T., Ghavalas, N. and Coutsis, J.G. (2010) First record of *Cacyreus marshalli* in Greece, and comments on the potential occurrence of *Zizeeria karsandra* on the Greek island of Crete (Lepidoptera: Lycaenidae). *Phegea* 38(3), 85.

Asher, J., Warren, M., Fox, R., Harding, P., Jeffcoate, G. and Jeffcoate, S. (2001) *The Millennium Atlas of Butterflies in Britain and Ireland*. Oxford University Press, Oxford, UK.

Atahan, A., Atahan, G., Gül, M. and Atahan, M. (2018) *Hatay'in Kelebekleri [Butterflies of Hatay]*. Privately published.

Atay, E. and Tatlı, M. (2019) Contribution to the Lepidoptera Fauna of Hatay (Turkey). *Turkish Journal of Agriculture – Food Science and Technology* 7(5), 719–723.

Ayberk, H. and Akkuzu, E. (2007) Contributions to the Knowledge of Papilionoidea (Lepidoptera: Rhopalocera) Fauna from northern Cyprus. *Pakistan Journal of Biological Sciences* 10(11), 1845–1849.

Baidya, S., Basu, D.N., Roy, S. and Roy, A.B. (2018) Occurrence of interspecific mating between two species of *Danaus* Kluk, 1780 (Lepidoptera: Nymphalidae) in nature. *Psyche: A Journal of Entomology*, Article ID 3059017, 1–5. doi: 10.1155/2018/3059017

Baker, R.R. (1968) A possible method of evolution of the migratory habit in butterflies. *Philosophical Transactions of the Royal Society of London* 253, 309–341.

Baker, R.R. (1978) *The Evolutionary Ecology of Animal Migration*. Hodder & Stoughton, London.

Bálint, Z., Yammine, W. and Katona, G. (2016) Butterfly and skipper records from Lebanon (Lepidoptera: Papilionoidea). *Folia Entomologica Hungarica* 77, 105–118.

Barbudo, D.J., González, J.R. and Haeger, J.F. (1988) *Capparis spinosa* (Capparidaceae): an oviposition substrate for *Lampides boeticus* Linnaeus, in southern Spain (Lepidoptera: Lycaenidae). *Nota Lepidopterologica* 10(4), 218–223.

Barton, I., Rosewarne, K. and John, E. (2013) *Chilo partellus* (Swinhoe, 1884) (Lepidoptera: Pyraloidea: Crambidae) a stem borer on Cyprus. *Entomologist's Record and Journal of Variation* 125, 181–185.

Başbay, O. (2013) İstilacı Kelebek [Invasive Butterfly]. *National Geographic Turkey* Aralık, 32.

Başbay, O. and John, E. (2021) A review of current range expansion of *Cacyreus marshalli* Butler, 1898 (Lepidoptera: Lycaenidae, Polyommatinae) in western Turkey. *Entomologist's Gazette* 72, 59–63.

Başbay, O., Salimeh, M. and John, E. (2020a) Preliminary note on the range expansion of *Papilio demoleus* (Lepidoptera: Papilionidae) in south-eastern Turkey, and on the first documented arrival of this invasive species at the Mediterranean coast of Syria. *Phegea* 48(1), 20–21.

Başbay, O., Salimeh, M. and John, E. (2020b) Westward Palaearctic range expansion of *Papilio demoleus* Linnaeus, 1758 (Lepidoptera: Papilionidae) and its arrival at Mediterranean coastal regions of Turkey and Syria. *Entomologist's Gazette* 71, 257–272.

Baytaş, A. (2007) *A Field Guide to the Butterflies of Turkey*. NTV Yayınları, Istanbul, Turkey.

Beale, C. and Sprenger, A. (2006) *Butterflies of the Aammiq area Lebanon*. A Rocha, Beirut, Lebanon.

Benyamini, D. (1993) The butterflies of Mt Hermon (Lepidoptera: Rhopalocera and Hesperiidae). *Linneana Belgica* XIV(4), 167–204.

Benyamini, D. (2002a) *A Field Guide to the Butterflies of Israel* (including butterflies of Mount Hermon, Sinai and Jordan). Keter Publishing House, Jerusalem, Israel [in Hebrew].

Benyamini, D. (2002b) The butterflies of Dana Reserve, Edom, Jordan (Lepidoptera: Papilionoidea et Hesperioidea) Part 2. *Linneana Belgica*, XVIII, 265–272.

Benyamini, D. (2005) Euphorbiaceae – a new host-plant family for *Carcharodus alceae* Esper, 1780 (Lepidoptera: Hesperiidae) and a discussion on the use of Euphorbiaceae by butterfly larvae (Papilionoidea, Hesperioidea) in the world. *Nota lepidopterologica* 28(2), 75–92.

Benyamini, D. (2008) Is *Euchloe falloui* Allard, 1867 (Pieridae) the butterfly with the longest diapause? *Nota lepidopterologica* 31(2), 293–295.

Benyamini, D. (2021a) *Hypolimnas misippus* (Linnaeus, 1767) (Lepidoptera: Nymphalidae) in Cyprus and other records in the Eastern Mediterranean, an overlooked record and east Mediterranean butterfly migration corridors. *Entomologist's Gazette* 72, 243–252.

Benyamini, D. (2021b) *Butterflies of the Levant* (and nearby areas). Vol. III Nymphalidae. 4D MicroRobotics Ltd, Beit Arye, Israel.

Benyamini, D. and John, E. (2020) *Butterflies of the Levant* (and nearby areas). Vol. II Papilionidae, Pieridae and Hesperiidae. 4D MicroRobotics Ltd, Beit Arye, Israel.

Benyamini, D. and Kurman, Sh. (2020) The First Records of *Cacyreus marshalli* Butler, 1898 from Israel, the east Mediterranean, Western Asia and Alexandria, Egypt. *Atalanta* 51(1/2), 3–6.

Benyamini, D., Bruschini, C., Serra, G. and John. E. (2007) First confirmed records of *Papilio demoleus* (Linnaeus, 1758) in Syria, with comments on the species appearance in Turkey and the Dominican Republic. *News of the Israeli Lepidopterists Society* 24(2), 4–11.

Benyamini, D., Aristophanous, A., Aristophanous, M. and John, E. (2018) The biology of the Cyprus endemic blue *Glaucopsyche paphos* Chapman, 1920 (Lepidoptera: Lycaenidae, Polyommatinae). *Entomologist's Gazette* 69, 151–165.

Bethune-Baker, G.T. (1913) Comparative notes on *Chilades galba*, Led., and *phiala*, Gr. Gr. *Transactions of the Entomological Society of London* 1913, 201–205.

Bhakare, M. and Smetacek, P. (2010) Two instances of inter-generic mating by Lycaenidae (Lepidoptera) in Maharashtra, India. *Journal of Research on the Lepidoptera* 43, 23–25.

Birch, R.J., Markl, G. and Gottschalk, T.K. (2021) Aestivation as a response to climate change: the Great Banded Grayling *Brintesia circe* in Central Europe. *Ecological Entomology.* Available at: https://onlinelibrary.wiley.com/doi/epdf/10.1111/een.13081 (accessed 2 October 2021).

Biricik, M. (2011) First record of *Junonia orithya* (Linnaeus, 1758) (Lepidoptera: Nymphalidae) in Turkey. *Zoology in the Middle East* 53, 130–132.

Boggs, C.L. and Dau, B. (2004) Resource specialization in puddling Lepidoptera. *Environmental Entomology* 33, 1020–1024.

BOLDSYSTEMS (2014) *Zizeeria karsandra*. Available at: http://www.boldsystems.org/index.php/Taxbrowser_Taxonpage?taxid=259026 (accessed 30 September 2013).

Bolotov, I.N., Mizin, I.A., Zheludkova, A.A., Askenova, O.V., Kolosova, Y.S. *et al.* (2021) Long-distance dispersal of migrant butterflies to the Arctic Ocean islands, with a record of *Nymphalis xanthomelas* at the northern edge of Novaya Zemlya (76.95°N). *Nota lepidopterologica* 44, 73–90.

Boppré, M. (1986) Insects pharmacophagously utilizing defensive plant chemicals (pyrrolizidine alkaloids). *Naturwissenschaften* 73, 17–26.

Bowles, N. (2013) An observation of apparent dispersive movement in Long-tailed Blue *Lampides boeticus* (L.) (Lep.: Lycaenidae). *Entomologist's Record and Journal of Variation* 125, 193–194.

Bozano, G.C. (1990) A contribution to the knowledge of the butterfly fauna of east Jordan. *Nota lepidopterologica* 13, 2–7.

Braby, M.F. (2016) Migration records of butterflies (Lepidoptera: Papilionidae, Hesperiidae, Pieridae, Nymphalidae) in the 'Top End' of the Northern Territory. *Australian Entomologist* 43(3), 151–160.

Braby, M.F., Farias Quipildor, G.E., Vane-Wright, R.I. and Lohman, D.J. (2015) Morphological and molecular evidence supports recognition of *Danaus petilia* (Stoll, 1790) (Lepidoptera: Nymphalidae) as a species distinct from *D. chrysippus* (Linnaeus, 1758). *Systematics and Biodiversity* 13(4), 386–402.

Bretherton, R.F. (1954) A week's butterfly collecting in Cyprus. *Entomologist* 87, 207–211.

Bretherton, R.F. (1974) Butterflies in Cyprus, June 1973. *Entomologist's Record and Journal of Variation* 86, 1–5.

Bucknill, J.A.S. (1911) The Butterflies and Moths of Cyprus (Part II). List of Butterflies known to occur in Cyprus. *English School Magazine, Nicosia*, Appendix, 3–14.

Bury, J. (2014) Rediscovery of *Charaxes jasius jasius* (LINNAEUS, 1767) (Lepidoptera: Nymphalidae) in Lebanon. *CESA News* 97, 13–19.

Callier, V. and Nijhout, H.F. (2011) Control of body size by oxygen supply reveals size-dependent and size-independent mechanisms of molting and metamorphosis. *PNAS* 108(35),

14664–14669. Available at: www.pnas.org/cgi/doi/10.1073/pnas.1106556108 (accessed 20 September 2021).

CBRS (2021) Cyprus Butterfly Recording Scheme [maintained by the first author]. Available at: http://www.cyprusbutterflies.co.uk (accessed 1 December 2021).

Chapman, T.A. (1920) An undescribed Lycaenid Butterfly from Cyprus, *Glaucopsyche paphos*, sp. n. (Lycaenidae). *Transactions of the Royal Entomological Society of London* 1920, 166.

Chowdhury, S., Fuller, R.A., Dingle, H., Chapman, J.W. and Zalucki, M.P. (2021) Migration in butterflies: a global overview. *Biological Reviews* 96(4), 1462–1483. doi: 10.1111/brv.12714

Christofides, Y. (2017) *Illustrated Flora of Cyprus*. Self-published.

Colom, P., Traveset, A., Carreras, D. and Stefanescu, C. (2021) Spatio-temporal responses of butterflies to global warming on a Mediterranean island over two decades *Ecological Entomology* 46(2), 262–272.

Comay, O., Yehuda, O.B., Benyamini, D., Schwartz-Tzachor, R., Pe'er, I. *et al.* (2020) Analysis of monitoring data where butterflies fly year-round. *Ecological Applications* 30(8), e02196, 1–17. Available at: https://esajournals.onlinelibrary.wiley.com/doi/full/10.1002/eap.2196 (accessed 2 October 2021).

Comay, O., Yehuda, O.B., Schwartz-Tzachor, R., Benyamini, D., Pe'er, I. *et al.* (2021) Environmental controls on butterfly occurrence and species richness in Israel: The importance of temperature over rainfall. *Ecology and Evolution* 11, 12035–12050. Available at: https://onlinelibrary.wiley.com/doi/pdf/10.1002/ece3.7969 (accessed 2 October 2021).

Corbet, A.S. (1949) The Linnaean name of Indo-Australian Rhopalocera. Part 7. Summary of determinations. *Proceedings of the Royal Entomological Society of London (B)* 18, 191–199.

Coutsis, J.G. (2012) A comparison between the female genitalia of *Gegenes pumilio* and those of *G. nostrodamus* (Lepidoptera: Hesperioidea). *Phegea* 40(1), 39–40.

Coutsis, J.G. and Benyamini, D. (2021) About the recent transfer of the species-group taxon *floccifera* and its closest relatives from the genus *Carcharodus* to the genus *Muschampia* (Lepidoptera: Hesperiidae, Pyrginae). *Phegea* 49(2), 61–66.

Coutsis, J.G. and Pamperis, L.N. (2012) Intergeneric, homosexual butterfly pairings (Lepidoptera: Nymphalidae, Satyrinae). *Entomologist's Gazette* 63, 146.

Cuadrado, M. (2017) The year-round phenology of the migratory Red Admiral *Vanessa atalanta* (Lepidoptera: Nymphalidae) in a Mediterranean area in southern Spain. *European Journal of Entomology* 114, 517–525.

Cuvelier, S. and Mølgaard, M.S. (2012) Butterflies and Skippers in the Dodecanese Islands (Greece): new data and an update on their distribution (Lepidoptera: Hesperioidea and Papilionoidea). *Phegea* 40(3), 66–80.

Dapporto, L. and Dennis, R.L.H. (2009) Conservation biogeography of large Mediterranean islands. Butterfly impoverishment, conservation priorities and inferences for an ecological 'island paradigm'. *Ecography* 32, 169–179.

de Freina, J.J. (1991) Untersuchungen zu Artrecht, Verbreitung und Systematik von *Lasiommata paramegaera* (Hübner, 1824) und *Lasiommata megera* (Linnaeus, 1758) (Lep. Satyridae). *Mitteilungen der Münchner Entomologischen Gesellschaft* 81, 165–173.

de Freina, J.J. (2014) Zoogeographische, ökologische und taxonomische Studie zu den Zwillingsarten *Chilades trochylus* Freyer, 1845 und *Chilades putli* Kollar, 1848 (Lepidoptera: Lycaenidae, Lycaeninae, Polyommatini). *Nachrichten des entomologischen Vereins Apollo*, N.F. 34(4), 145–160.

de Jong, R. (2016) Reconstructing a 55-million-year-old butterfly (Lepidoptera, Hesperiidae). *European Journal of Entomology* 113, 423–428.

de Jong, R. (2017) Fossil butterflies, calibration points and the molecular clock (Lepidoptera: Papilionoidea). *Zootaxa* 4270(1), 001–063.

de Jong, R. and Coutsis, J.G. (2017) A re-appraisal of *Gegenes* Hübner, 1819 (Lepidoptera, Hesperiidae) based on male and female genitalia, with the description of a new genus, *Afrogegenes. Tijdschrift voor Entomologie* 160, 1–21.

de Lattin, G. (1949) Über die Artfrage in der *Hipparchia semele* L.-Gruppe. *Entomologische Zeitschrift* 59, 113–118, 124–126, 131–132.

de Lattin, G. (1952) Two new subspecies of *Hipparchia semele* Linnaeus. *Entomologist's Record and Journal of Variation* 64, 336–337.

de Lesse, H. (1960) Spéciation et variation chromosomique chez les Lépidoptères Rhopalocères. *Annales des Sciences Naturelles (Series* 12) 2, 1–223.

de Lesse, H. (1967) Formules chromosomiques de Lépidoptères Rhopalocères d'Afrique du Nord. *Bulletin de la Société entomologique de France* 72, 20–25.

de Vos, R. (1999) Trekvlinders in 1998 (negenenvijftigste jaarverslag) (Lepidoptera). *Entomologische Berichten, Amsterdam* 59(12), 173–185.

de Worms, C.G.M. (1967) Three weeks in Cyprus, April–May 1967. *The Entomologist's Record and Journal of Variation* 79, 245–249.

Dening, R.C. (1985) Butterflies in northern Cyprus in early June, 1981. *Entomologist's Record and Journal of Variation* 97, 92–93.

Dening, R.C. (1987) Butterflies in northern Cyprus. *Entomologist's Record and Journal of Variation* 99, 226–227.

Dennis, R.L.H. (1991) *Kirinia roxelana* Cramer, 1777 (Lepidoptera: Nymphalidae, Satyrinae) oviposition on an arboreal substrate. *Entomologist's Gazette* 42, 239–241.

Dennis, R.L.H. (1995) *Euchloe ausonia* (Hübner) (Lepidoptera: Pieridae) oviposition on *Brassica nigra* (L.) Koch (Cruciferae): big immature plants are preferred. *Entomologist's Gazette* 46, 253–255.

Dennis, R.L.H. (1997) An inflated conservation load for European butterflies: increases in rarity and endemism accompany increases in species richness. *Journal of Insect Conservation* 1, 43–62.

Dennis, R.L.H. (2010) *A Resource-based Habitat view for Conservation: Butterflies in the British landscape.* Wiley-Blackwell, Oxford, UK.

Dennis, R.L.H. (2020) *Butterfly Biology Systems: Connections and Interactions in Life History and Behaviour*. CABI, Wallingford, UK.

Dennis, R.L.H. and Hardy, P.B. (2018) *British and Irish Butterflies. An Island Perspective*. CABI, Wallingford, UK.

Dennis, R.L.H. and Schmitt, T. (2009) Faunal structures, phylogeography and historical interference. In: Settele, J., Shreeve, T., Konvička, M. and Van Dyck, H. (eds) *Ecology of Butterflies in Europe*. Cambridge University Press, Cambridge, UK, pp. 250–280.

Dennis, R.L.H. and Shreeve, T.G. (1988) Hostplant-habitat structure and the evolution of butterfly mate-locating behaviour. *Zoological Journal of the Linnean Society* 94, 301–318.

Dennis, R.L.H. and Thomas, C.D. (2000) Bias in butterfly distribution maps: the influence of hotspots and recorder's home range. *Journal of Insect Conservation* 4, 73–77.

Dennis, R.L.H., Shreeve, T.G., Olivier, A. and Coutsis, J.G. (2000) Contemporary geography dominates butterfly diversity gradients within the Aegean archipelago (Lepidoptera: Papilionoidea, Hesperioidea). *Journal of Biogeography* 27, 1365–1383.

Dennis, R.L.H., Olivier, A., Coutsis, J.G. and Shreeve, T.G. (2001) Butterflies on islands in the Aegean Archipelago: predicting numbers of species and incidence of species using geographical variables. *Entomologist's Gazette* 52, 3–39.

Dennis, R.L.H., Dapporto, L., Shreeve, T.G., John, E., Coutsis, J.G. *et al.* (2008). Butterflies of European islands: the implications of the geography and ecology of rarity and endemicity for conservation. *Journal of Insect Conservation* 12, 205–236.

Descimon, H. and Mallet, J. (2009) Bad species. In: Settele, J., Shreeve, T., Konvička, M. and Van Dyck, H. (eds) *Ecology of Butterflies in Europe*. Cambridge University Press, Cambridge, UK, pp. 219–249.

Dincă, V., Dapporto, L., Somervuo, P., Vodă, R., Cuvelier, S. *et al.* (2021) High resolution DNA barcode library for European butterflies reveals continental patterns of mitochondrial genetic diversity. *Communications Biology* 4, 315. doi: 10.1038/s42003-021-01834-7

Donoso, I., Stefanescu, C., Martínez-Abraín, A. and Travaset, A. (2016) Phenological asynchrony in plant–butterfly interactions associated with climate: a community-wide perspective. *Oikos* 125(10), 1434–1444.

Döring, E. and Hoffmann, M.H. (2004) Distribution analysis of monophagous butterflies (Lepidoptera) and their host plants in the western Palaearctic. *Schlechtendalia* 12, 85–93.

Eckweiler, W. and Bozano, G.C. (2011) *Guide to the Butterflies of the Palaearctic Region*. Satyrinae Part IV. Omnes Artes, Milano, Lombardy, Italy.

Edwards, M., Varnava, A., Stavrinides, M. and John, E. (2016) Hymenoptera. In: Sparrow, D.J. and John, E. (eds) *An Introduction to the Wildlife of Cyprus*. Terra Cypria, Limassol, Cyprus, pp. 411–441.

Eid, E., Katbeh-Bader, A., Al Otoom, M. and Othman, Y. (2009) Contribution to the Entomofauna of Dibeen Forest Reserve in Jordan. *Cesa News* 49, 19–41.

Eitschberger, U. and Stamer, P. (1990) *Cacyreus marshalli* Butler, 1898, eine neue Tagfalterart für sie europäische Fauna? (Lepidoptera, Lycaenidae). *Atalanta* 21(1/2), 101–108.

Eitschberger, U., Reinhardt, R. and Steiniger, H. (1991) *Wanderfalter in Europa (Lepidoptera). Atalanta* 22(1), 1–67 + XVI plates.

Ellison, R.E. and Wiltshire, E.P. (1939) The Lepidoptera of the Lebanon: with notes on their season and distribution. *Transactions of the Royal Entomological Society of London* 88(1), 1–57.

Erhardt, A. and Mevi-Schütz, J. (2009) Adult food resources in butterflies. In: Settele, J., Shreeve, T., Konvička, M. and Van Dyck, H. (eds) *Ecology of Butterflies in Europe*. Cambridge University Press. Cambridge, UK, pp. 9–16.

Evans, W.H. (1937) *A catalogue of the African Hesperiidae, indicating the classification and nomenclature adopted in the British Museum*. British Museum (Natural History), London.

Feltwell, J.S.E. (2008) An outbreak of *Libythea celtis* Laicharting (Lepidoptera : Libytheidae) in the Basses-Cévennes, France. *British Journal of Entomology and Natural History* 21(2), 57–60.

Feltwell, J.S.E. and Vane-Wright, R.I. (1982) Classification of *Pieris brassicae*. In: Feltwell, J. (ed.) *Large White Butterfly the Biology, Biochemistry and Physiology of Pieris brassicae (Linnaeus). Series Entomologica* 18. The Hague, Boston, London, pp. 1–11.

Fernández-Palacios, J.M., Kreft, H., Irl, S.D.H., Norder, S., Ah-Peng, C. *et al.* (2021) Scientists' warning – The outstanding biodiversity of islands is in peril. *Global Ecology and Conservation* 31, e01847. doi: 10.1016/j.gecco.2021.e01847

Feulner, G.R., Roobas, B., Carlisle, T. and Meyer, H. (2014) First UAE and Arabian records of *Chilades pandava*, the Cycad Cupid butterfly, an introduced Oriental species (Lepidoptera: Lycaenidae) hosted by the ornamental sago plant *Cycas revolute*. *Tribulus* 22, 48–56.

Fiedler, K. (2021) The ant associates of Lycaenidae butterfly caterpillars – revisited. *Nota Lepidopterologica* 44, 159–174.

Fric, Z., Dickinson, R., Fetouh, G., Larsen, T.B., Schön, W. and Wiemers, M. (2014) First record of the cycad blue, *Chilades pandava*, in Egypt – a new invasive butterfly species in the Mediterranean region and on the African continent (Lepidoptera: Lycaenidae). *African Entomology* 22(2), 315–319.

Fric, Z., Marešová, J., Kadlec, T., Tropek, R., Pyrcz, T.W. and Wiemers, M. (2019) World travellers: phylogeny and biogeography of the butterfly genus *Leptotes* (Lepidoptera: Lycaenidae). *Systematic Entomology* 44, 652–665.

Fuchs, J. (1996) Zwei Wochen Schmetterlingsbeobachtungen auf Zypern (4.-18.6.1995) (Lep., Rhopalocera). *Berichte des Kreises Nürnberger Entomologen Galathea* 12/2, 57–62.

Galanos, C.J. (2014) First records of *Pararge aegeria* and *Cacyreus marshalli*, and a verification of *Muschampia proto* from the Greek Island of Rhodes. First records of *Cacyreus marshalli* and *Gegenes* sp. from the Greek Island of Tilos; Dodecanese Complex S.E. Aegean (Lepidoptera: Hesperioidea and Papilionoidea). *Phegea* 42(4), 74–77.

Galanos, C.J. (2020) Bionomics of *Freyeria trochylus* (Freyer, 1844) and *Zizeeria karsandra* (Moore, 1865) (Lepidoptera, Lycaenidae) on Rodos Island, Greece. *Nota Lepidopterologica* 43, 139–150.

García-Barros, E. (1988) Delayed ovarian maturation in the butterfly *Hipparchia semele* as a possible response to summer drought. *Ecological Entomology* 13, 391–398.

García-Barros, E. (2000) Notas sobre la biología de los adultos de *Pandoriana pandora* (Dennis and Schiffermüller, 1775) en la España central (Lepidoptera: Nymphalidae). *SHILAP Revista de lepidopterologia* 28(109), 97–102.

GBIF (2021) Global Biodiversity Information Facility. Available at: https://www.gbif.org (accessed 28 July 2021).

Geier, T. and Schurian, K. (2014) Bemerkungen zur Larvalökologie des Segelfalters (*Iphiclides podalirius*), insbesondere zu regionalen Nahrungspflanzenpräferenzen und zum „Wackelgang“ älterer Raupen (Lepidoptera: Papilionidae). *Nachrichten des entomologischen Vereins Apollo*, N.F. 34(4), 201–209.

Georghiou, G.P. (1977) *The Insects and Mites of Cyprus*. Benaki Phytopathological Institute, Athens, Greece.

Gilbert, G. and Zalat, S. (2007) *Butterflies of Egypt. Atlas, Red Data Listing & Conservation.* Egyptian Environmental Affairs Agency, Nature Conservation Sector, Cairo.

Gordon, I.J. (1987) Natural selection for rare and mimetic colour pattern combinations in wild populations of the diadem butterfly, *Hypolimnas misippus* L. *Biological Journal of the Linnean Society* 31, 1–23.

Graham-Smith, G.S. and Graham-Smith, W. (1930) *Pieris brassicae*, L., with special reference to aberrations from Aberdeenshire. *Entomologist's Record and Journal of Variation* 42, 17–22.

Graves, P.P. (1907) Notes on Egyptian and Syrian butterflies. *The Entomologist's Record and Journal of Variation* 19, 64–68.

Graves, P.P. (1910) A contribution to the fauna of Syria. *The Entomologist's Record and Journal of Variation* 22, 271–273.

Graves, P.P. (1911) A contribution to the fauna of Syria. *The Entomologist's Record and Journal of Variation* 23, 31–36.

Graves, P.P. (1928) *Epinephele cypricola* sp. nov. *The Entomologist* 61, 5–7.

Greatorex-Davies, J.N., Brereton, T.B., Roy, D.B., Middlebrook, I. and Cruickshank, K.L. (2007) *United Kingdom Butterfly Monitoring Scheme report for 2006*. Centre for Ecology and Hydrology, Monks Wood, Huntingdonshire, UK.

Grill, A., de Vos, R. and van Arkel, J. (2004) The shape of endemics: Notes on male and female genitalia in the genus *Maniola* (Schrank, 1801), (Lepidoptera, Nymphalidae, Satyrinae) *Contributions to Zoology* 73(4), 293–304.

Grill, A., Gkiokia, E. and Alvarez, N. (2006) Evolutionary history and patterns of differentiation among European *Maniola* butterflies (Lepidoptera: Satyrinae). *European Journal of Entomology* 103, 613–618.

Grill, A., Cerny, A. and Fiedler, K. (2013) Hot summers, long life: egg laying strategies of *Maniola* butterflies are affected by geographic provenance rather than adult diet. *Contributions to Zoology* 82(1), 27–36.

Grimaldi, D. and Engel, M.S. (2005) *Evolution of the insects.* Cambridge University Press, Cambridge, New York, Melbourne.

Guillemard, F.H.H. (1888) VIII – Ornithological Notes of a Tour in Cyprus in 1887. *Ibis* 30(1), 94–124.

Hackman, W. (1952) On the insect fauna of Cyprus. Results of the expedition of 1939 by Harald, Håkan and P. H. Lindberg. IX Lepidoptera. *Societas Scientiarum Fennica Commentationes Biologicae* 13(8), 1–9.

Hadjisterkotis, E. (2001) The Cyprus mouflon, a threatened species in a biodiversity 'hotspot' area. In: Nahlik, A. and Walter, U. (eds) *Proceedings of the Third International Symposium on Mouflon*, Sopron, Hungary, pp. 71–81.

Hadjisterkotis, E. (2008) *Review of biodiversity research results from Cyprus that directly contribute to the sustainable use of biodiversity in Europe*. Unit of Environmental Studies, Ministry of the Interior, Technical Report, Lefkosía, Cyprus.

Haeler, E., Fiedler, K. and Grill, A. (2014) What prolongs a butterfly's life? Trade-offs between dormancy, fecundity and body size. *PLoS ONE* 9(11) e111955. doi: 10.1371/journal.pone.0111955

Hardy, P.B. (1994) Butterflies in Crete. *The Entomologist's Record and Journal of Variation* 106, 203–211.

Heath, A. (1997) A review of African genera of the tribe Aphnaeini (Lepidoptera; Lycaenidae). *Metamorphosis, Occasional Supplement* 2, 1–60.

Hecht, M. (2021) The Lemon Swallowtail reached my garden in Kibbutz Afik in the southern Golan Heights [translated from Hebrew]. *News of the Israeli Lepidopterists Society* 38(2), 4 [in Hebrew].

Heikkilä, M., Kaila, L., Mutanen, M., Peña, C. and Wahlberg, N. (2012) Cretaceous origin and repeated tertiary diversification of the redefined butterflies. *Proceedings of the Royal Society B* 279, 1093–1099.

Hemming, A.F. (1929) Revision of the *baton* group of the genus *Turanana* Bethune-Baker, with an account of an unrecognized species, *T. vicrama* Moore. *The Entomologist* 62, 27–34, 60–64, 84–89.

Henning, S.F. (1990) Courtship and mating of butterflies. *The Entomologist's Record and Journal of Variation* 102, 205–213.

Hermann, S.L., Blackledge, C., Haan, N.L., Myers, A.T. and Landis, D.A. (2019) Predators of monarch butterfly eggs and neonate larvae are more diverse than previously recognised. *Nature.com Scientific Reports* 9, 14304. doi: 10.1038/s41598-019-50737-5

Hesselbarth, G. (1983) Beitrag zur Biologie von Ypthima asterope KLUG (Lepidoptera: Satyridae). *Nachrichten des entomologischen Vereins Apollo, Frankfurt* N.F. 4(1), 7–14.

Hesselbarth, G., van Oorschot, H. and Wagener, S. (1995) *Die Tagfalter der Türkei unter Berücksichtigung der angrenzenden Länder*. Selbstverlag Sigbert Wagener, Bocholt, Borken, Germany.

Higgins, L.G. (1973) Crete in late June, 1973. *Entomologist's Record and Journal of Variation* 99, 209–222.

Higgins, L.G. and Riley, N.D. (1970) *The Butterflies of Britain and Europe*. Collins, London.

Higgins, L.G. and Riley, N.D. (1983) *The Butterflies of Britain and Europe*. HarperCollins, London.

Hofmann, P. and Rose, K. (1987) Zur Rhopaloceren-Fauna Zyperns. *Nachrichten des entomologischen Vereins Apollo*, *Frankfurt* N.F. 7(4), 133–141.

Holik, O. (1949) Ueber die Gattung *Satyrus* L. *Zeitschrift der Wiener Entomologischen Gesellschaft* 34, 98–105.

Hołyński, R.B. (2010) Taxonomy and the mediocrity of DNA barcoding – some remarks on PACKER *et al.* 2009: DNA Barcoding and the Mediocrity of Morphology. *Arthropod Systematics and Phylogeny* 68(1), 143–150.

Humphreys, R.K. and Ruxton, G.D. (2018) A review of thanatosis (death feigning) as an anti-predator behaviour. *Behavioral Ecology and Sociobiology* 72, 22. doi: 10.1007/s00265-017-2436-8

Hutchinson, J.T. and Cobham, C.D. (1909) *A Handbook of Cyprus, 6th Issue*. Edward Stanford, London.

Hutsebaut, J., Leertouwer, H.L. and Stavenga, D.G. (2020) Polymorphism of *Colias croceus* from the Azores caused by differential pterin expression in the wing scales. *Journal of Insect Physiology* 127, 104–114.

ICN (2011) Invertebrate Conservation News. *Amateur Entomologists' Society*. No. 66.

Igarashi, S. (1979) *Papilionidae and their early stages* Vol. I + II. Kodansha, Tokyo, Japan.

Iordanou, N. and Charalambous, P. (1998) *Chemical control of the citrus leafminer, Phyllocnistis citrella (Stainton) (Lepidoptera, Phyllocnistidae), in Cyprus*. Agricultural Research Institute, Nicosia Technical Bulletin 197, Lefkosía, Cyprus.

Islam, M.S., Yasmin, M. and Islam, A.T.M.F. (2019) Studies on the biology and population abundance of lemon butterfly, *Papilio demoleus* L. (Papilionidae: Lepidoptera). *Bangladesh Journal of Entomology* 29, 77–90.

IUCN Red List (2013) *Cyprus' biodiversity at risk*. Available at: https://www.iucn.org/sites/dev/files/content/documents/cyprus_biodiversity_at_risk_fact_sheet_may_2013.pdf (accessed 9 April 2021).

Jahnavi, M., Ramakrishna Rao, A. and Sarada, G. (2018) Biology and morphology of citrus butterfly *Papilio demoleus* Linnaeus (Lepidoptera: Papilionidae) on acid lime. *Journal of Entomology and Zoology Studies* 6(1), 1556–1561.

John, E. (2000) *Butterflies of Cyprus 1998 (records of a year's sightings).* Amateur Entomologists' Society Pamphlet No. 15.

John, E. (2001a) Migration of *Vanessa cardui* (L.) (Lep.: Nymphalidae) through Cyprus, March 2001. *Entomologist's Record and Journal of Variation* 113, 269–281.

John, E. (2001b) Autumn butterfly records from Cyprus, 2000. *Bulletin of the Amateur Entomologists' Society* 60, 84–96.

John, E. (2005) Cyprus: a brief overview of its butterflies and an introduction to the island's Butterfly Recording Scheme. *Antenna* 29, 33–37.

John, E. (2007) Butterflies of Cyprus: an update to the AES booklet (John 2000) with additional comments on endemism and threatened species, plus changes to the island's butterfly checklist. *Bulletin of the Amateur Entomologists' Society* 66, 140–150.

John, E. (2013) *Glaucopsyche paphos* (Chapman, 1920) (Paphos Blue) – a second brood? *'cypricola' Newsletter of the Cyprus Butterfly Study Group* 1, 5.

John, E. and Coutsis, J.G. (2022) Observations suggesting extended pupal diapause in *Pontia chloridice* (Hübner, [1813]) (Lepidoptera: Pieridae, Pierinae) in Cyprus and Greece. *Phegea* 50(1), 4–8.

John, E. and Dennis, R.L.H. (2019) Post-aestivation puddling by female *Chazara briseis larnacana* (Oberthür, 1909) and *Maniola cypricola* (Graves, 1928) (Lepidoptera: Nymphalidae, Satyrinae): influences and implications. *Entomologist's Gazette* 70, 97–109.

John, E. and Makris, C. (2001) *Libythea celtis* (Laicharting 1782) (Lepidoptera: Libytheidae): Presence of a breeding colony in the Tróodos Mountains, Cyprus. *Entomologist's Gazette* 52, 173–180.

John, E. and Mérit, X. (2008) *Chilades galba* (Lederer, 1855) and *Aporia crataegi* (Linnaeus, 1758): significant contributions to the Cyprus Butterfly Recording Scheme, April/May 2008. *Atalanta* 39(1–4), 165–171.

John, E. and Parker, R. (2002) Dispersal of *Hipparchia cypriensis* (Holik, 1949) (Lepidoptera: Nymphalidae, Satyrinae) in Cyprus, with notes on its ecology and life-history. *Entomologist's Gazette* 53, 3–18.

John, E. and Parker, R. (2003) The Butterfly Year in Cyprus. *Sanctuary* 32, 14–18.

John, E. and Skule, B. (2016) Lepidoptera. In: Sparrow, D.J. and John, E. (eds) *An Introduction to the Wildlife of Cyprus*. Terra Cypria, Limassol, Cyprus, pp. 268–385.

John, E. and Tamson, S. (2011) Predation of an African Grass Blue *Zizeeria karsandra* (Moore, 1865) by the crab spider *Thomisus onustus* (Walckenaer, 1805) in Cyprus. *Bulletin of the Amateur Entomologists' Society* 70, 224–225.

John, E. and Tennent, W.J. (2012) Marine (seawater) puddling by butterflies: is the sea an under-utilised sodium resource? *Entomologist's Gazette* 63, 135–145.

John, E. and Vane-Wright, R.I. (2019) *Danaus chrysippus chrysippus* (Linnaeus, 1758) ff. 'alcippus' and 'alcippoides' (Lepidoptera: Nymphalidae, Danainae) from Cyprus. *Entomologist's Gazette* 70, 135–146.

John, E., Coutsis, J.G. and Makris, C. (2006) A review of records for *Colias erate* (Esper, [1805]) (Lep.: Papilionoidea, Pieridae) in Cyprus: were they all yellow forms of *Colias croceus* (Geoffroy, 1785)? *Entomologist's Gazette* 57, 3–12.

John, E., Russell, P., Hall, D. and Christofides, Y. (2008a) Notes on the life history, ecology and distribution of *Pontia chloridice* (Hübner, 1808) (Lepidoptera: Pieridae), and a first record of *Hyposoter ebenitor* (Aubert, 1972) (Hymenoptera: Ichneumonidae) in Cyprus. *Entomologist's Gazette* 59, 209–226.

John, E., Cottle, N., McArthur, A. and Makris, C. (2008b) Eastern Mediterranean migrations of the Small White *Pieris rapae* (Linnaeus, 1758) (Lepidoptera: Pieridae): observations from Cyprus 2001 and 2007. *Entomologist's Gazette* 59, 71–78.

John, E., Gascoigne-Pees, M. and Larsen, T.B. (2010) *Ypthima asterope* (Klug, 1832) (Lepidoptera: Nymphalidae, Satyrinae): its biogeography, lifecycle, ecology and present status in Cyprus, with additional notes from Rhodes and the eastern Mediterranean. *Entomologist's Gazette* 61, 1–22.

John, E., Haines, D.H. and Haines, H.M. (2011) *Chazara persephone* (Hübner, 1803) (Lepidoptera: Nymphalidae, Satyrinae) – has the species retained a secretive presence in Cyprus for one hundred years? *Entomologist's Gazette* 62, 79–82.

John, E., Makris, C. and Christofides, Y. (2013a) *Cleome iberica* DC.: a new host-plant for *Pontia chloridice* (Hübner, [1813]) (Lepidoptera: Pieridae) in the Levant and its influence on the butterfly's phenology in Cyprus. *Entomologist's Gazette* 64, 19–26.

John, E., Wiemers, M., Makris, C. and Russell, P. (2013b) The *Pontia edusa* (Fabricius, 1777) / *Pontia daplidice* (Linnaeus, 1758) complex: confirmation of the presence of *Pontia daplidice* (Lepidoptera: Pieridae) in Cyprus, and of *Cleome iberica* DC. as a new host-plant for this species in the Levant. *Entomologist's Gazette* 64, 69–78.

John, E., Sparrow, D.J. and Sparrow, R. (2015a) Migration of Lepidoptera in the vicinity of Cyprus, eastern Mediterranean, in April 2014. *Entomologist's Gazette* 66, 255–262.

John, E., Stefanescu, C., Honey, M.R., Crawford, M. and Taylor, D. (2015b) Ceremonial releases of *Danaus plexippus* (Linnaeus, 1758) (Lepidoptera: Nymphalidae, Danainae) in the Iberian Peninsula, the Balearic Islands and Cyprus: implications for biogeography, potential for colonisation and a provisional listing of Asclepiadoideae from these regions. *Entomologist's Gazette* 66, 141–156.

John, E., Başbay, O. and Seven, E. (2018a) *Pontia glauconome* Klug, 1829 (Lepidoptera: Pieridae, Pierinae): a preliminary note on the species' first recorded presence in Turkey. *Entomologist's Gazette* 69, 64.

John, E., Thomas, B., Başbay, O., Cebeci, Z., and Coutsis, J.G. (2018b) *Cacyreus marshalli* Butler, 1898 (Lepidoptera: Lycaenidae, Polyommatinae): a preliminary note on its appearance in Crete and on range expansion in Greece and coastal areas of northwestern Turkey. *Entomologist's Gazette* 69, 18.

John, E., Hardman, M. and Smith, M. (2019a) How important are olfactory cues for host-plant detection by migrating *Danaus chrysippus* (Linnaeus, 1758) (Lepidoptera: Nymphalidae, Danainae) in Cyprus? *Entomologist's Gazette* 70, 223–238.

John, E., Hawkes, W.L.S. and Walliker, E.J. (2019b) A review of Mediterranean records of *Catopsilia florella* (Fabricius, 1775) (Lepidoptera: Pieridae, Coliadinae), with notes on the spring 2019 arrival in Cyprus of this Afrotropical migrant. *Phegea* 47(3), 80–86.

John, E., Başbay, O., Seven, E. and Kaymaz, N. (2020) *Pontia glauconome* (Klug, 1829) (Lepidoptera: Pieridae, Pierinae) in south-eastern Turkey: confirmation of breeding populations, with notes on the biology of early stages and on a species of the larval parasitoid *Hyposoter* Förster, 1869 (Hymenoptera: Ichneumonidae, Campopleginae). *Entomologist's Gazette* 71, 27–44.

John, E., Salimeh, M., Başbay, O. and Wiemers, M. (2021a) Determination by molecular analysis of the presence of nominate *Papilio demoleus* Linnaeus, 1758 (Lepidoptera: Papilionidae) in Mediterranean Syria and of confirmed range expansion of this swallowtail species into Lebanon. *Entomologist's Gazette* 72, 137–146.

John, E., Bağlar, H., Başbay, O. and Salimeh, M. (2021b) First appearance in Cyprus of *Papilio demoleus* Linnaeus, 1758 (Lepidoptera: Papilionidae), as it continues its predicted westward spread in the Palearctic region. *Entomologist's Gazette* 72, 257–264.

John, E., Bağlar, H., Başbay, O., Konstantinou, G., Salimeh, M. & Wiemers, M. (2022a) Confirmation of nominotypical subspecies *Papilio demoleus demoleus* Linnaeus, 1758 (Lepidoptera: Papilionidae) in Cyprus and of potential establishment. *Entomologist's Gazette* 73, 117–128.

John, E., Başbay, O., Salimeh, M. and Bağlar, H. (2022b) Where next? The seemingly inexorable spread of *Papilio demoleus* Linnaeus, 1758 (Lime Swallowtail) (Lepidoptera: Papilionidae) in countries of the eastern Mediterranean and its arrival in Cyprus in 2021. *Bulletin of the Amateur Entomologists' Society* 80 (545), in press.

Johnson, C.G. (1969) *Migration and Dispersal of Insects by Flight*. Methuen & Co., London.

Jugovic, J. and Kržič, A. (2019) Behavior and oviposition preferences of a black-veined white, *Aporia crataegi* (Lepidoptera: Pieridae). *Journal of Entomological and Acarological Research* 51, 8108 (50–59).

Jutzeler, D. and Lafranchis, T. (2005) Biologie et écologie d'*Hyponephele lupina* (Costa, 1836) en Grèce. Comparaison avec *Hyponephele lycaon* (Rottemberg, 1775) *(Lepidoptera: Nymphalidae, Satyrinae)*. *Linneana Belgica* XX, 35–44.

Jutzeler, D., Lafranchis, T. and Wiemers, M. (2005) Comparaison des états pre-imaginaux de *Kirinia roxelana* (Cramer, 1777) de Grèce et d'*Esperarge climene* (Esper, 1783) d'Iran complétée par des indications sur la taxinomie, l'écologie et la distribution et enrichie d'observations récentes des deux espèces faites en Grèce (*Lepidoptera*: *Nymphalidae*). *Lineana Belgica* XX, 51–62.

Jutzeler, D., Russell, P. and Eitschberger, U. (2019) Reassessment of the taxonomic rank of *Hipparchia cypriensis* (Holik, 1949) from Cyprus with reference to *H. pellucida* (Stauder, 1923) from Georgia based of (*sic*) their pre-imaginal stages (Lepidoptera, Satyrinae). *Atalanta* 50(1-4), 118–139.

Karsholt, O. and Razowski, J.O. (1996) *The Lepidoptera of Europe. A Distributional Checklist*. Apollo Books, Stenstrup, Denmark.

Katbeh-Bader, A., Amr, Z.S. and Isma'el, S. (2003) The Butterflies of Jordan. *Journal of Research on the Lepidoptera* 37, 11–26.

Katbeh-Bader, A., Amr, Z.S., Abu Baker, M. and Mahasneh, A. (2004) The Butterflies of Jordan. *Denisia* 14, 265–281.

Kattulas, M. and Koutsaftikis, A. (1977) Systematische, ökologische und ethologische Untersuchungen der Lepidopterenfauna der Insel Kastelloriso. 1 Teil. *Biologia gallohellen* 7, 151–161.

Kawahara, A.Y. (2006) Biology of the snout butterflies (Nymphalidae, Libytheinae) Part 1: *Libythea* Fabricius. *Transactions of the Lepidopterological Society of Japan* 57(1), 13–33.

Kawahara, A.Y. and Breinholt, J.W. (2014) Phylogenomics provides strong evidence for relationships of butterflies and moths. *Proceedings of the Royal Society B* 281, 20140970. doi: 10.1098/rspb.2014.0970

Keynäs, K. and Mikkola, K. (1970) *Pontia daplidice – Euchloë chloridice* migration. *Circular of the Finnish Lepidopterological Society* 6/70, 5–6 [in Finnish].

Kitching, R.L. and Luke, B. (1985) The myrmecophilous organs of the larvae of some British Lycaenidae (Lepidoptera): a comparative study. *Journal of Natural History* 19(2), 259–276.

Koçak, A.Ö. (1981) Critical check-list of European Papilionoidea (Lepidoptera). *Priamus* 1(2), 50.

Köstler, W. (2001) Zur Macrolepidopterenfauna der Südtürkei im Monat November (Insecta, Lepidoptera). *Atalanta* 32(1/2), 233–244.

Köstler, W. (2002) Neuere Beobachtungen zur Biologie und Morphologie der Präimaginalstadien von *Pontia chloridice* (Hübner, [1813]) (Lepidoptera: Pieridae). *Entomologische Zeitschrift* 112, 213–217.

Kotiaho, J.S., Kaitala, V., Komonen, A. and Päivinen, J. (2005) Predicting the risk of extinction from shared ecological characteristics. *PNAS* 102(6), 1963–1967. Available at: https://www.pnas.org/content/102/6/1963 (accessed 20 September 2021).

Kreuzinger, A.J., Fiedler, K., Letsch, H. and Grill, A. (2015) Tracing the radiation of *Maniola* (Nymphalidae) butterflies: new insights from phylogeography hint at one single incompletely differentiated species complex. *Ecology and Evolution* 5(1), 46–58.

Kristensen, N.P., Scoble, M.J. and Karsholt, O. (2007) Lepidoptera phylogeny and systematics: the state of inventorying moth and butterfly diversity. *Zootaxa* 1668, 699–747. Available at: https://www.biotaxa.org/Zootaxa/article/view/zootaxa.1668.1.30 (accessed 20 September 2021).

Ksantini, I., Jardak, T. and Zeghal, N. (2011) First report on *Virachola livia* Klug. (Lepidoptera: Lycaenidae) and its effects on different pomegranate varieties in Tunisia. *Bulletin OEPP/EPPO Bulletin* 41(2), 178–182.

Kudrna, O. (1975) On a hitherto undescribed European species of the genus *Hipparchia* Fabricius (Lep., Satyridae). *Entomologist's Gazette* 26, 197–207.

Kudrna, O. (1977) *A Revision of the Genus Hipparchia Fabricius*. E.W. Classey, Faringdon, UK.

Kudrna, O. (1986) *Butterflies of Europe, Vol. 8. Aspects of the conservation of European Butterflies*. AULA-Verlag GmbH, Wiesbaden, Hesse, Germany.

Kudrna, O. (1996) Mapping European butterflies; handbook for recorders. *Oedippus* 12, 1–60.

Kudrna, O. (2002) *The Distribution Atlas of European Butterflies*. Oedippus 20. Gesellschaft für Schmetterlingsschutz e. V. in cooperation with Apollo Books, Stenstrup, Denmark.

Kudrna, O. (2013) European butterflies, global warming and predicting future – science or business? *Entomologische Zeitschrift* 123(3), 103–113.

Kudrna, O. (2019) *Distribution of Butterflies and Skippers in Europe.* Společnost pro Ochranu Motýlů (SOM), South Bohemia, Czech Republic.

Kudrna, O., Harpke, A., Lux, K., Pennerstorfer, J., Schweiger, O. *et al.* (2011) *Distribution Atlas of Butterflies in Europe*. Gesellschaft für Schmetterlingsschutz e. V., Halle, Saxony-Anhalt, Germany.

Kudrna, O., Pennerstorfer, J. and Lux, K. (2015) *Distribution Atlas of European Butterflies and Skippers*. Wissenschaftlicher Verlag Peks e.K., Schwanfeld, Schweinfurt, Bavaria, Germany.

Lafranchis, T. (2004) *Butterflies of Europe*. Diatheo, Paris, France.

Lafranchis, T. (2005) Biologie, écologie et repartition de *Gegenes pumilio* (Hoffmansegg, 1804) en Grèce (*Lepidoptera*, *Hesperiidae*). *Linneana Belgica* XX(4), 116–120.

Lafranchis, T. (2019) Notes on the biology of some butterflies in Greece (Lepidoptera: Papilionoidea). *Entomologist's Gazette* 70, 113–134.

Larsen, T.B. (1974) *Butterflies of Lebanon*. National Council for Scientific Research, Beirut, Lebanon.

Larsen, T.B. (1975) Provisional notes on butterfly migration in Lebanon. *Atalanta* 6, 62–74.

Larsen, T.B. (1976) The importance of migration to the butterfly faunas of Lebanon, east Jordan and Egypt (Lepidoptera, Rhopalocera). *Notulae Entomologicae* 56, 73–83.

Larsen, T.B. (1977) Recent expansion in the range of *Aporia crataegi* (Linnaeus) in east Jordan. (Pieridae). *Nota lepidopterologica* 1(1), 19–21.
Larsen T.B. (1978) The first record of *Catopsilia florella* F. from Lebanon for more than fifty years. *Atalanta* 9, 253–254.
Larsen, T.B. (1980) The two subspecies of *Polyommatus icarus* (Rottemburg) in Lebanon; a probable case of multiple invasion (Lep.; Lycaenidae). *Nota lepidopterologica* 2(4), 147–153.
Larsen, T.B. (1982a) The importance of migration to the butterfly fauna of Arabia (Lep., Rhopalocera). *Atalanta Bd*. XIII 4, 248–259.
Larsen, T.B. (1982b) *Gegenes pumilio* Hoffmannsegg, 1804; a review with cytological evidence that two species are involved (Hesperiidae). *Nota lepidopterologica* 5(2–3), 103–110.
Larsen, T.B. (1983) Insects of Saudi Arabia. Lepidoptera, Rhopalocera. A monograph of the butterflies of the Arabian Peninsula. *Fauna of Saudi Arabia* 5, 333–478.
Larsen, T.B. (1984a) *Butterflies of Saudi Arabia and its Neighbours*. Stacey International, London.
Larsen, T.B. (1984b) A tropical migrant butterfly new to the eastern Mediterranean *Junonia orithya here* LANG (Lepidoptera, Nymphalidae). *Atalanta* 15, 101–102.
Larsen, T.B. (1986) Tropical Butterflies of the Mediterranean. *Nota lepidopterologica* 9(1–2), 63–67.
Larsen, T.B. (1988) The anatomy of a major butterfly migration in southern India. *Atalanta* 18, 267–281.
Larsen, T.B. (1990) *Butterflies of Egypt.* Apollo Books, Svendborg, Funen, Denmark.
Larsen, T.B. (1991) The art of feigning death – thanatosis in *Euploea* (Danainae) and other aposematic butterflies. *Entomologists' Record and Journal of Variation* 103, 263–266.
Larsen, T.B. (1992) A migration of *Catopsilia florella* (F.) in Botswana; morphological differences between migratory and non-migratory populations (Lepidoptera: Pieridae). *Tropical Lepidoptera* 3, 2–11.
Larsen, T.B. (1994) Butterfly biodiversity and conservation in the Afrotropical region. In: Pullin, A.S. (ed) *Ecology and Conservation of Butterflies*. Chapman and Hall, London, pp. 290–303.
Larsen, T.B. (1996) *The Butterflies of Kenya and their Natural History*. Oxford University Press, Oxford, New York, Tokyo.
Larsen, T.B. (1997) Mass hilltopping of earwigs on the Troodos summit in Cyprus (Dermaptera). *Entomologists' Record and Journal of Variation* 109, 89–90.
Larsen, T.B. (2005) *Butterflies of West Africa*. Apollo Books, Stenstrup, Denmark.
Larsen, T.B. (2015) The tropical butterflies of the Levant: a biogeographical and ecological perspective. *News of the Israeli Lepidopterists Society* 32, 8–36. [Presentation given at 'Butterflies of the Middle East – 1st International Congress', 29–31 May 2008, Jerusalem, Israel. Proceedings not published.]
Larsen, T.B. and Larsen, K. (1980) *Butterflies of Oman*. Bartholomew Books, Edinburgh, UK.
Larsen, T.B. and Nakamura, I. (1983) The butterflies of east Jordan. *Entomologist's Gazette* 34, 135–208.
Larsen, T.B. and Pittaway, A.R. (1982) Notes on the ecology, biology and taxonomy of *Apharitis acamas* (Klug) Lep.; Lycaenidae). *Entomologist's Gazette* 33, 163–168.
Lawson, N., Vane-Wright, R.I. and Boppré, M. (2021) The puzzle of monarch butterflies (*Danaus plexippus*) and their association with plants containing pyrrolizidine alkaloids. *Ecological Entomology* 46, 999–1005.
Lederer, J. (1855) Beitrag zur Schmetterlings Fauna von Cypern, Beirut und einem Teil Kleinasiens. *Verhandlungen der Zoologisch-Botanischen Vereins in Wien* 5, 177–254.
Lepidoptera Libanotica (2019) Available at: https://butterflies.spnl.org/luthrodes-pandava-lebanon-record/ (accessed 6 April 2021).

Lindstedt, C., Murphy, L. and Mappes, J. (2019) Antipredator strategies of pupae: how to avoid predation in an immobile life stage? *Philosophical Transactions Royal Society B* 374, 20190069. doi: 10.1098/rstb.2019.0069

Lorković, Z. (1971) *Gegenes nostrodamus* F. and *G. pumilio* Hffgg. on the eastern Adriatic coast. *Acta entomologica Jugoslavica* 7, 56.

Lukhtanov, V.A. and Pazhenkova, A. (2021) The taxa of the *Hyponephele lycaon – H. lupina* species complex (Lepidoptera, Nymphalidae, Satyrinae): deep DNA barcode divergence despite morphological similarity. *Folia Biologica (Kraków)* 69(1), 11–21.

Lukhtanov, V.A., Pazhenkova, A. and Novikova, V. (2016) Mitochondrial chromosome as a marker of animal migratory routes: DNA barcoding revealed Asian (non-African) origin of a tropical migrant butterfly *Junonia orithya* in south Israel. *Comparative Cytogenetics* 10(4), 671–677.

Lushai, G., Smith, D.A.S., Gordon, I.J., Goulson, D., Allen, J.A. and Maclean, N. (2003) Incomplete sexual isolation in sympatry between subspecies of the butterfly *Danaus chrysippus* (L.) and the creation of a hybrid zone. *Heredity* 90, 236–246.

Maes, D., Verovnik, R., Wiemers, M., Brosens, D., Beshkov, S. *et al.* (2019) Integrating national Red Lists for prioritising conservation actions for European butterflies. *Journal of Insect Conservation* 23, 301–330.

Magni, D. and Caudullo, G. (2016) *Celtis australis* in Europe: distribution, habitat, usage and threats. In: San-Miguel-Ayanz, J., de Rigo, D., Caudullo, G., Houston Durrant, T. and Mauri, A. (eds) *European Atlas of Forest Tree Species*. Publications Office of the European Union, Luxembourg, p. 80. Available at: https://ies-ows.jrc.ec.europa.eu/efdac/download/Atlas/pdf/Celtis_australis.pdf (accessed 21 September 2021).

Makris, C. (2003) *Butterflies of Cyprus*. Bank of Cyprus Cultural Foundation, Nicosia, Cyprus.

Mallet, J. (2007) Subspecies, Semispecies, Superspecies. In: Levin, S.A. (ed) *Encyclopaedia of Biodiversity*. Elsevier, Oxford, UK, Vol. 2, pp. 1–5.

Manil, L. (1990) Les Rhopalocères de Chypre. *Linneana Belgica* XII(8), 313–391.

Markogianni, V., Tzirkalli, E., Gücel, S., Dimitriou, E. and Zogaris, S. (2014) Remote sensing application for identifying wetland sites on Cyprus: Problems and prospects. *Proceedings SPIE Volume 9229, Second International Conference on Remote Sensing and Geoinformation of the Environment (RSCy2014)*. doi: 10.1117/12.2066144

Martin, J. and Russell, P. (2013) Increasing numbers of records of *Danaus chrysippus* (Linnaeus, 1758) and *Pelopidas thrax* (Hübner, 1821) (Lepidoptera: Nymphalidae, Danainae; Hesperiidae) in the Eastern Aegean islands. *Entomologist's Gazette* 64, 85–93.

McArthur, A. and John, E. (2019) *Pieris rapae* (Linnaeus, 1758) (Lepidoptera: Pieridae, Pierinae): a further note on trans-Mediterranean spring migrations from Cyprus. *Entomologist's Gazette* 70, 244.

McCullagh, B.S., Alexiuk, M.R., Payment, J.E., Hamilton, R.V., Lalonde, M. *et al.* (2020) It's a moth! It's a butterfly! It's the complete mitochondrial genome of the American moth-butterfly *Macrosoma conifera* (Warren, 1897) (Insecta: Lepidoptera: Hedylidae)! *Mitochondrial DNA Part B* 5(3), 3615–3617.

Merckx, T. and Van Dyck, H. (2005) Mate location behaviour of the butterfly *Pararge aegeria* in woodland and fragmented landscapes. *Animal Behaviour* 70, 411–416.

Mérit, X. and Mérit, V. (2007) Rediscovery of *Hypolimnas misippus* (Linnaeus, 1764) as a unique female in Lebanon during spring 2003 (Lepidoptera, Nymphalidae). *Atalanta* 38, 319–320.

Merrill, R.M., Gutiérrez, D., Lewis, O.T., Gutiérrez, J., Díez, S.B. and Wilson, R.J. (2008) Combined effects of climate and biotic interactions on the elevational range of a phytophagous insect. *Journal of Animal Ecology* 77, 145–155.

Middleton-Welling, J., Dapporto, L., García-Barros, E., Wiemers, M., Nowicki, P. *et al.* (2020) A new comprehensive trait database of European and Maghreb butterflies, Papilionoidea. *Scientific Data* 7, 351. doi: 10.1038/s41597-020-00697-7

Montagud-Alario, S. and García-Alamá, J.A. (2010) *Mariposas diurnas de la comunitat Valenciana: (Papilionoidea and Hesperioidea)*. Collección Biodiversidad, 17. Conselleria de Medi Ambient, Aigua, Urbanisme i Habitatge, Generalitat Valenciana, Valencia.

Müller, G.C., Kravchenko, V.D., Phillips, A., Shwarz-Tzachor, R. and Benyamini, D. (2005) The first record of the Eremic Lycaenid *Deudorix livia* (Klug, 1834) in Europe. *Atalanta* 36(1/2), 109–112.

Myers, N., Mittermeier, R.A., Mittermeier, C.G., da Fonseca, G.A.B. and Kent, J. (2000) Biodiversity hotspots for conservation priorities. *Nature* 403, 853–858.

Nazari, V. and Carbonell, F. (2018) *Allancastria cerisyi* (Godart, 1824). Version 29 March 2018 (under construction). Available at: http://tolweb.org/Allancastria_cerisy/65422/2018.03.29 in The Tree of Life Web Project, http://tolweb.org/ (accessed 21 September 2021).

Nazari, V. and Sperling, F.A.H. (2007) Mitochondrial DNA divergence and phylogeography in western Palaearctic Parnassiinae (Lepidoptera: Papilionidae): How many species are there? *Insect Systematics and Evolution* 38, 121–138.

Nazari, V., Zakharov, E.V. and Sperling, F.A.H. (2007) Phylogeny, historical biogeography, and taxonomic ranking of Parnassiinae (Lepidoptera, Papilionidae) based on morphology and seven genes. *Molecular Phylogenetics and Evolution* 42, 131–156.

Nekrutenko, Yu. P. (1984) The blue butterflies of the USSR fauna, assigned to the genus *Chilades* (Lepidoptera, Lycaenidae). *Vestnik Zoologii* 1984(3), 29–40 [in Russian].

Nieukerken, E.J. van, Kaila, L., Kitching, I.J., Kristensen, N.P., Lees, D.C. *et al.* (2011) Order Lepidoptera Linnaeus, 1758. In: Zhang, Z.-Q. (ed.) *Animal Biodiversity: an outline of higher-level classification and survey of taxonomic richness.* Magnolia Press, Auckland, New Zealand. *Zootaxa* 3148, pp. 212–221.

Numa, C., van Swaay, C., Wynhoff, I., Wiemers, M., Barrios, V. *et al.* (2016) *The Status and Distribution of Mediterranean butterflies*. IUCN, Malaga, Spain.

Oberthür, C. (1909) *Satyrus briseis larnacana* ssp. nova. *Études de Lépidoptérologie Comparée*. Fascicule III. Rennes imprimerie, Rennes, Brittany, France, pp. 263–265.

Ochuodho, J.O. and Modi, A.T. (2007) Light-induced transient dormancy in *Cleome gynandra* L. seeds. *African Journal of Agricultural Research* 2(1), 587–591.

Olivier, A. (1990) Notes on the butterflies of the Greek island of Kastellórizo (Lepidoptera : Hesperioidea and Papilionoidea). *Phegea* 18(4), 169–190.

Olivier, A. (1993) *The Butterflies of the Greek island of Ródos: Taxonomy, Faunistics, Ecology and Phenology* with a tentative synthesis on the biogeography of the butterflies of Kríti (Crete), Kárpathos, Ródos, the Eastern Aegean islands and Kípros (Cyprus) (Lepidoptera: Hesperioidea and Papilionoidea). Vlaamse Vereniging voor Entomologie, Antwerp, Belgium.

Olivier, A. (1996) Notes on the taxonomic status and supposed biogeographic affinity of the *Pseudochazara anthelea* (Hübner, [1824]) populations from Kípros (Cyprus) and from the Greek island of Kós (Lepidoptera: Nymphalidae Satyrinae). *Phegea* 24(1), 5–12.

Olivier, A. and Coutsis, J.G. (1997) A revision of the superspecies *Hipparchia azorina* and of the *Hipparchia aristaeus* group (Nymphalidae : Satyrinae). *Nota Lepidopterologica* 20(3/4), 150–292.

Owen, D.F. (1971) *Tropical Butterflies*. Clarendon Press, Oxford, UK.

Özden, Ö. and Hodgson, D.J. (2011) Butterflies (Lepidoptera) highlight the ecological value of shrubland and grassland mosaics in Cypriot garrigue ecosystems. *European Journal of Entomology* 108, 431–437.

Pamperis, L.N. (2009) *The Butterflies of Greece* (2nd edn revised and enlarged). Hellenic Society for the Protection of Nature, Editions Pamperis, Athens, Greece.

Paradiso, F., Martelli, F., Cerrato, C., Ghidotti, S., Viterbi, R. *et al.* (2019) From Africa to the Alps: risk assessment on an invasion by *Cacyreus marshalli* (Butler, 1898). *Journal of Insect Conservation* 23, 279–288. doi: 10.1007/s10841-018-00124-8

Parker, R. (1983) The Butterflies of Cyprus. *Entomologist's Gazette* 34, 17–53.

Parker, R. and John, E. (2003) Fight behaviour of the Cyprus Grayling, *Hipparchia cypriensis* in the Troodos Mountains. *Bulletin of the Amateur Entomologists' Society* 62, 27–30.

Parker, R. and John, E. (2016) The spread of the Geranium Bronze, *Cacyreus marshalli*: a summary of recent extensions of its range around the Mediterranean, two new records and a UK sighting. *Bulletin of the Amateur Entomologists' Society* 75, 33–36.

Penney, D. (2012) *Field Guide to Wildlife of The Gambia*. 2nd edition. Siri Scientific Press, Manchester, UK.

Petrakis, P.V. and Legakis, A. (2005) Insect migration and dispersal with emphasis on Mediterranean ecosystems. In: Elewa, A.M.T. (ed.) *Migration of Organisms: Climate. Geography. Ecology*. Springer-Verlag, Berlin and Heidelberg, Germany, pp. 85–126.

Pierce, N.E., Braby, M.F., Heath, A., Lohman, D.J., Mathew, J. *et al.* (2002) The ecology and evolution of ant association in the Lycaenidae (Lepidoptera). *Annual Review of Entomology* 47, 733–771.

Pittaway, A.R. (1980) Butterflies (Lepidoptera) of Qatar, April-June, 1979. *Entomologist's Gazette* 31, 103–111.

Pittaway, A.R. (1981) Further notes on the butterflies and hawkmoths (Lepidoptera) of Eastern Saudi Arabia. *Entomologist's Gazette* 32, 27–35.

Pittaway, A.R. (1985) Lepidoptera: Rhopalocera of Western Saudi Arabia. *Fauna of Saudi Arabia* 7, 172–197.

Polak, S., Schein, V., Valič, P., Verovnik, R., Čelik, T. *et al.* (2009) *Metulji Notranjske in Primorske: slikovni priročnik za določanje dnevnih metuljv v naravi*. Postojna: Notranjski muzej, Slovenia.

Pollard, E. and Yates, T.J. (1993) *Monitoring Butterflies for Ecology and Conservation*. Chapman and Hall, London.

Pyrcz, T.W., Lorenc-Brudecka, J. and Zubek, A. (2015) New informative characters in the female genitalia in assessing the phylogeny of Palearctic Satyrina. Poster presentation: XIXth European Congress of Lepidopterology. 27 September – 02 October 2015, Radebeul, Germany, p. 118.

Rabl, C. and Rabl, D. (2015) Die Einwanderung von *Libythea celtis* (Laicharting, 1782) (Lepidoptera: Nymphalidae) nach Österreich. *Beiträge zur Entomofaunistik* 16, 3–8.

Reavey, D. (1993) Why body size matters to caterpillars. In: Stamp, N.E. and Casey, T. M. (eds) *Caterpillars – Ecological and Evolutionary Constraints on Foraging*. Chapman & Hall, New York and London, pp. 248–279.

Rebel, H. (1916) Über die Lepidopterenfauna Cyperns. XXVI. *Jahresbericht des Wiener Entomologischen Vereins* 26(1915), 93–110.

Rebel, H. (1924) Lepidopterologische Nachträge zu einigen ostmediterranen Insularfaunen. *Jahresbericht des Wiener entomologischen Vereins* 30, 37–49.

Rebel, H. (1939) Zur Lepidopterenfauna Cyperns. *Mitteilungen der Münchner entomologischen Gesellschaft* 29, 487–515.

Riley, N.D. (1921) Some undescribed Rhopalocera from Mesopotamia and N.W. Persia; and other notes. *Annals and Magazine of Natural History* 8, 590–600.

Riley, N.D. (1925) The species usually referred to *Cigaritis* Boisduval (Lepidoptera; Lycaenidae). *Novitates Zoologicae* 32, 70–95.

Rödder, D., Schmitt, T., Gros, P., Ulrich, W. and Habel, J.C. (2021) Climate change drives mountain butterflies towards the summits. *Scientific Reports* 11,14382. doi: 10.1038/s41598-021-93826-0

Rowlings, M. and Cuvelier, S. (2018) *Zizeeria karsandra* (Lepidoptera: Lycaenidae) recorded from Crete (Greece): observations, distribution and habitats. *Phegea* 46(4), 126–131.

Russell, P.R. (2013) Intergeneric pairings involving *Maniola jurtina* L. (Lepidoptera: Nymphalidae: Satyrinae): are sight and smell overridden by this butterfly's sex drive? *Entomologist's Gazette* 64, 11–17.

Rutowski, R. (1991) The evolution of male mate-location behaviour in butterflies. *The American Naturalist* 138(5), 1121–1139.
Sahoo, R.J., Lohman, D.J., Wahlberg, N., Müller, C.J., Brattström, O. *et al.* (2018) Evolution of *Hypolimnas* butterflies (Nymphalidae): Out-of-Africa origin and *Wolbachia*-mediated introgression. *Molecular Phylogenetics and Evolution* 123, 50–58.
Saitoh, K. (1984) Chromosomes of three butterfly taxa from Yemen. *Tyo to Ga* 34(4), 167–170.
Sala, G. and Bollino, M. (1997) A contribution to the knowledge of the Papilionidae of Kriti Island (Lepidoptera, Papilionidae). *Atalanta* 28(1/2), 35–41.
Salata, S., Georgiadis, C. and Borowiec, L. (2019) Invasive ant species Hymenoptera: Formicidae) of Greece and Cyprus. *North-western Journal of Zoology* 15(1), 13–23.
Salmon, M.A. (2000) *The Aurelian Legacy, British Butterflies and their collectors.* Harley Books, Colchester, UK.
Samodurow, W.A., Korolew, W.A. and Tschikolowez, W.W. (2001) Eine Übersicht über die Satyriden der Gattung *Hyponephele* Muschamp, 1915. VII. Die Arten *Hyponephele lycaon* (Rottemburg, 1775), *H. pasimelas* (Staudinger, 1886), *H. lycaonoides* D. Weiss, 1978, *H. przhewalskyi* Dubatolov, Sergeev and Zhdanko, 1994, *H. dzhungarico* Samodurov, 1996, *H. galtscha* (Grum-Grshimailo, 1893) und *H. lupina* (Costa, 1836) (Lepidoptera, Satyridae). *Atalanta* 32(1/2), 111–186.
Sanetra, M. and Fiedler, K. (1996) Behaviour and morphology of an aphytophagous lycaenid caterpillar: *Cigaritis* (*Apharitis*) *acamas* Klug, 1834 (Lepidoptera: Lycaenidae). *Nota lepidopterologica* 18(1), 57–76.
Sbordoni, V., Cesaroni, D., Coutsis, J.G. and Bozano, G.C. (2018) *Guide to the Butterflies of the Palaearctic Region*. Satyrinae Part V. Omnes Artes, Milano, Lombardy, Italy.
Schaider, P. (1976) *Libythea celtis* (Laicharting, 1782) ein Wanderfalter? (Lep. Libytheidae). *Atalanta* 7, 19–22.
Schneider, F. (1962) Dispersal and migration. *Annual Review of Entomology* 7, 223–242.
Schurian, K.G. (1995) Die Futterpflanze von *Danaus chrysippus* L. in der Südtürkei (Lepidoptera: Nymphalidae, Danainae). *Nachrichten des Entomologischen Vereins Apollo*, N.F. 15(4), 505–506.
Schurian, K.G., Grandisch, H. and Mark, H-G. (1993) Beobachtungen zur Biologie und Ökologie von *Danaus chrysippus* L. in der Südtürkei (Lepidoptera: Nymphalidae, Danainae). *Nachrichten des entomologischen Vereins Apollo*, N.F. 13(3a), 343–350.
Seizmair, M. (2013) Bemerkenswerte Rhopaloceren-Nachweise der Herbstfauna des Libanon (Lepidoptera, Rhopalocera). *Atalanta* 4(1–4), 89–92.
Shaw, M., Stefanescu, C. and Van Nouhuys, S. (2009) Parasitoids of European Butterflies. In: Settele, J., Shreeve, T., Konvička, M. and Van Dyck, H. (eds) *Ecology of Butterflies in Europe*. Cambridge University Press, Cambridge, UK, pp. 130–156.
Shields, O. (1967) Hill-topping. An ecological study of summit congregation behavior of butterflies on a southern California hill. *Journal of Research on the Lepidoptera* 6, 69–178.
Shields, O. (1974) Toward a theory of butterfly migration. *Journal of Research on the Lepidoptera* 13(4), 217–238.
Shlalo, A. and Kasis, W. (2016) Biological study of almond butterfly *Aporia crataegi* in central Syria. *Arab Journal of Plant Protection* 34(2), 82–94 [in Arabic].
Shreeve, T.G. and Dennis, R.L.H. (2009) Cross-water transfer by common butterflies (Lepidoptera: Rhopalocera): clear examples of ignoring the highway. *Entomologist's Gazette* 60, 107–109.
Shreeve, T.G., Dennis, R.L.H. and Wakeham-Dawson, A. (2006) Phylogenetic, habitat, and behavioural aspects of possum behaviour in European lepidoptera. *Journal of Research on the Lepidoptera* 39, 80–85.
Simes, J.A. (1920) *Gonepteryx cleopatra*, L. Has it more than one brood per annum? *Entomologist's Record and Journal of Variation* 32, 197–199.

Simonsen, T.J., de Jong, R., Heikkilä, M. and Kaila, L. (2012) Butterfly morphology in a molecular age – does it still matter in butterfly systematics? *Arthropod Structure and Development* 41, 307–322.

Smart, P. (1991) *The Illustrated Encyclopedia of the Butterfly World*. Tiger Books International, London.

Smith, D.A.S. (2014) *African Queens and their Kin – A Darwinian Odyssey*. Brambleby Books, Taunton, UK.

Smith, D.A.S. (2021) The status of the African Queen butterfly, *Danaus chrysippus* (Linnaeus, 1758) (Lepidoptera: Nymphalidae, Danainae), in the Canary Islands, including an irruption on Fuerteventura. *Entomologist's Gazette* 72, 209–220.

Smith, D.A.S. and Gordon, I.J. (1987) The genetics of the butterfly *Hypolimnas misippus* (L.): the classification of phenotypes and the inheritance of forms *misippus* and *inaria*. *Heredity* 59, 467–475.

Snegovaya, N.Y. and Petrov, V. A. (2019) A catalogue of butterflies (Lepidoptera, Rhopalocera) of Azerbaijan. *Acta Biologica Sibirica* 5(3), 62–117.

Solonkin, I.A., Shkurikhin, A.O., Oslina, T.S. and Zakharova, E.Y. (2021) Changes in the body size of black-veined white, *Aporia crataegi* (Lepidoptera: Pieridae), recorded in a natural population in response to different spring weather conditions and at different phases of an outbreak. *European Journal of Entomology* 118, 214–224.

Soyhan, T., Başer, S. and Nazari, V. (2013) First record of *Cacyreus marshalli* Butler, 1898 (Lycaenidae) from Turkey. *Nota lepidopterologica* 36(2), 189–190.

Sparrow, D.J. and John, E. (2016) Geography. In: Sparrow, D.J. and John, E. (eds) *An Introduction to the Wildlife of Cyprus*. Terra Cypria, Limassol, Cyprus, pp. 3–7.

Spieth, H.R. and Schwarzer, E. (2001) Aestivation in *Pieris brassicae* (Lepidoptera: Pieridae): Implications for parasitism. *European Journal of Entomology* 98, 171–176.

Spieth, H.R., Pörschmann, U. and Teiwes, C. (2011) The occurrence of summer diapause in the large white butterfly *Pieris brassicae* (Lepidoptera: Pieridae): a geographical perspective. *European Journal of Entomology* 108, 377–384.

Spitsyn, V.M., Kondakov, A.V., Tomilova, A.A., Karamkhudoeva, M. and Potapov, G.S. (2019) A taxonomic revision of the butterfly genera *Esperarge* and *Kirinia* (Lepidoptera: Nymphalidae: Satyrinae). *Ecologica Montenegrina* 26, 118–126.

Staudinger, O. (1878) Lepidopteren-Fauna Kleinasien's. *Horae Societatis Entomologicae Rossicae* XIV, 176–482.

Staudinger, O. (1881) Lepidopteren-Fauna Kleinasien's. Nachträge. *Horae Societatis Entomologicae Rossicae* XVI, 65–135.

Stefanescu, C. (2001) The nature of migration in the red admiral butterfly *Vanessa atalanta*: evidence from the population ecology in its southern range. *Ecological Entomology* 26, 525–536.

Stefanescu, C. (2006) The Nettle-tree Butterfly *Libythea celtis*, the only European 'snout'. *Cynthia* 6, VI–VII.

Stefanescu, C., Pintureau, B., Tschorsnig, H.-P. and Pujade-Villar, J. (2003) The parasitoid complex of the butterfly *Iphiclides podalirius feisthamelii* (Lepidoptera: Papilionidae) in north-east Spain. *Journal of Natural History* 37, 379–396.

Stefanescu, C., Jubany, J. and Dantart, J. (2006) Egg-laying by the butterfly *Iphiclides podalirius* (Lepidoptera, Papilionidae) on alien plants: a broadening of host range or oviposition mistakes? *Animal Biodiversity and Conservation* 29(1), 83–90.

Stefanescu, C., Páramo, F., Åkesson, S., Alarcón, M., Ávila, A. *et al.* (2012) Multi-generational long-distance migration of insects: studying the painted lady butterfly in the Western Palaearctic. *Ecography* 35, 001–014. doi: 10.1111/j.1600-0587.2012.07738.x

Stefanescu, C., Soto, D.X., Talavera, G., Vila, R., and Hobson, K.A. (2016) Long-distance autumn migration across the Sahara by painted lady butterflies: exploiting resource pulses in the tropical savannah. *Biology Letters* 12, 20160561. doi: 10.1098/rsbl.2016.0561

Stichel, H. (1907) Neue Unterarten von *Zerynthia cerisyi* God. *Entomologische Zeitschrift* 29, 178.

Talavera, G., Lukhtanov, V.A., Pierce, N.E. and Vila, R. (2012) Establishing criteria for higher-level classification using molecular data: the systematics of *Polyommatus* blue butterflies (Lepidoptera, Lycaenidae). *Cladistics* (2012), 1–27.

Talavera, G., Bataille, C., Benyamini, D., Gascoigne-Pees, M. and Vila, R. (2018) Afrotropical Painted Lady butterflies recolonize the Mediterranean in early spring. *Biology Letters* 14, 20180274. doi: 10.1098/rsbl.2018.0274

Talhouk, A.S. (1954) A list of insects found on plants of economic importance in Syria. *Bulletin de la Société entomologique d'Égypte* XXXVIII, 305–309.

Tea, K.Y-T., Wei, J.S., Gan, C.W. and Lohman, D.J. (2020) Beachgoing butterflies: marine puddling on black sand beaches of Tangkoko Batuangus Nature Reserve, North Sulawesi. *Journal of the Lepidopterists' Society* 74(2), 127–131.

Tea, K.Y-T., Wei, J.S., Beaver, E.P. and Lohman, D.J. (2021) Kleptopharmacophagy? Milkweed butterflies scratch and imbibe from Apocynaceae-feeding caterpillars. *Ecology* 102, Issue 12. doi: 10.1002/ECY.3532

Tennent, W.J. (1985) Some Records of Spring Butterflies in Cyprus in 1984. *Entomologist's Gazette* 36,105–109.

Tennent, W.J. (1995) Field observations of the 'hill-topping' phenomenon in north-west Africa – and an introduction to 'ravining' (Lep: Rhopalocera). *The Entomologist's Record and Journal of Variation* 107, 57–67.

Tennent, W.J. (1996) *The Butterflies of Morocco, Algeria and Tunisia*. Gem Publishing, Wallingford, UK.

Tennent, W.J., Aguiar, A.M.F., Healey, D. and Ferreira, A. (2013) Recent records of *Hypolimnas misippus* (Linnaeus, 1764) (Lepidoptera: Nymphalidae) from Madeira, Portugal, with a resumé of historical records from Macaronesia. *Entomologist's Gazette* 64, 101–108.

Thomas, J. and Lewington, R. (1991) *The Butterflies of Britain and Ireland*. Dorling Kindersley, London.

Thomson, G. (1976) Le genre *Maniola* (Lepidoptera : Satyridae). Notes sur les genitalia males et femelles. *Linneana Belgica* VI, 126–142.

Thomson, G. (1987) *Enzyme variation at morphological boundaries in Maniola and related genera (Lepidoptera: Nymphalidae: Satyrinae).* Doctor of Philosophy Thesis. University of Stirling, Scotland.

Thomson, G. (1992) Egg surface morphology of Manioline butterflies. *Atalanta* 23(1/2):195–214.

Thomson, G. (2011) *The Meadow Brown Butterflies: a study in genetics, morphology and evolution*. Privately published.

Tiple, A., Agashe, D., Khurad, A.M. and Kunte, K. (2009) Population dynamics and seasonal polyphenism of *Chilades pandava* butterfly (Lycaenidae) in central India. *Current Science* 97(12), 1774–1779.

Tolman, T. (1992) The larval host-plant of *Pontia chloridice* (Hübner, [1813]) in Greece (Lepidoptera : Pieridae). *Phegea* 20(3), 109–113.

Tolman, T. and Lewington, R. (1997) *Butterflies of Britain & Europe*. HarperCollins, London.

Toper Kaygin, A., Yildiz, Y. and Avci, M. (2006) Lepidoptera fauna in Bartin province, in western Black Sea region of Turkey. *African Journal of Agricultural Research* 4(9), 815–822.

Trakel (undated) Kelebekler. Available at: http://www.trakel.org/kelebekler/ (accessed 20 June 2021).

Tshikolovets, V.V. (2011) *Butterflies of Europe and the Mediterranean Area*. Tshikolovets Publications, Pardubice, Czech Republic.

Tshikolovets, V., Naderi, A. and Eckweiler, W. (2014) *The Butterflies of Iran and Iraq*. Tshikolovets Publications, Pardubice, Czech Republic.

Turner, H.J. (1920) The Butterflies of Cyprus. *Transactions of the Entomological Society of London* 1920, 170–207.

Tzirkalli, E., Kadis, C., Halley, J.M., Vogiatzakis, I., Wilson, R.J. *et al.* (2019) Conservation ecology of butterflies on Cyprus in the context of Natura 2000. *Biodiversity and Conservation* 28(7), 1759–1782.

van Oorschot, H. and Coutsis, J.G. (2014) *The genus Melitaea Fabricius, 1807(Lepidoptera: Nymphalidae, Nymphalinae) Taxonomy and systematics with special reference to the male genitalia.* Tshikolovets Publications, Pardubice, Czech Republic.

Van Swaay, C.A.M. and Warren, M.S. (1999) *Red Data Book of European Butterflies (Rhopalocera).* Nature and Environment, No. 99. Council of Europe Publishing, Strasbourg, Grand Est, France.

Van Swaay, C., Warren, M. and Loïs, G. (2006) Biotope use and trends of European butterflies. *Journal of Insect Conservation* 10, 189–209.

Van Swaay, C., Cuttelod, A., Collins, S., Maes, D., López Munguira, M. *et al.* (2010a) *European Red List of Butterflies.* Publications Office of the European Union, Luxembourg.

Van Swaay, C.A.M., Harpke, A., Van Strien, A., Fontaine, B., Stefanescu, C. *et al.* (2010b) *The Impact of Climate Change on Butterfly Communities 1990-2009.* Report VS2010.025, Butterfly Conservation Europe and De Vlinderstichting, Wageningen, Gelderland, The Netherlands.

Van Swaay, C., Maes, D., Collins, C., Munguira, M.L., Šašić, M. *et al.* (2011) Applying IUCN criteria to invertebrates: how red is the Red List of European butterflies? *Biological Conservation* 144, 470–478.

Vane-Wright, R.I. (2004) Butterflies at that awkward age. *Nature* 428, 477–480.

Vane-Wright, R.I. (2020) *Euploea dorippus* Klug, 1845: species, semispecies, subspecies, junior subjective synonym of *Danaus chrysippus chrysippus* (Linnaeus), and/or form – what does Klug's *dorippus* represent? (Lepidoptera: Nymphalidae, Danainae). *Nota lepidopterologica* 43, 117–138. doi: 10.3897/nl.43.47936

Vane-Wright, R.I. and John, E. (2019) A note on *Papilio aegyptius* Schreber, 1759, and its synonymy with *Danaus chrysippus chrysippus* (Linnaeus, 1758) (Lepidoptera: Nymphalidae, Danainae). *Entomologist's Gazette* 70, 147–154.

Verity, R. (1905) *Papilio machaon* ab. *cypria. Rhopalocera Palaearctica, iconographie et description des Papillons Diurnes de la région paléarctique. Papilionidae et Pieridae.* R. Verity, Florence, Italy.

Verity, R. (1919) Seasonal polymorphism and races of some European Grypocera and Rhopalocera. *Entomologist's Record and Journal of Variation* 31, 43–48, 87–89.

Verovnik, R. and Wiemers, M. (2010) *European Butterflies – Taxonomy* (in association with Balletto, E., Coutsis, J., Karsholt, O. *et al.*). Butterfly Conservation Europe, Wageningen, Gelderland, The Netherlands.

Verovnik, R. and Wiemers, M. (2016) Species delimitation in the Grayling genus *Pseudochazara* (Lepidoptera, Nymphalidae, Satyrinae) supported by DNA barcodes. *ZooKeys* 600: 131–154. doi: 10.3897/zookeys.600.7798

Vodă, R., Dapporto, L., Dincă, V., Shreeve, T.G., Khaldi, M. *et al.* (2016) Historical and contemporary factors generate unique butterfly communities on islands. *Scientific Reports* 6, 28828. Available at: https://www.ncbi.nlm.nih.gov/pmc/articles/PMC4926222/pdf/srep28828.pdf (accessed 23 December 2021).

Vogiatzakis, I.N., Manolaki, P., Zomeni, M. and Zotos, S. (2016) Habitats. In: Sparrow, D.J. and John, E. (eds) *An Introduction to the Wildlife of Cyprus.* Terra Cypria, Limassol, Cyprus, pp. 24–39.

Wagstaffe, A. (2016) Geology. In: Sparrow, D.J. and John, E. (eds) *An Introduction to the Wildlife of Cyprus.* Terra Cypria, Limassol, Cyprus, pp. 12–23.

Wahlberg, N., Braby, M.F., Brower, A.V.Z., de Jong, R., Lee, M.M. *et al.* (2005) Synergistic effects of combining morphological and molecular data in resolving the phylogeny of butterflies and skippers. *Proceedings of the Royal Society B-Biological Sciences* 272, 1577–1586. doi: 10.1098/rspb.2005.3124

Wakeham-Dawson, A. and Dennis, R.L.H. (2001) A quantitative description of the male genitalia of 23 taxa of *Pseudochazara* de Lesse, 1951 (Lepidoptera: Nymphalidae, Satyrinae). *Entomologist's Gazette* 52, 227–250.

Wakeham-Dawson, A., Parker, R., John, E. and Dennis, R.L.H. (2003) Comparison of the male genitalia and androconia of *Pseudochazara anthelea acamanthis* (Rebel, 1916) from Cyprus, *Pseudochazara anthelea anthelea* (Hübner, [1924]) from mainland Turkey and *Pseudochazara anthelea amalthea* (Frivaldszky, 1845) from mainland Greece (Nymphalidae, Satyrinae). *Nota lepidopterologica* 25(4), 251–263.

Waring, P. and Thomas, R.C. (1994) Butterflies and moths of south-east Majorca. *The Bulletin of the Amateur Entomologists' Society* 53, 121–125.

Weller, S., Yakovlev, V.R., Zolotuhin, V.V., Zwick, A. (2011) Order Lepidoptera Linnaeus, 1758. In: Zhang, Z.-Q. (ed.) *Animal Biodiversity: An Outline of Higher-level Classification and Survey of Taxonomic Richness*. Monograph Zootaxa 3148. Magnolia Press, Auckland, New Zealand, pp. 212–221.

Werner, U. and Buszko, J. (2005) Detecting biodiversity hotspots using species–area and endemics–area relationships: the case of butterflies. *Biodiversity and Conservation* 14, 1977–1988.

Wickman, P.-O. (2009) Mating behaviour in butterflies. In: Settele, J., Shreeve, T., Konvićka, M. and Van Dyck, H. (eds) *Ecology of Butterflies in Europe*. Cambridge University Press, Cambridge, UK, pp. 17–28.

Wiemers, M. (1983) Tagfalterbeobachtungen in Mittelgriechenland im August 1982. *Nachrichten des entomologischen Vereins Apollo*, N.F. 4(2), 25–58.

Wiemers, M. (2007) Invasion of the Lime Swallowtail *Papilio demoleus* L. into SE Asia. *Berichte des Naturwissenschaftlich-Medizinischen Verein Innsbruck, Suppl.* 17, 268–269.

Wiemers, M., Stradomsky, B.V. and Vodolazhsky, D.I. (2010) A molecular phylogeny of *Polyommatus* s. str. and *Plebicula* based on mitochondrial *COI* and nuclear *ITS2* sequences (Lepidoptera: Lycaenidae). *European Journal of Entomology* 107, 325–336.

Wiemers, M., Balletto, E., Dincă, V., Fric, Z.F., Lamas, G. *et al.* (2018) An updated checklist of the European Butterflies (Lepidoptera, Papilionoidea). *ZooKeys* 811, 9–45. doi: 10.3897/zookeys.811.28712

Wiemers, M., Chazot, N., Wheat, C.W., Schweiger, O. and Wahlberg, N. (2020) A complete time-calibrated multi-gene phylogeny of the European butterflies. *ZooKeys* 938: 97–124. doi: 10.3897/zookeys.938.50878

Wiklund, C. and Forsberg, J. (1991) Sexual size dimorphism in relation to female polygamy and protandry in butterflies: a comparative study of Swedish Pieridae and Satyridae. *Oikos* 60(3), 373–381.

Williams, C.B. (1930) *The Migration of Butterflies*. Oliver & Boyd, Edinburgh, Scotland.

Williams, C.B. (1958) *Insect Migration*. Collins, London.

Williams, C.B. (1970) The migrations of the Painted Lady butterfly, *Vanessa cardui* (Nymphalidae), with special reference to North America. *Journal of the Lepidopterists' Society* 24, 157–175.

Williams, C.B., Cockbill, G.F., Gibbs, M.E. and Downes, J.A. (1942) Studies in the migration of Lepidoptera. *Transactions of the Royal Entomological Society of London* 92(1), 101–283.

Wiltshire, E.P. (1948) Middle East Lepidoptera, IV: Two New Forms or Species and Thirty-five New Records from Cyprus. *Entomologist's Record and Journal of Variation* 60, 79–87.

Wiltshire, E.P. (1949) Some more new records from Cyprus. *Entomologist's Record and Journal of Variation* 61, 73–78.

Wiltshire, E.P. (1951) Further new Records of Lepidoptera from Cyprus, Iraq and Persia (Iran). *Entomologist's Record and Journal of Variation* 63, 1–6.

Wiltshire, E.P. (1957) *The Lepidoptera of Iraq*. Revised and enlarged edn. Nicholas Kaye Ltd, London.

Wimmer, J. (1985) Beitrag zur Macrolepidopterenfauna von Zypern. *Jahresbericht Steyrer Entomologenrunde* 19, 54–61.

Wing, S.L. and Boucher, L.D. (1998) Ecological aspects of the cretaceous flowering plant radiation. *Annual Review of Earth and Planetary Sciences* 26, 379–421.

Wu, L.-W., Hsu, Y.-F. and Yen, S.-H. (2010) Elucidating origins of the Cycad Blue (*Chilades pandava*): a threat to cycad plants worldwide, with a discussion on the evolution of Cycas-feeding behaviour. PhD Thesis. National Taiwan Normal University, Taipei, Taiwan.

Yunus, M. and Munir, M. (1972) Host plants and host preference of lemon butterfly, *Papilio demoleus* Linn. caterpillars. *Pakistan Journal of Zoology* 4(2), 231–232.

Yusifova, N. and Rákosy, L. (2019) New records of the Lime Swallowtail *Papilio demoleus* Linnaeus, 1758 (Lepidoptera: Papilionidae) in Azerbaijan. *Entomologica romanica* 23, 23–26.

Zarikian, A.N. and Ghrejyan, T. (2018) High altitude Papilionoidea (Lepidoptera) of Al-Lazzab reserve in Syria. *Indian Journal of Entomology* 80(4), 1227–1235.

Zhang, J., Brockmann., E., Cong, Q., Shen, J. and Grishin, N. (2020) A genomic perspective on the taxonomy of the subtribe Carcharodina (Lepidoptera: Hesperiidae: Carcharodini). *Zootaxa* 4748(1), 182–194.

Zobar, D. and Genc, H. (2008) Biology of the Queen of Spain Fritillary, *Issoria lathonia* (*Lepidoptera: Nymphalidae*). *Florida Entomologist* 91(2), 237–240.

Zografou, K., Grill, A., Wilson, R.J., Halley, J.M., Adamidis, G.C. and Kati, V. (2020) Butterfly phenology in Mediterranean mountains using space-for-time substitution. *Ecology and Evolution* 10, 928–939.

Zorkot, H.A. (2016). *A Field Guide to the Butterflies of Lebanon and the Middle East*. Society for the Protection of Nature in Lebanon, Beirut, Lebanon.

General Index

Page numbers in *italics* refer to illustrations.

Index of Common Lepidoptera Names

Page numbers in **bold** refer to the main section on each species, including illustrations. Other illustrations are given in *italics*.

Index of Scientific Lepidoptera Names

Page numbers in **bold** refer to the main section on each species, including illustrations. Other illustrations are given in *italics*.